METAL–POLYMER NANOCOMPOSITES

METAL–POLYMER NANOCOMPOSITES

Edited by

Luigi Nicolais
Gianfranco Carotenuto
Institute of Composite
and Biomedical Materials
National Research Council Naples, Italy

WILEY-INTERSCIENCE

A JOHN WILEY & SONS, INC., PUBLICATION

Published by John Wiley & Sons, Inc., Hoboken, New Jersey.
Published simultaneously in Canada.

Library of Congress Cataloging-in-Pub lication Data:
Metal-polymer nanocomposites / edited by Luigi Nicolais and Gianfranco Carotenuto.
p. cm.
Includes bibliographical references and index.
ISBN 0-471-47131-3 (Cloth)
1. Nanostructured materials. 2. Metallic composites. 3. Polymeric composites.
I. Nicolais, Luigi. II. Carotenuto, Gianfranco.
TA418.9.N35M525 2005
620.1′6–dc22

2004009418

Printed in the United States of America

10 9 8 7 6 5 4 3 2 1

CONTENTS

PREFACE

As advanced technologies are expanding, the need for novel functional materials significantly increases. Nowadays, materials with a special combination of properties (e.g., magnetic–transparent, conductive–transparent, catalytic–magnetic, etc.) are strictly required. Materials based on nano-sized metals will surely represent an adequate solution to many present and future technological demands, since they exhibit both novel properties (e.g., plasmon resonance, superparamagnetism, etc.) and unique properties combinations.

Nano-sized metals have special characteristics that can be exploited for a number of advanced functional applications. However, the difficult handling of these incredibly small objects (1–30 nm) has represented a strong limitation to their use. Manipulations of single nanoscopic objects by surface tunneling microscopy (STM), spontaneous self-assembly, and dielectrophoresis are the only available approaches for building functional devices using nano-sized metals. In addition, most of nano-sized metals are very instable: They can aggregate because of the high surface free energy and can be oxidized-contaminated by air, moisture, SO_2, and so on. The embedding of nanoscopic metals into dielectric matrices represents a valid solution to the manipulation and stabilization problems. In the functional field, polymers are particularly interesting as an embedding phase, since they may have a variety of characteristics: They can be an electrical and thermal insulator or conductor, may have a hydrophobic or hydrophilic nature, can be mechanically hard, plastic, or rubbery, and so on. Finally, polymer-embedding is the easiest and most convenient way for nanostructured metals' stabilization, handling, and use. Polymer-embedded nanostructures are frequently termed nanocomposites because of their biphasic nature.

The fundamental knowledge on the preparation and nature of metal–polymer nanocomposites has a long history that is connected to the names of many famous scientists. The oldest technique for the preparation of metal–polymer nanocomposites that can be found in the literature was described in an abstract that appeared in 1835. In an aqueous solution, a gold salt was reduced in the presence of gum arabic, and subsequently a nanocomposite material was obtained in the form of a purple solid simply by co-precipitation with ethanol. Around 1900, widely forgotten reports describe the preparation of polymer nanocomposites with uniaxially oriented inorganic particles and their remarkable optical properties. Dichroic plants and animal fibrils (e.g., linen, cotton, spruce, or chitin, amongst others) were prepared by impregnation with solutions of silver nitrate, silver acetate, or gold chloride, followed by reduction of the corresponding metal ions under the action of light. Dichroic films were also obtained using gold chloride-treated gelatin that was subsequently drawn, dried, and finally exposed to light. Similar results were obtained when gelatin was mixed with colloidal gold before drying and drawing. In 1904, Zsigmondy (Nobel Laureate in Chemistry, 1925) reported that nanocomposites of colloidal gold and gelatin reversibly changed the color from blue to red upon swelling with water. In order to explain the mechanism of nanocomposite color change, they suggested that the material absorption must also be influenced by the interparticle distance. In addition, around the same time, the colors of gold particles embedded in dielectric matrices was the subject of detailed theoretical analyses by Maxwell Garnett, who explained the color shifts upon variation of particle size and volume fraction in a medium. During the following three decades, dichroic fibers were prepared with many different metals (i.e., Pd, Pt, Cu, Ag, Au, Hg, etc.). The dichroism was found to depend strongly on the employed element, and optical spectra of dichroic nanocomposites, made of stretched poly(vinyl alcohol) films containing gold, silver, or mercury, were presented in 1946. It was assumed already in the early reports that dichroism was originated by the linear arrangement of small particles or by polycrystalline rod-like particles located in the uniaxially oriented spaces present in the fibers. An electron micrograph published in 1951 showed that tellurium needles were present inside a dichroic film made of stretched poly(vinyl alcohol). In 1910, Kolbe proved that dichroic nanocomposite samples based on gold contained the metal indeed in its zero-valence state. Such affirmation was confirmed a few years later by X-ray scattering; in particular it was shown that zero-valence silver and gold were present in the respective nanocomposites made with oriented ramie fibers, and the ring-like interference patterns of the metal crystallites showed that the individual primary crystallites were not oriented. Based on Scherrer's equation, which was developed in this period, the average particle diameter of silver and gold crystallites was determined in fibers of ramie, hemp, bamboo, silk, wool, viscose, and cellulose acetate to be between 5 and 14 nm.

Metals undergo the most considerable property change by size reduction, and their composites with polymers are very interesting for functional applications. The new properties observed in nano-sized metals (mesoscopic metals) are produced by quantum-size effects (i.e., electron confinement and surface effect). These properties are size-dependent and can be simply tuned by changing the dimension. Since the same element may show different sets of properties by size variation, a Three-dimensional Periodic Table of elements has been proposed. Confinement effects arise in nano-sized metal domains since conduction electrons are allowed to move in a very small space, which is comparable to their De Broglie wavelength; consequently their states are quantized just like in the atoms, and these systems are termed *artificial atoms*. Surface effects are produced because with a decrease in size, matter consists more and more of surface atoms than of inner atoms. As a result, the matter properties slowly switch from that determined by the characteristics of inner atoms to that belonging to surface atoms. In addition, the surface nature of a nano-sized object significantly differs from that of a massive object. Atoms on the surface of a massive crystalline solid are principally located on basal planes, but they transform almost completely in edge and corner atoms with a decrease in size. Because of the very low coordination number, edge and corner atoms are highly chemically reactive, supercatalytically active, highly polarizable, and so on, in comparison with atoms on basal planes.

Because of quantum-size effects, mesoscopic metals show a set of properties completely different from that of their massive counterpart. Particularly interesting is: the size-dependent ferromagnetism and the superparamagnetism characterizing all metals (included diamagnetic metals like silver); the chromatism observed with silver, gold, and copper metals due to plasmon absorption; the photo- and thermoluminescence; and the supercatalytic effect (hyperfine catalysts are characterized by an extraordinarily higher catalytic activity and a different selectivity compared to corresponding fine powders). In addition, because of the band-structure disappearance, metals become thermally and electrically insulators at very small sizes. They are highly chemically reactive (heterogeneous reactions become stoichiometric and new reaction schemes are possible, for example: nano-sized noble metals are very reactive), are super absorbent, and show completely different thermodynamic parameters (for example, they melt at much lower temperatures). Many of these unique chemical–physics characteristics of nano-sized metals leave unmodified after embedding in polymers (e.g., optical, magnetic, dielectric, and thermal transport properties), and therefore they can be used to provide special functionalities to polymers.

A limited number of methods have been developed for the preparation of metal–polymer nanocomposites. Usually, such techniques consist of highly specific approaches, which can be classified as *in situ* and *ex situ* methods. In the *in situ* methods, two steps are needed: First, the monomer is polymerized in solution, with metal ions introduced before or after polymerization. Then metal ions in the polymer matrix are reduced chemically, thermally, or by UV irradiation. In the *ex situ* processes, the metal nanoparticles are chemically synthesized, and their surface is organically passivated. The derivatized nanoparticles are dispersed into a polymer solution or liquid monomer that is then polymerized.

For the comprehension of mechanisms involved in the appearance of novel properties in polymer-embedded metal nanostructures, their characterization represents the fundamental starting point. The microstructural characterization of nanofillers and nanocomposite materials is performed mainly by transmission electron microscopy (TEM), large-angle X-ray diffraction (XRD), and optical spectroscopy (UV–Vis). These three techniques are very effective in determining particle morphology, crystal structure, composition, and particle size.

Of the many techniques that have been used to study the structure of metal–polymer nanocomposites, transmission electron microscopy has undoubtedly been the most useful. This technique is currently used to probe the internal morphology of nanocomposites. High-quality images can be obtained because of the presence in the sample of regions that do not allow high-voltage electron beam passage (i.e., the metallic domains) and a region perfectly transparent to the electron beam (i.e., the polymeric matrix). High-resolution transmission electron microscopy (HRTEM) allows morphological investigations with resolution of 0.1 nm, and thus this technique makes it possible to accurately image nanoparticle sizes, shape, and atomic lattice.

Large-angle X-ray powder diffraction (XRD) has been one of the most versatile techniques utilized for the structural characterization of nanocrystalline metal powders. The modern improvements in electronics, computers, and X-ray sources have allowed XRD to become an indispensable tool for identifying nanocrystalline phases as well as crystal size and crystal strain. The comparison of the crystallite size obtained by the XRD diffractogram using the Scherrer formula with the grain size obtained from the TEM image allows us to establish if the nanoparticles have a mono- or polycrystalline nature.

Most metal clusters are characterized by the surface plasmon resonance, which is an oscillation of the surface plasma electrons induced by the electromagnetic field; consequently, their microstructure can be indirectly investigated by optical spectroscopy (UV–Vis spectroscopy). The characteristics of this absorption (shape, intensity, position, etc.) are strictly related to the nature, structure, topology, and so on, of the cluster system. In fact, the absorption frequency is a fingerprint of the particular metal, the eventual peak splitting reflects aggregation phenomena, the intensity of the peak is related to the particle size, the absorption wavelength is related to the particle shape, the shift of the absorption with increasing of temperature is indicative of a cluster melting, and so on. For bimetallic particles, information about inner structure (intermetallic or core/shell) and composition can be obtained from the absorption frequency. Differently from *off-line* techniques (e.g., TEM, XRD), this method allows *on-line* and *in situ* cluster sizing and monitoring of morphological evolution of the system. This method has been used also in the study of cluster nucleation and growth mechanisms.

This outline of the principal characterization techniques for nanocomposite materials is far from being complete. Advances in Raman spectroscopy, energy dispersive spectroscopy, infrared spectroscopy, and many other techniques are of considerable importance as well. In fact, the success that nanostructured materials are having in the last few years is strictly related to the advanced characterization techniques that are available today.

Applications of metal–polymer nanocomposites have already been made in different technological fields; however, the use of a much larger number of devices based on these materials can be predicted for the near future.

Because of the plasmon surface absorption band, atomic clusters of metals can be used as pigments for optical plastics. The color of the resulting nanocomposites is light-fast and intensive; in addition, these materials are perfectly transparent, since the cluster size is much lower than light wavelength. Gold, silver, and copper can be used for color filter application. Also, UV absorbers can be made, for example, by using Pd clusters. The plasmon surface absorption frequency is modulated by making intermetallic particles (e.g., Pd/Ag, Au/Ag) of adequate composition.

Polymeric films containing uniaxially oriented pearl-necklace type of nanoparticle arrays exhibit a polarization-dependent and tunable color. The color of these systems is very bright and can change, strongly modifying the light polarization direction. These materials are obtained by dispersing metal nanoparticles in polymeric thin films and subsequently reorganizing the dispersed phase into pearl-necklace arrays by solid-state drawing at a temperature below the polymer melting point. The formation of these arrays in the films is the cause of a strong polarization-direction-dependent color that can be used in the fabrication of liquid-crystal color display and special electrooptical devices.

Surface plasmon resonance has been used to produce a wide variety of optical sensors—that is, systems that are able to change their color in the presence of specific analytes. These devices can be used as sensors for immunoassay, gas, and liquid.

Metals are characterized by ultrahigh/low refractive indices and therefore can be used to modify the refractive index of optical plastics. Ultrahigh/low refractive index optical nanocomposites can be used in the waveguide technology (e.g., planar waveguides and optical fibers).

Plastics doped by atomic clusters of ferromagnetic metals show magneto-optical properties (i.e., when subject to a strong magnetic field, they can rotate the vibration plane of a plane-polarized light), and therefore they can be used as Faraday rotators. These devices have a number of important optical applications (e.g., magnetooptic modulators, optical isolators, optical shutters, etc.).

Nano-sized metals (e.g., gold, silver) have attracted much interest because of the nonlinear optical polarizability, which is caused by the quantum confinement of the metal electron cloud. When irradiated with light above a certain threshold power, the optical polarizability deviates from the usual linear dependence on that power. By incorporating these particles into a clear polymeric matrix, nonlinear optical devices can be made in a readily processable form. These materials are used to prepare a number of devices for photonics and electrooptics.

Finally, polymer-embedding represents a simple but effective way to use mesoscopic properties of nano-sized metals. A large variety of advanced functional devices can be based on this simple material class. In the last few years, a number of pioneering techniques have been developed for preparing metal–polymer nanocomposite materials. In particular, the *in situ* techniques based on the thermolysis of special organic metal precursors seems to be a very promising approach, principally for the possibility to produce metal–polymer nanocomposites on a large scale by techniques already available for thermoplastic polymer hot-processing.

Luigi Nicolais
Gianfranco Carotenuto

Naples
March 2004

CONTRIBUTORS

V. I. Belotelov Institute of General Physics, RAS, Moscow, Russia; and M. V. Lomonosov Moscow State University, Faculty of Physics, Physics of Oscillations Department, Moscow, Russia

L. M. Bronstein Department of Chemistry, Indiana University, Bloomington, Indiana

G. Carotenuto Institute of Composite and Biomedical Materials, National Research Council, Naples, Italy

W. Caseri Department of Materials, Institute of Polymers, ETH Zentrum, Zürich, Switzerland

G. I. Dzhardimalieva Institute of Problems of Chemical Physics, Russian Academy of Sciences, Chernogolovka Moscow Region, Russia

G. N. Gerasimov Karpov Institute of Physical Chemistry (Russian State Scientific Center), Moscow, Russia

A. Heilmann Fraunhofer Institute for Mechanics of Materials, Halle (Saale), Germany

G. U. Kulkarni Chemistry and Physics of Materials Unit, Jawaharlal Nehru Center for Advanced Scientific Research, Bangalore, India

B. Martorana Fiat Research Center, Orbassano, Italy

L. Nicolais Institute of Composite and Biomedical Materials, National Research Council, Naples, Italy

P. Perlo Fiat Research Center, Orbassano, Italy

A. D. Pomogailo Institute of Problems of Chemical Physics, Russian Academy of Sciences, Chernogolcvka Moscow Region, Russia

C. N. R. Rao Chemistry and Physics of Materials Unit, Jawaharlal Nehru Center for Advanced Scientific Research, Bangalore, India

A. S. Rozenberg Institute of Problems of Chemical Physics, Russian Academy of Sciences, Chernogolovka Moscow Region, Russia

A. L. Stepanov Kazan Physical–Technical Institute of the Russian Academy of Sciences, Kazan, Russian Federation; and Institute for Experimental Physics and Erwin Schrödinger Institute for Nanoscale Research, Karl-Franzens–University Graz, Graz, Austria

P. J. Thomas Chemistry and Physics of Materials Unit, Jawaharlal Nehru Center for Advanced Scientific Research, Bangalore, India

L. I. Trakhtenberg Karpov Institute of Physical Chemistry (Russian State Scientific Center), Moscow, Russia

A. K. Zvezdin Institute of General Physics, RAS, Moscow, Russia; and Fiat Research Center, Orbassano, Italy

1

PHYSICAL AND CHEMICAL PROPERTIES OF NANO-SIZED METAL PARTICLES

C. N. R. Rao, G. U. Kulkarni, and P. J. Thomas

Chemistry and Physics of Materials Unit, Jawaharlal Nehru Center for Advanced Scientific Research, Bangalore, India

1. INTRODUCTION

A nanocluster is a tiny chunk of the bulk measuring a few nanometers with a finite number of atoms in it. Nano-sized metals with sizes in the range of 1–50 nm are considered important and are obtainable as sols—a dispersion of a solid in a liquid. Metal sols possess fascinating colors and have long been used as dyes and catalysts. That such dyes indeed consist of tiny metal chunks was established as early as 1857 by Faraday [1, 2]. Modern techniques of synthesis enable one to obtain sols of metals that can be dried and redissolved like water colors. The nano-sized clusters display a remarkable tendency to remain single-crystalline and hence are also called nanocrystals. In addition, nanocrystals possess a high surface area: A great fraction of atoms in a nanocrystal is on its surface [3].

An added dimension to research on nanocrystals is their size-dependent properties. The electronic, magnetic, and optical properties of a nanocrystal depend on its size [3]. In small nanocrystals, the electronic energy levels are

Metal-Polymer Nanocomposites, Edited by Luigi Nicolais and Gianfranco Carotenuto
ISBN 0-471-47131-3

not continuous as in the bulk but are discrete, due to the confinement of the electron wavefunction to the physical dimensions of the particles [4]. This phenomenon is called *quantum confinement*; therefore, nanocrystals are also known as quantum dots. In other words, a small nanocrystal could be a very bad conductor, although nanocrystals are tiny silhouettes of the conducting bulk. Likewise, a tiny nanocrystal of a ferromagnet can be paramagnetic in nature. In several respects, small nanocrystals behave like molecules. The nanocrystals can be discretely charged with electrons with characteristic charging energies. This means that a nanocrystal carrying an extra electron can exhibit properties different from those of a neutral species.

The shrinking dimensions of the current microelectronic devices and the realization that current lithographic processes cannot extend to the nanoworld [5] have lent tremendous thrust to research aimed at ordering nanocrystals into functional networks [6–11]. The nanocrystals akin to covalent systems self-assemble into ordered arrays in one, two, and three dimensions under the right conditions. Lattices of nanocrystals consist of interacting nanocrystals and may exhibit novel properties arising out of such interactions. Thus, the ability to engineer such assemblies extends the reach of current lithographic techniques and holds promise for a new generation of electronics of the nanoworld [6]. In this context, synthesis and programmed assembly of nanocrystals assume significance.

In this chapter we discuss the physical and chemical properties of metal nanocrystals, with emphasis on size-dependent properties. The ability of nanocrystals to form mesoscopic organizations in one, two, and three dimensions is also examined. Collective properties of nanocrystal organization are presented.

2. SYNTHETIC STRATEGIES

2.1. General Methods

Chemical synthesis of sols of metals results in nanoparticles embedded in a layer of ligands or stabilizing agents, which prevent the aggregation of particles. The stabilizing agents employed include surfactants such as long-chain thiols or amines or polymeric ligands such as polyvinylpyrrolidone (PVP). Reduction of metal salts dissolved in appropriate solvents produces small metal particles of varying size distributions [12–15]. A variety of reducing agents have been employed for reduction. These include electrides, alcohols, glycols, metal borohydrides, and certain specialized reagents such as tetrakis(hydroxymethyl) phosphonium chloride.

Successful nanocrystals synthesis has also been carried out employing soft templates such as the water pool in a reverse micelle, the interface of two phases. Reverse micellar methods have been successfully utilized in the preparation of Ag, Au, Co, Pt, and Co nanocrystals [16, 17]. The synthesis of nanocrystals at the air–water interface as in Langmuir–Blodgett films or at a liquid–liquid interface is currently attracting wide attention [7, 18, 19]. It has been shown recently that films of metal, semiconductor, and oxide nanocrystals can be prepared using a water–toluene interface [20, 21]. A typical film of Au nanocrystals is shown in Figure 1.1. Traditionally, clusters of controlled sizes have been generated by abalation of a metal target in vacuum followed by mass selection of the plume to yield cluster beams [22, 23]. Such cluster beams could be subject to *in situ* studies or be directed on to solid substrates. In order to obtain nanocrystals in solution, Harfenist et al. [24] steered a mass-

Figure 1.1. Nanocrystalline film of Au formed at the toluene–water interface (**middle**). Gold is introduced as a toluene solution of $Au(PPh_3)Cl$ while partially hydrolyzed THPC [tetrakis(hydromethyl) phosphonium chloride] in water acts as a reducing agent. The film is obtained when the two layers are allowed to stand for several hours. When dodecanethiol is added to the toluene layer, the film breaks up forming an organosol (**left**) while mercaptoundecanoic acid added to water produces a hydrosol (**right**). Shown below are the corresponding TEM images showing nanocrystals. The scale bars correspond to 50 nm.

selected Ag cluster beam through a toluene solution of thiol and capped the vacuum-prepared particles.

Colloids of alloys have been made by the chemical reduction of the appropriate salt mixture in the solution phase. In the case of semiconductor nanocrystals, a mixture of salts is subject to controlled precipitation. Thus, Ag–Pd and Cu–Pd colloids of varying composition have been prepared by alcohol reduction of mixtures of silver nitrate or copper oxide with palladium oxide [25]. Fe–Pt alloy nanocrystals have been made by thermal decomposition of the Fe and Pt acetylacetonates in high-boiling organic solvents [26]. Au–Ag alloy nanocrystals have been made by co-reduction of silver nitrate and chloroauric acid with sodium borohydride [27, 28]. Au–Ag alloying and segregation has been brought about by the use of lasers on Au–Ag-layered particles [29, 30].

2.2. Size Control

The successful synthesis of nanocrystals involves three steps: nucleation, growth, and termination by the capping agent or ligand [12–14]. Though the reaction temperature and reagent concentrations provide a rudimentary control of the three steps, it is often impossible to independently control them and so the obtained nanocrystals usually exhibit a distribution in size. Typically, the distribution is log-normal with a standard deviation of 10% [14]. Given the fact that properties of the nanocrystals are size-dependent, it is significant to be able to synthesize nanocrystals of precise dimensions with minimal size distributions. This can be accomplished to a limited extent by size-selective precipitation either by centrifugation or by use of a miscible solvent–nonsolvent liquid mixture to precipitate nanocrystals. Schmid [31] and Zamaraev and co-workers [32] succeeded in preparing truly monodisperse nanocrystals that they called "cluster compounds." These cluster compounds are like macromolecules with a core containing metal–metal bonds, yet they are obtainable in definite stoichiometries, with typical examples being $[Pt_{38}(CO)_{44}H_2]^{2-}$ and $Au_{55}(PPh_3)_{12}Cl_6$. The enhanced stability of Au_{55} was recently demonstrated clearly by Boyen et al. [33], who exposed a series of Au_n nanocrystals to oxidation and found that Au_{55} does not get oxidized under conditions that oxidize bulk Au. These nanocrystals are bequeathed with special stability because they consist of a magic number of metal atoms that enable the complete closure of successive shells of atoms in a cubic close-packed arrangement. The magic numbers 13, 55, 147, 309, and 561 correspond to the closure of 1, 2, 3, 4, and 5 shells, respectively [34]. A schematic illustration of magic nuclearity nanocrystals is shown in Figure 1.2. Since the breakthrough, several magic nuclearity nanocrystals have been prepared including PVP-stabilized Pd_{561} nanocrystals [35]. Figure 1.3, shows scanning tunneling and transmission electron microscopic (TEM) images of polymer-protected Pd_{561} nanocrystals.

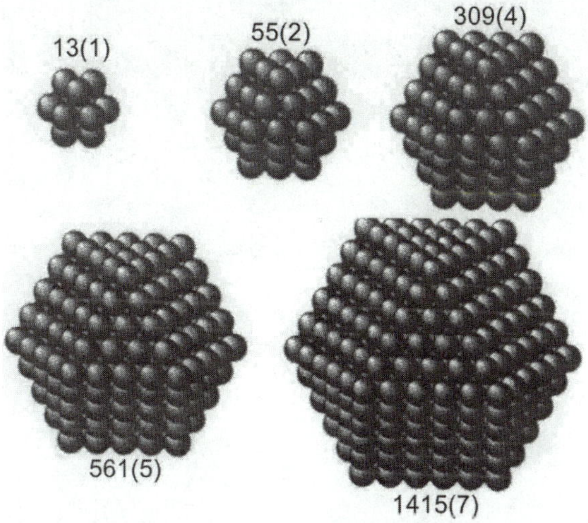

Figure 1.2. Metal nanocrystals in closed-shell configurations with a magic number of atoms. The number of shells is indicated in brackets.

2.3. Shape Control

Since the properties of the nanocrystals follow from the confinement of the electrons to the physical dimensions of the nanocrystals, it would be interesting to vary the shape of the nanocrystals and study the effect of confinement of electrons in such artificial shapes [36]. For example, it is predicted that light emitted from a nanorod would be linearly polarized along the growth axis [37]. Such predictions have led to the revival of interest in synthetic strategies yielding nonspherical nanocrystals. Conventional methods such as those due to Turkevich et al. [13] yield, in addition to spherical particles, a mixture of shapes—triangular, teardrop, and so on—which was then thought of as undesirable. Today, smarter synthetic schemes have been designed which yield selectively nanocrystals in the form of rods, elongated spheres, cubes, and hexagons.

2.4. Tailoring the Ligand Shell

Nanocrystals in their native form are dominated by the surface species, and the capping agents employed play a role in determining the property of the nanocrystals [38]. Hence, in addition to controlling the size and the shape of the nanocrystals, it is also necessary to tailor its surface with the right capping agent. In addition to traditional capping agents that include ions, surfactants,

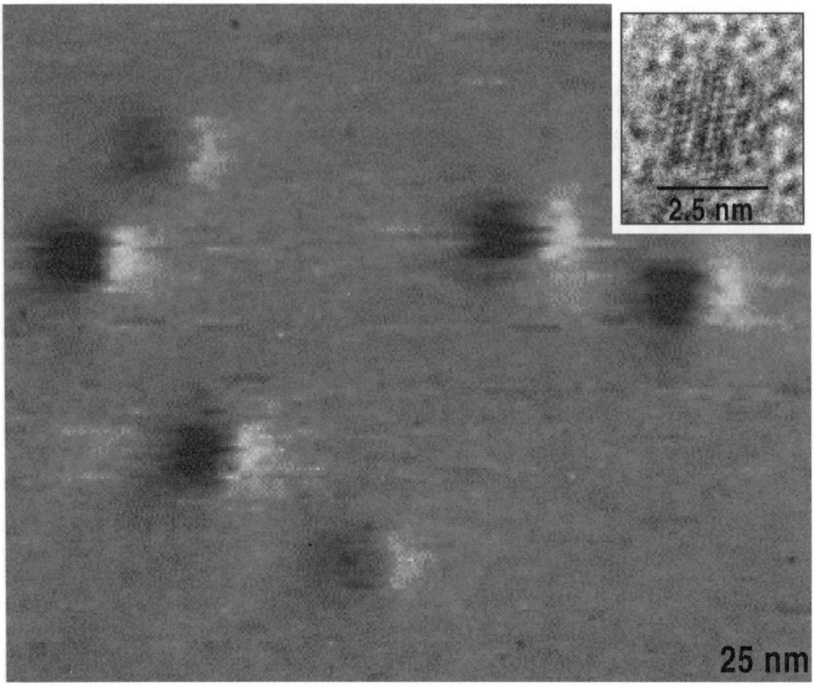

Figure 1.3. Scanning tunneling microscopy image of polymer coated Pd_{561} nanocrystals. The nanocrystals are seen as fluffy balls against plane background of the graphite substrate. The inset shows a high-resolution electron micrograph (HRTEM) of an individual nanocrystal. We see the characteristic 11 [111] fringes in the icosahedral shape measuring 2.5 nm. The diameter estimated from STM is ~3.4 nm, with the difference being due to the ligand shell.

and polymers, a new genre of ligands—dendrimers, hydrogen-bonding fragments of protein, DNA, and dyes—with pendent thiol groups as well as silica layers have been used as capping agents [39–41]. In some cases, a layer of a noble metal is used as a buffer between the core nanocrystal and the ligand shell [42]. Thus, a layer of gold lends special stability to Fe nanocrystals and helps to prevent oxidation and to preserve the magnetic properties of Fe.

Of special interest with regard to tailoring the ligand shells are reactions that enable the total replacement of one set of ligands with another [43–47]. These reactions also typically enable the transfer of nanocrystals from one phase to another. A novel method of thiol-derivatizing hydrosols of metal sols has been developed by Sarathy et al. [44, 45]. The procedure involves mixing vigorously a hydrosol containing metal particles of the desired size distribution with a toluene solution of an alkane thiol in the presence of a strong acid or reducing agent. The completion of the derivatization is marked by a vivid interchange of the colors from the aqueous layer to the hydrocarbon layer. The advantage of this method is that well-characterized metal particles can be easily thiol-derivatized in a nonaqueous medium. A variety of hydrosols of Au, Ag, and Pt have been thiolized by this procedure.

3. PHYSICAL PROPERTIES OF NANOCRYSTALS

The electronic structure of a nanocrystal critically depends on its very size. For small particles, the electronic energy levels are not continuous as in bulk materials, but discrete, due to the confinement of the electron wavefunction because of the physical dimensions of the particles (see Figure 1.4). The average electronic energy level spacing of successive quantum levels, δ, known as the so-called Kubo gap, is given by

$$\delta = \frac{4E_F}{3n} \tag{1}$$

where E_F is the Fermi energy of the bulk material and n is total number of valence electrons in the nanocrystal. Thus, for an individual silver nanoparticle of 3-nm diameter containing approximately 1000 silver atoms, the value

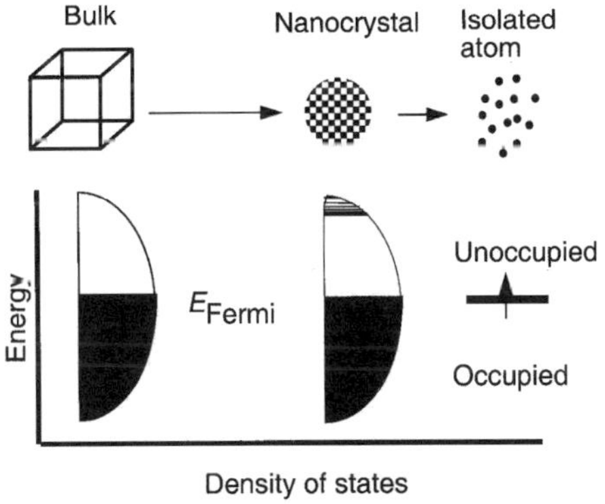

Figure 1.4. Density of states for metal nanocrystals. The density of states is discrete at the band edges. The Fermi level is at the center of a band in a metal, and hence kT exceeds the electronic energy level spacing even at room temperatures and small sizes.

of δ would be 5–10 meV. Since the thermal energy at room temperature, kT, is ~25 meV, a 3-nm particle would be metallic ($kT > \delta$). However, at low temperatures the level spacings, especially in small particles, may become comparable to kT, rendering them nonmetallic [4]. Because of the presence of the Kubo gap in individual nanoparticles, properties such as electrical conductivity and magnetic susceptibility exhibit quantum size effects. The resultant discreteness of energy levels also brings about fundamental changes in the characteristic spectral features of the nanoparticles, especially those related to the valence band.

Extensive investigations of metal nanocrystals of various sizes obtained, for example, by the deposition of metals on amorphized graphite and other substrates, as a result of X-ray photoelectron spectroscopy and related techniques [48–50], have yielded valuable information on their electronic structure. An important result from these experiments is that as the metal particle size decreases, the core-level binding energy of metals such as Au, Ag, Pd, Ni, and Cu increases sharply. This is shown in the case of Pd in Figure 1.5, where the binding energy increases by over 1 eV at small size. The variation in the binding energy is negligible at large coverages or particle size, since the binding energies are close to those of the bulk, macroscopic metals. The increase in the core-level binding energy in small particles occurs due to the poor screening of the core hole and is a manifestation of the size-induced metal–nonmetal transition in nanocrystals. Further evidence for the occurrence of such a metal–nonmetal transition driven by the size of the individual particle is provided by other electron spectroscopic techniques such as UPS and BIS. All these measurements

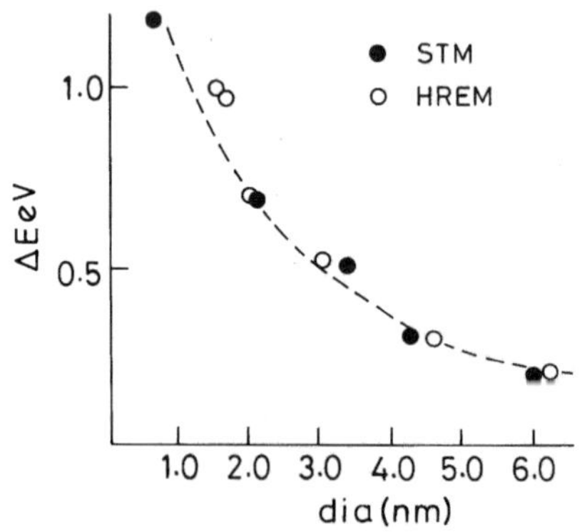

Figure 1.5. Variation of the shift, δE, in the core-level binding energy (relative to the bulk metal value) of Pd with the nanoparticle diameter. The diameters were obtained from HREM and STM images. (Reproduced with permission from reference 50.)

indicate that an electronic gap manifests itself for a nanoparticle with a diameter of 1–2 nm and possessing 300 ± 100 atoms.

Photoelectron spectroscopic measurements [51] on mass-selected Hg_n nanoparticles (n ~ 3–250) in the gas phase reveal that the characteristic HOMO–LUMO (s–p) energy gap decreases gradually from 3.5 eV for $n = 3$ to 0.2 eV for n ~ 250. The band gap closure is predicted at n ~ 400. The metal–nonmetal transition in gaseous Hg nanoparticles was examined by Rademann et al. [52] by measuring the ionization energies (IE). For $n < 13$, the dependence of IE on n suggested a different type of bonding. A small Hg particle with atoms in the $6s^2 6p^0$ configuration held together by relatively weak van der Waals forces, is essentially nonmetallic. As the nanoparticle grows in size, the atomic $6s$ and $6p$ levels broaden into bands, and an insulator metal transition appears to occur driven by the physical dimensions of the individual particle. Note that this is the same element, Hg, behaving as either a metal or a nonmetal, depending upon its physical size!

The electronic absorption spectra of metal nanocrystals in the visible region are dominated by the plasmon bands. This absorption is due to the collective excitation of the itinerant electron gas on the particle surface and is characteristic of a nanocrystal of a given size. In metal colloids, surface plasmon excitations impart characteristic colors to the metal sols, with the beautiful wine-red color of gold sols being well known [53–56]. The dependence of the plasmon peak on the dielectric constant of the surrounding medium and the diameter of the nanocrystal was predicted theoretically by Mie and others at the turn of the last century [57–60]. The dependence of the absorption band of thiol-capped Au nanocrystals on solvent refractive index was recently verified by Templeton et al. [61]. Link and El-Sayed [53, 54] found that the absorption band splits into longitudinal and transverse bands in Au nano-rods.

Direct information on the gap states in nanocrystals of metals and semiconductors is obtained by scanning tunneling spectroscopy (STS). This technique provides the desired sensitivity and spatial resolution, making it possible to carry out tunneling spectroscopic measurements on individual particles. A systematic STS study of Pd, Ag, Cd, and Au nanoparticles of varying sizes deposited on a graphite substrate has been carried out under ultrahigh vacuum conditions, after having characterized the nanoparticles by XPS and STM [62]. The I–V spectra of bigger particles were featureless while those of the small particles (<1 nm) showed well-defined peaks on either side of zero bias due to the presence of a gap (see Figure 1.6). It is seen that small particles of ~1-nm diameter are in fact nonmetallic! From the various studies discussed hitherto, it appears that the size-induced metal-insulator transition in metal nanocrystals occurs in the range of 1- to 2-nm diameter or 300 ± 100 atoms.

Theoretical calculations of the electronic structure of metal nanocrystals throw light on the size-induced changes in the electronic structure. Rosenblit

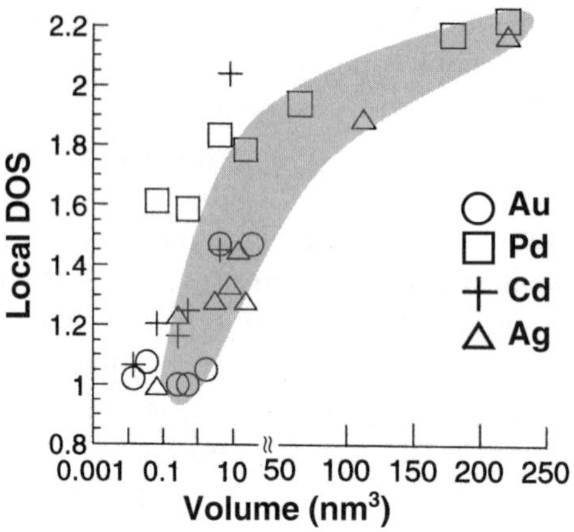

Figure 1.6. Variation of the nonmetallic band gap with nanocrystal volume for various metal nanocrystals, deposited on a graphite surface. The band gaps were estimated by means of scanning tunneling spectroscopy. (Reproduced with permission from reference 62.)

and Jortner [63] calculated the electronic structure of a model metal cluster and predicted electron localization to occur in a cluster of diameter 0.6 nm. A molecular orbital calculation on an Au_{13} cluster [64] in icosahedral and cuboctahedral structures shows that the icosahedral structure undergoes Jahn–Teller distortion while the cuboctahedral structure does not distort. The onset of the metallic state is barely discernible in the Au_{13} cluster. Relativistic density functional calculations of gold clusters [65] with $n = 6$–147 show that the average interatomic distance increases with the nuclearity of the cluster. The HOMO–LUMO electronic gap decreases with particle size from 1.8 eV for Au_6 (0.5-nm diameter) to 0.3 eV for Au_{147} (2-nm diameter). *Ab initio* molecular dynamics simulations of aluminum clusters [66] with $n = 2$, 6, 12, 13, 55, and 147 reveal that the minimum energy structures of Al_{13} and Al_{55} are distorted icosahedra whereas Al_{147} is a near-cuboctahedron. The HOMO–LUMO gap increases from 0.5 eV for Al_2 to 2 eV for Al_{13}; the gap is around 0.25 eV for Au_{55} and decreases to 0.1 eV for Au_{147}. The convergence of the cluster properties toward those of the corresponding bulk materials with increase in size is noteworthy.

In a bulk metal, the energy required to add or remove an electron is its work function. In a molecule, the corresponding energies, electron affinity and ionization potential, respectively, are, however, nonequivalent. Because nanocrystals are intermediary, the two energies differ only to a small extent [67], with the difference being the charging energy, *U*. This is a Coulombic energy and is different from electronic energy gap. Furthermore, Coulombic states can be similar for both semiconductor and metallic nanocrystals, unlike the electronic states. A manifestation of single electron charging is the Coulomb staircase behavior observed in the tunneling spectra [68] when a nanocrystal covered with an insulating ligand shell is held between two tunnel junctions. A typical staircase along with its theoretical fit is shown in Figure 1.7a. Such measurements have also been carried out on Pd and Au nanocrystals in the size range, 1.5 ± 6.5 nm [69]. The charging energies follow a scaling law [70] of the form $U = A + B/d$, where *A* and *B* are constants characteristic of the metal and *d* is the particle diameter (see Figure 1.7b).

Magnetic properties of nanoparticles of transition metals such as Co and Ni show marked variations with size. It is well known that in the nanometric domain, the coercivity of the particles tends to zero [71]. Thus, the nanocrystals behave as superparamagnets with no associated coercivity or retentivity. The blocking temperature that marks the onset of this superparamagnetism also increases with the nanocrystal size. Furthermore, the magnetic moment per atom is seen to increase as the size of a particle decreases [72].

4. CHEMICAL PROPERTIES PROPERTIES OF NANOCRYSTALS

The surface area of nanocrystals increases markedly with the decrease in size. Thus, a small metal nanocrystal of 1-nm diameter will have 100% of its atoms on the surface. A nanocrystal of 10-nm diameter, on the other hand, would have about 15% of its atoms on the surface. A small nanocrystal with a higher surface area would be expected to be more reactive. Furthermore, the qualitative change in the electronic structure arising due to quantum confinement in small nanocrystals will also bestow unusual catalytic properties on these particles, totally different from those of the bulk metal. We illustrate these important aspects with a few examples from the recent literature. A low-temperature study [73] of the interaction of elemental O_2 with Ag nanocrystals of various sizes (Figure 1.8) has revealed the capability of smaller nanocrystals to dissociate dioxygen to atomic oxygen species. On bulk Ag, the adsorbed oxygen species at 80 K is predominantly O^{2-}. This interaction of O_2 with Ag dependent on its particle size is remarkable. Another important example is the reaction of H_2S with Ni nanocrystals giving rise to S^{2-} species, with nanocrystals of different sizes exhibiting different temperature profiles (see Figure 1.9). Unlike bulk nickel, small nanocrystals show less dependence in their catalytic activity on ambient temperature.

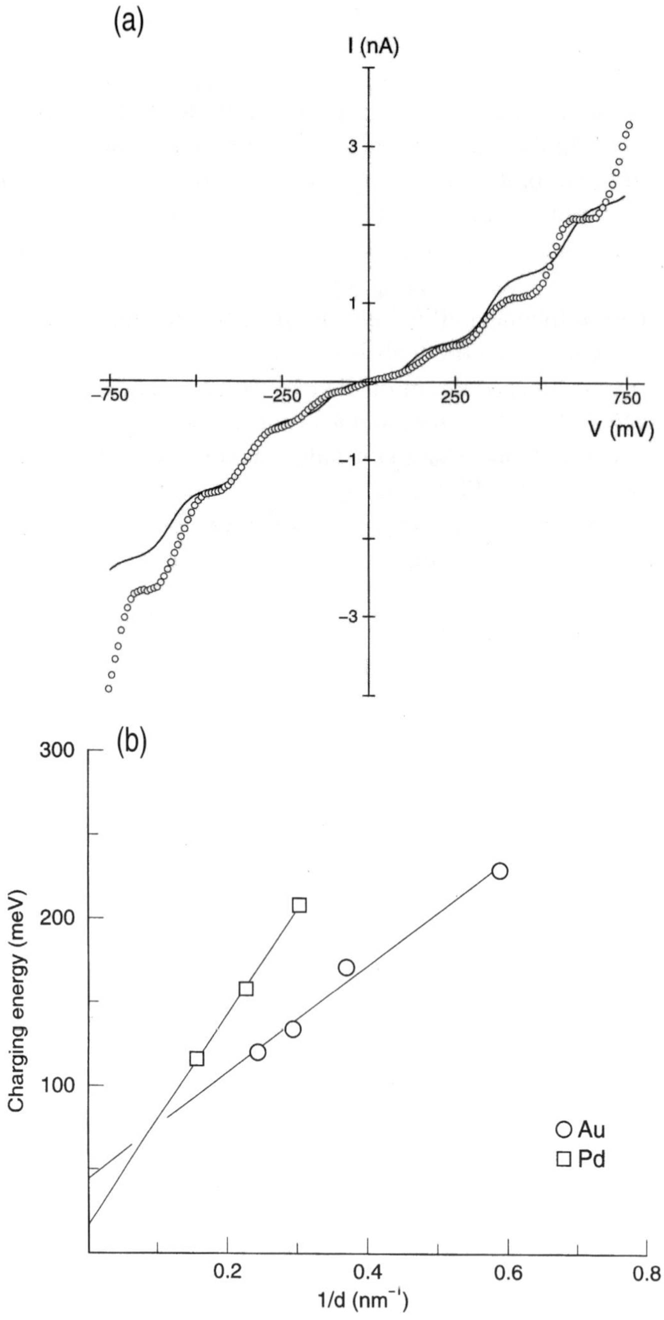

Figure 1.7. (**a**) *I–V* characteristics of an isolated 3.3-nm Pd nanocrystal (dotted line) and the theoretical fit (solid line) obtained at 300 K using a semiclassical model according to which the observed capacitance (C) can be resolved into two components C_1 and C_2, and the resistance (R) can be resolved into R_1 and R_2, such that $C = C_1 + C_2$ and $R = R_1 + R_2$. For $C_1 \ll C_2$ and $R_1 \ll R_2$, the model predicts steps in the measured current to occur at critical voltages, $V_c = n_c e/C + (q_0 + e/2)/C$, where q_0 is the residual charge. (**b**) Variation of the charging energies of Pd and Au nanocrystals with inverse diameters (d). (Reproduced with permission from reference 69.)

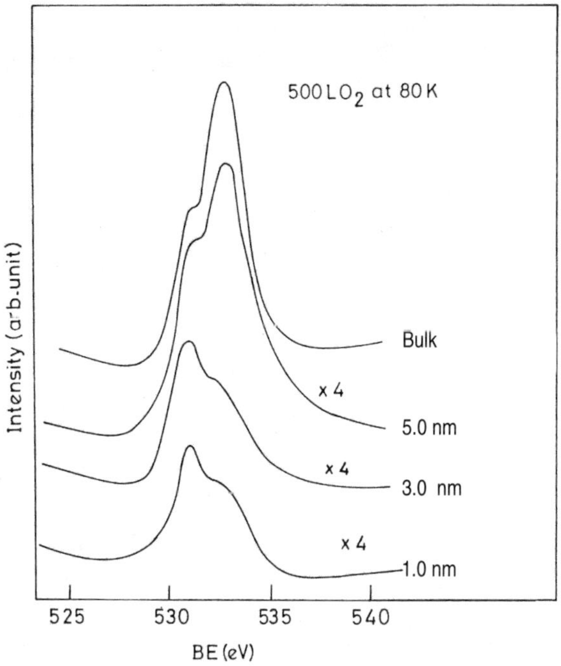

Figure 1.8. Change in the O(1*s*) spectra of Ag clusters exposed to 500 L O_2 at 80 K. The diameters of the clusters were estimated from metal coverage. The lower binding energy peak at 531 eV corresponds to O^- while that at 533 eV arises due to molecular oxygen. (Reproduced with permission from reference 74.)

The ability of Cu, Pd, Pt, and Ni nanoparticles to absorb CO has been thoroughly investigated. Carbon monoxide from a bulk Cu surface desorbs above 250 K. Small Cu particles, however, retain CO up to much higher temperatures [74]. A similar observation has been made in the case of Pd particles [75]. The results obtained with Ni particles are more interesting. In addition to showing a trend similar to the above, small Ni particles are also capable of dissociating CO to form carbidic species on the particle surface (see Figure 1.10) [76]. This could be due to the Ni(3*d*) level in small clusters coming close to the antibonding energy level of CO(2*p**). Heiz and co-workers have studied the ability of a size-selected Pt cluster with nuclearity between 5 and 20 atoms to oxidize CO [77]. The small Pt clusters are all catalytically active and exhibit a different temperature and activity profile, depending on the nuclearity (see Figure 1.11). Simple molecular orbital and geometry-based considerations suffice to describe the catalytic activity in such small clusters [77].

Bulk Au is a noble metal. Goodman and co-workers [78], however, found that Au nanocrystals supported on a titania surface show a marked size effect

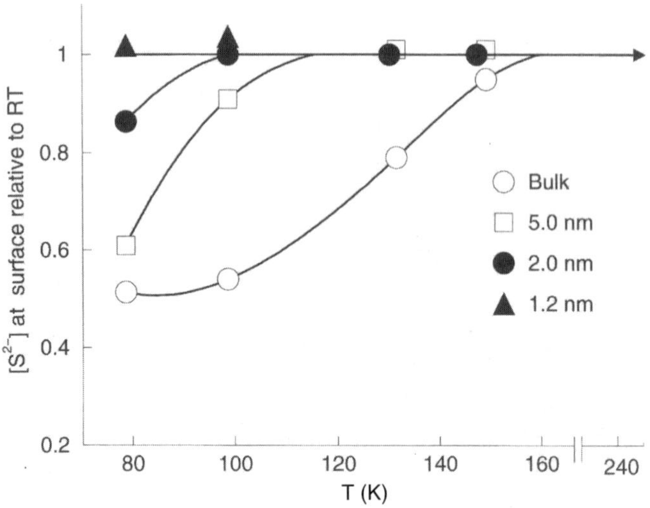

Figure 1.9. Variation of the normalized areas of the signal for the core-level transitions of S^{2-} with temperature for different sizes of Ni clusters deposited on graphite. (Reproduced with permission from reference 73.)

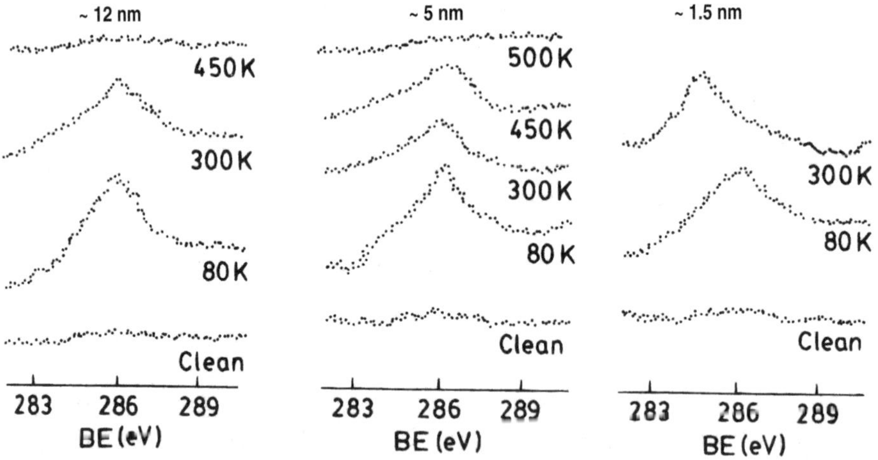

Figure 1.10. Change in the C(1*s*) spectra of CO adsorbed on Ni clusters with temperature. The feature at 286 eV corresponds to molecularly adsorbed CO while that at 284 eV arises due to the formation of carbidic species. (Reproduced with permission from reference 74.)

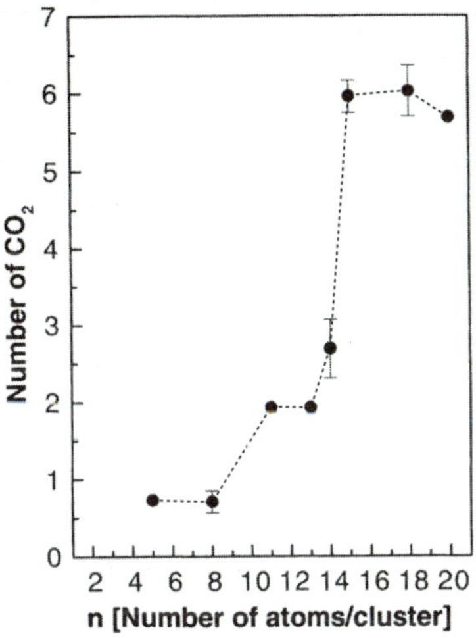

Figure 1.11. A plot of the number of catalytically produced CO_2 molecules against the nuclearity of Pt clusters. The CO_2 molecules produced by oxidation of CO are studied by means of temperature-programmed desorption mass spectrometry. (Reproduced with permission from reference 77.)

in their catalytic ability for CO oxidation reaction, with Au nanoparticles in the range of 3.5 nm exhibiting the maximum chemical reactivity. A metal-to-nonmetal transition was observed in the *I–V* spectra as the cluster size is decreased to below 3.5 nm^3 (consisting of ~300 atoms). This result is quite similar to that obtained with Pd particles supported on oxide substrate [79]. In another study of Au particles supported on a zinc oxide surface, smaller particles (<5 nm) exhibited a marked tendency to adsorb CO while those with diameters above 10 nm did not significantly adsorb CO [80]. Heiz and co-workers have found that small Au clusters (Au_n, $n \leq 20$) exhibit size-dependent catalytic activity above a nuclearity of 8 (see Figure 1.12) [81]. The increased activity of these metal particles is attributed to the charge transfer between the oxide support and the particle surface. It is possible that defects on the oxide surface also play a role in determining the catalytic activity of the nanocrystals.

5. PROGRAMMED ASSEMBLIES

Like molecular systems, nanocrystals capped with suitable ligands spontaneously assemble into ordered aggregates. That such self-assembly can occur through a variety of weak forces is being recognized. Cooperative assemblies of ligated metal and semiconductor and as well as of colloidal polymer spheres

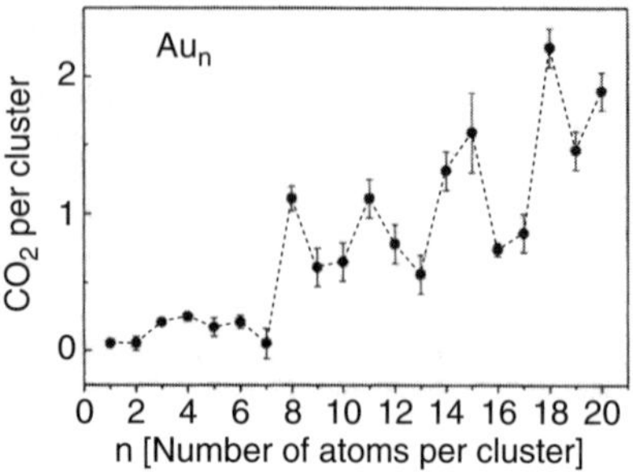

Figure 1.12. A plot of the number of catalytically produced CO_2 molecules against the nuclearity of Au clusters supported on a defect-rich MgO(100) surface. The CO_2 molecules produced by oxidation of CO are studied by means of temperature-programmed desorption mass spectrometry. (Reproduced with permission from reference 81.)

seem to occur through the mediation of electrostatic and capillary forces [67, 82, 83]. The forces that govern the nanocrystal assembly, however, are different in many ways. Surface tension for example, plays an important role [14] because in a nanocrystal a large fraction of atoms are present at the surface. Surfactant molecules that self-assemble on solid surfaces have proved to be the best means of obtaining ordered arrays of nanocrystals [67]. The way in which the nanocrystals organize themselves depends critically on the core diameter, the nature of the ligand, the substrate, and even the dispersive medium used [84]. Thiolized metal nanocrystals readily arrange into two-dimensional arrays on removal of the solvent [6]. Using suitable methods, they can also be put into one-dimensional organization in the form of strings or assembled in a stepwise fashion in a three-dimensional superlattice (see Figure 1.13).

5.1. One-Dimensional Arrangements

Hornayak et al. [85] used the ordered channels of porous alumina as templates to obtain linear arrangements of Au nanocrystals. By varying the pore size, the diameter of the nanowire could be controlled. A linear arrangement has also been obtained by coordinating Au particles (~1.4 nm) stabilized with phosphine ligands to single-stranded DNA oligonucleotide of the desired length and specific sequence [86, 87]. Pt nanocrystals in the form of ribbons has been obtained using a cholesteric liquid crystalline template [88]. Organization of particles in a one-dimensional lattice has met with limited success. Heath and co-workers [89] have fabricated wires of Ag nanocrystals by compressing a dispersion of Ag (4.5 nm) nanocrystals in toluene. The wires were one nanocrystal thick and a few nanocrystals wide, and they extended in length from 20 to 300 nm. The

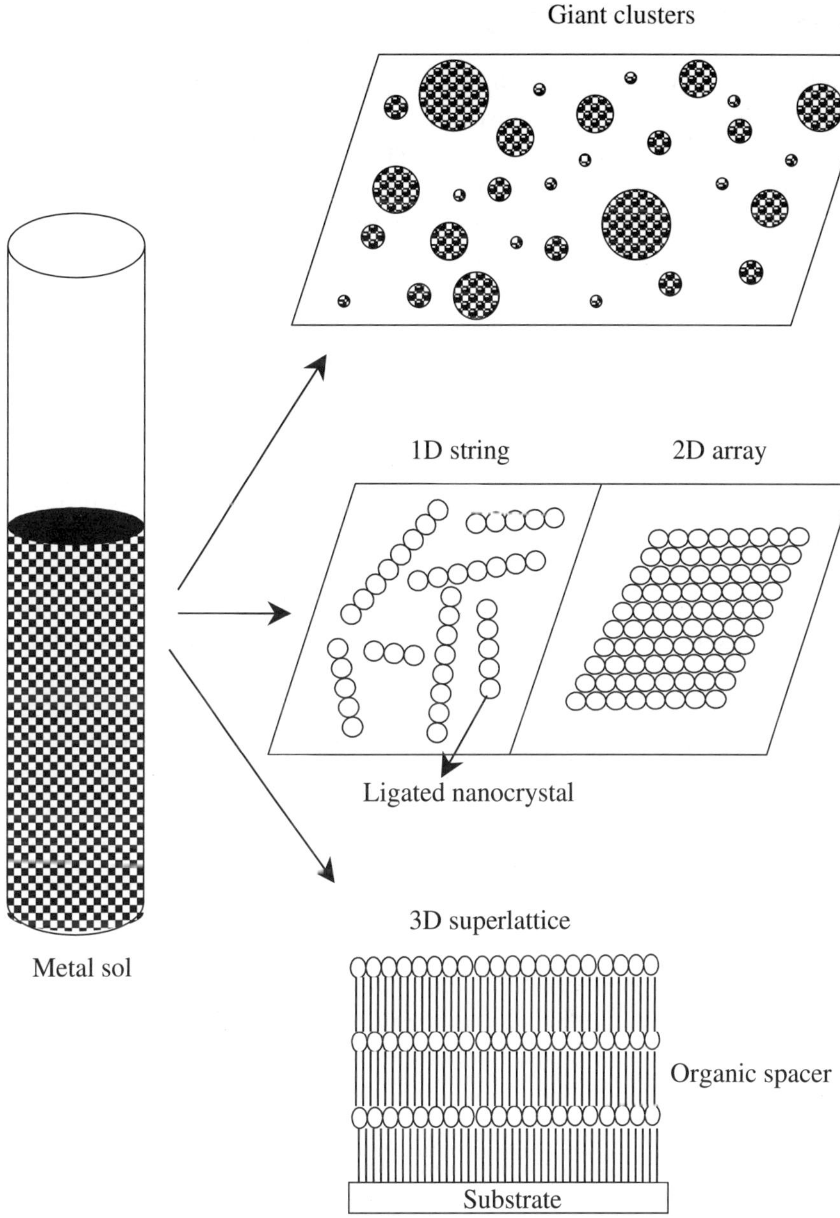

Figure 1.13. Schematic illustration of the various metal nanocrystal organizations.

interwire separation distance as well as the alignment of the wires could be controlled by compressing the film. Based on preliminary experimental observations, it has been suggested tobacco mosaic virus tubules could serve as templates for growth of one-dimensional lattice of quantum dots [90].

5.2. Two-Dimensional Arrays

Ligands based on long-chain thiols or phosphines have served as good candidates for assembling monodisperse nanocrystals on a flat substrate. Two-dimensional organizations of a variety of nanocrystals can be brought about by simply evaporating a drop of the sol on a flat substrate.

Gold organosols using alkane thiols as surfactants were first prepared by Schiffrin and co-workers [91] by phase transferring gold ions and carrying out reduction in the presence of thiols. Several workers have adopted this procedure to obtain thiolized metal nanocrystals [92–95]. Whetten et al. [93] centrifuged the organosol and separated out fractions containing nanocrystals of different mean sizes to prepare well-ordered two-dimensional arrays of size-selected Au nanocrystals. Harfenist et al. [24] found that that Ag nanocrystals prepared by using a cluster beam were stable in air and formed extended two-dimensional arrays. Fitzmaurice and co-workers [94] have obtained two-dimensional arrays of dodecanethiol-covered Ag nanocrystals. The Ag nanocrystals were prepared following the method of Schiffrin and co-workers [91]. Well-ordered arrays of magic nuclearity nanocrystals, Pd_{561} and Pd_{1415}, have been successfully obtained (see Figure 1.14) after replacing their polymer coating by alkanethiols, following the phase transfer method discussed previously [96]. Long-chain fatty acids have also been used for ligating and assembling metal nanocrystals. Colloidal dispersion of Co nanocrystals capped with fatty acids were found to self-assemble to yield hexagonally ordered arrays similar to those obtained with alkanethiols [97, 98]. Similarly, Ag nanocrystals capped with fatty acids of appropriate lengths yield cubic or hexagonal close-packed structures [99, 100]. Schmid et al. [101] have reported an ordered two-dimensional array of small Au_{55} nanocrystals (diameter ~1.4 nm) on a polymer film. At the other end of the size regime, big Au nanocrystals of 15- to 90-nm dimensions have also been organized into two-dimensional arrays [102]. Arrays of Au–Ag [27, 28] and Fe–Pt alloy nanocrystals [26] have been obtained. Magic nuclearity Pd_{561} nanocrystals have been exploited to make Pd–Ni core-shell particles with variable Ni loadings [103]. The nanocrystals so obtained possess a core-shell structure, where an Ni layer covers a Pd seed. The magic nuclearity Pd_{561} nanocrystals act as high-quality seeds and promote the formation of monodisperse Pd–Ni core-nanocrystals. Arrays of $Pd_{561}Ni_n$ (n up to 10,000 atoms) have been prepared after thiolizing the core-shell nanocrystals [104]. By a simple extension of this technique, arrays of triple-layer nanocrystals of the form $Pd_{561}Ni_{3000}Pd_{1500}$ were also obtained. Methods to organize nonspherical metal nanocrystals into two-dimensional arrays have met with very limited success. Thus, hexagonal Pt as well as elongated silver nanocrystals have been organized into ordered two-dimensional arrays [44, 105]. Interestingly, ordered two-dimensional lattices containing thiolized spherical Au particles of two

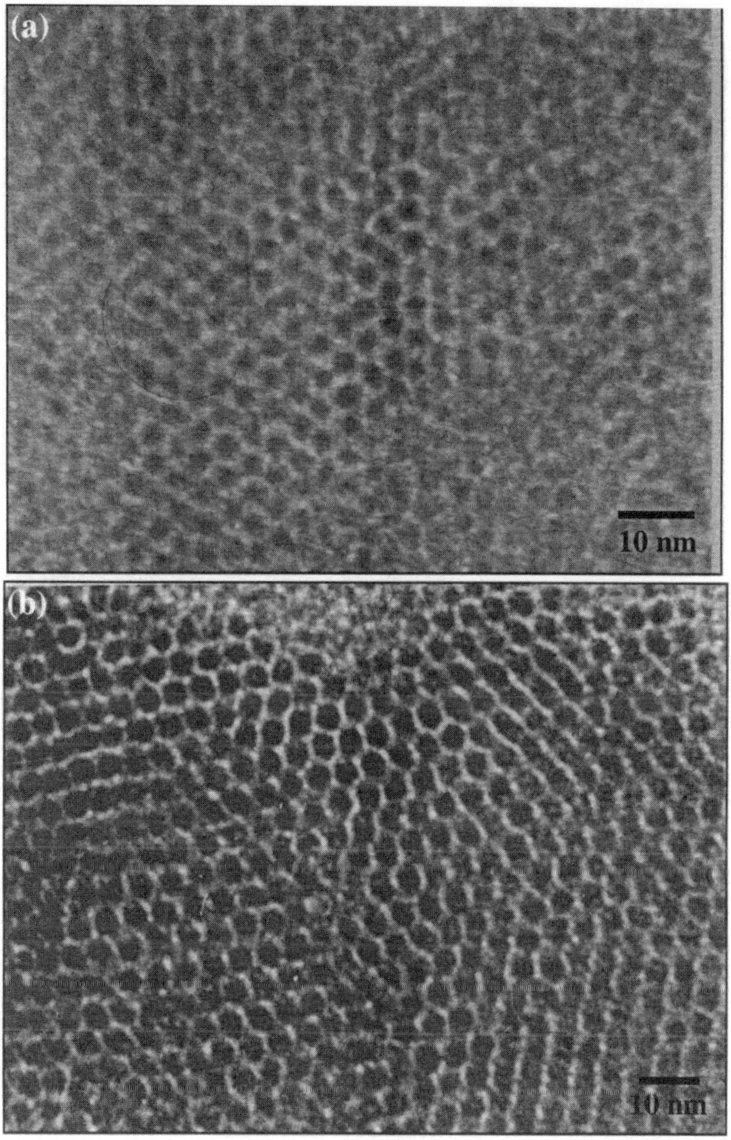

Figure 1.14. TEM micrographs showing hexagonal arrays of thiolized Pd nanocrystals. (**a**) Pd_{561} octanethiol. (**b**) Pd_{1415}, octanethiol. Organized arrays of these nanocrystals extend to lengths over several microns.

different sizes have been reported by Kiely et al. [95], who found that the nanocrystals of different radii follow the radius ratio rules formulated for alloying of different metals. Alloy arrays consisting of Au and Ag nanocrystals of different sizes have been made [106].

5.2.1. Stability and Phase Behavior of Two-Dimensional Arrays. The nanocrystal organizations mentioned above are mainly entropy-driven. The two lengths involved—the nanocrystal diameter (*d*) and the ligand chain length (*l*)—play an important role in deciding the nature of the organization—that is, its orderliness. It has been observed experimentally that for a given diameter of the nanocrystal, the packing changes swiftly as the length of the thiol ligand is increased. The stability diagram in terms of *d* and *l* shown in Figure 1.15 illustrates that extended close-packed organizations of nanocrystals are found for *d*/*l* values ~2. Although entropy-driven, the above cannot be treated as hard-sphere organizations. Based on a study of the effect of the solvent polarity on the self-assembly of ligated metal nanocrystals, Korgel et al. [84, 94] proposed a soft-sphere model taking the interparticle interaction into consideration. Accordingly, a ligated nanocrystal allows for penetration of the ligand shell up to its hard-sphere limit. In this model, the total potential energy, *E*, is considered to be a result of two types of forces between the nanocrystals:

$$E = E_{\text{steric}} + E_{vdW} \tag{2}$$

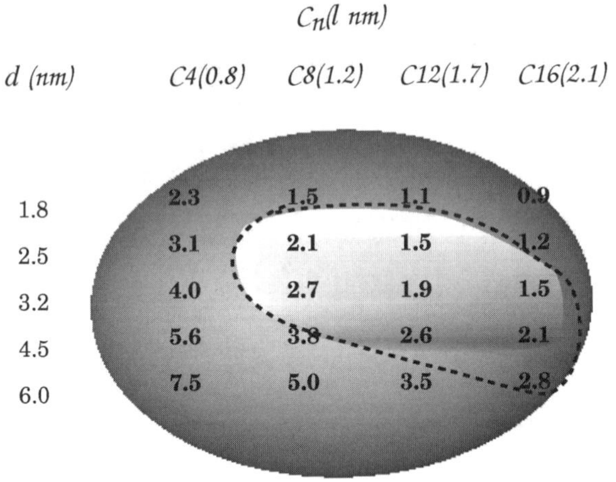

Figure 1.15. The *d*–*l* phase diagram for Pd nanocrystals thiolized with different alkanethiols. The mean diameter, *d*, was obtained from the TEM measurements on as-prepared sols. The length of the thiol, *l*, is estimated by assuming an all-trans conformation of the alkane chain. The thiol is indicated by the number of carbon atoms, C_n. The bright area in the middle encompasses systems that form close-packed organizations of nanocrystals. The surrounding darker area includes disordered or low-order arrangements of nanocrystals. The area enclosed by the dashed line is derived from calculations from the soft-sphere model.

$$E_{vdW} = \frac{A}{12}\left\{\frac{d^2}{\tau^2 - d^2} + \frac{d^2}{\tau^2} + 2\ln\left[\frac{\tau^2 - d^2}{t^2}\right]\right\} \tag{3}$$

$$E_{\text{steric}} = \frac{50dl^2}{(\tau - d)\pi\sigma_a^3} kTe^{-\pi(\tau-d)} \tag{4}$$

The van der Waals interaction due to the polarization of the metal cores constitutes the attractive term, and the steric interaction between the thiol molecules on the two surfaces forms the repulsive term, where τ is the interparticle distance. The Hamaker constant, A, for Pd nanocrystals, in toluene for instance, has been estimated to be 1.95 eV [107]. The calculated diameter of the area occupied by the thiol molecule (sa) on the particle surface is 4.3 Å [94]. The total energy is attractive over a range of interparticle distances, and the magnitude increases with fall in distance. There could be a range of interparticle distances where the attractive energy from the van der Waals term exceeds the repulsive energy due to the steric factor, giving rise to net stabilization of the two-particle system. Stabilization energies of 17 and 2 meV are obtained from the calculation for particles coated with octanethiol and dodecanethiol, respectively. When the stabilization energies have moderate values, comparable to the thermal energy of the nanocrystals, ordered organizations can be expected. If the d/l and hence the stabilization energy is not favorable, collapsed monolayers of nanocrystals or loosely packed structures are seen. Clearly, the interdigitation of thiol molecules plays a major role in attributing hardness to the ligated nanocrystal, which in turn decides the nature of the two-dimensional organization. A similar treatment should hold good for other nanocrystals.

5.3. Three-Dimensional Superlattices

Multilayer assemblies using monothiols (see Figure 1.13) are generally fragile and are not suited for use in functional devices. One of the means of obtaining robust structures involves multilayer deposition of nanocrystals and has been drawing a great deal of attention over the last few years, since they provide a convenient, low-cost means to prepare ultra-thin films of controlled thicknesses, suited for device applications. In a typical experiment, one end of a monolayer-forming bifunctional spacer is tethered to a flat substrates such as gold, aluminum, indium tin oxide, or glass, leaving the other end free to anchor nanocrystals [6, 7]. Subsequent layers can be introduced by dipping the substrate sequentially into the respective spacer molecule solution and the nanocrystal dispersion, with intermediate steps involving washing and drying. The formation of the multilayer assembly can be monitored using a variety of spectroscopy and microscopy tools. Thus, by employing Au substrates and dithiols as spacers, Sarathy et al. [108] have formed multilayer assemblies of various nanocrystals. Brust et al. [109] have reported the formation of multilayers of Au nanoparticles using dithiols. These workers have confirmed the layer-by-layer deposition of particle arrays by employing UV-vis spectroscopy and ellipsometry. Three-dimensional superlattices involving nanocrystals of different metals (e.g., Pt and Au) and of metals and semiconductors (e.g., Au and CdS) have also been prepared and characterized [108]. Such assemblies can be made with polyelectrolytes such as poly(diallyldimethylammonium chloride) (PDDA), polyethyleneimine (PEI) [110, 111], and poly(allylamine hydrochloride) (PAH), as well as with polymers such as poly-phenylenevinylene (PPV) [112, 113].

5.4. Giant Nanocrystals

It has been proposed that self-similarity in metal nanocrystal organization would manifest in the form of a giant cluster whose shape and size are direct consequences of the nanocrystals themselves [114]. The invariance of the shell effects in metal nanocrystals with scaling is shown schematically in Figure 1.16.

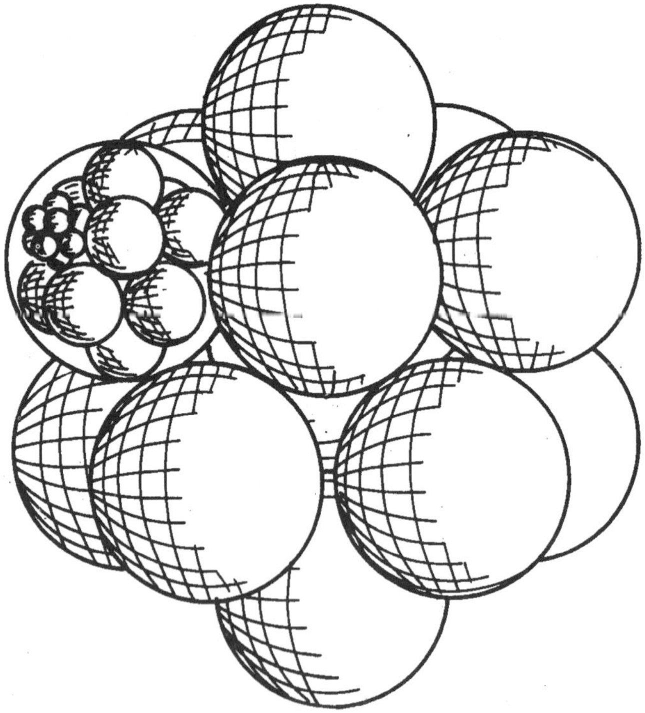

Figure 1.16. Self-similarity. Schematic illustration of the formation of a cluster of metal nanocrystals (super cluster) and a cluster of superclusters. The size effects operating in nanocrystals could be invariant to scaling. (Reproduced with permission from reference 114.)

Thus, Pd_{561} nanocrystals would be expected to self-aggregate into a giant cluster of the type $(Pd_{561})_{561}$ under suitable conditions. The monodisperse nature of the nanocrystals is thought to be important in assisting the self-aggregation process. Formation of such clusters was observed in the mass spectra of magic nuclearity Au_{55} nanocrystals. Secondary ion mass spectrometry indicated the presence of species with large m/z values, and these were attributed to $(Au_{13})_{55}$ giant clusters [115]. However, the giant clusters so obtained have not been isolated or imaged. One such observation was made in the case of Pd_{561} nanocrystals where the PVP-covered nanocrystals aggregated to form giant clusters [116]. The TEM image in Figure 1.17 is revealing. There are regions where the nanocrystals are densely packed in the form of giant aggregates with estimated nanocrystal nuclearities corresponding to various magic numbers. It is possible that the

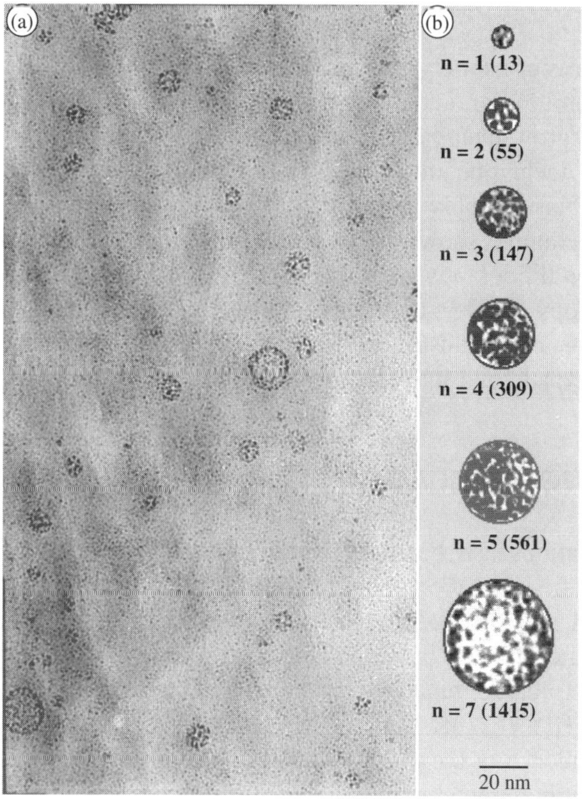

Figure 1.17. **(a)** TEM micrograph showing the giant clusters comprising Pd_{561} nanocrystals. Sample for TEM was prepared by the slow evaporation of a PVP–Pd_{561} hydrosol. **(b)** Giant clusters enclosed in circles whose diameters correspond to magic numbers. The n and the values in the parentheses indicate the number of nanocrystals and closed shells, respectively.

formation of the giant clusters is facilitated by the polymer shell that encases them. Unlike in the case of Pd nanocrystals coated with alkanethiols, which self-assemble to form ordered arrays, the polymer shell effectively magnifies the facets of the metallic core, thereby aiding a giant assembly of the nanocrystals.

The tendency of monodisperse nanocrystals to arrange into ordered three-dimensional arrays extending to a few microns has been noticed [117]. Careful tuning of crystallization conditions have yielded crystallites of micrometer dimensions consisting of Au_{55} nanocrystals and Fe–Pt alloy nanocrystals (4.5 nm) [118, 119]. However, it was observed that the nanocrystal arrangement in all the above crystallites was polymorphous. It is believed that such crystallites, consisting of ordered nanocrystals, could prove to be the best candidates to study the collective properties of an ensemble of nanocrystals.

5.5. Nanocrystal Patterning

Creating patterns of nanocrystals on surfaces has attracted wide attention. Such patterned substrates can act as templates to grow nanowires and can serve as etch masks to grow nanopillars and quantum dots [120–122]. Other than the layer-by-layer technique mentioned before, simple techniques such as spin coating have been employed to create a nanocrystalline pattern on surfaces [123]. In the example shown in Figure 1.18, a direct write lithographic technique—dip pen lithography [124, 125]—relies on a cantilever used for atomic force microscopy (AFM) to write on a substrate to create patterns of Au nanocrystals on mica substrates. Thus, nanocrystals of metals and semiconductors can be patterned into rectangles and lines of varying aspect ratios.

6. EMERGING APPLICATIONS

Several applications have been envisaged for nanocrystals, ranging from simple dyes to magnetic-resonance-imaging contrast agents [126], as components of single-electron and nanoscopic electronic circuitry [10, 127], magnetic media [97], as ingredients in catalyst and sensors, and so on. All the above applications seek to exploit the tunability provided by size-dependent properties of the nanocrystals [3].

The dependence of the plasmon band on the dielectric constant of the surrounding medium in metal nanocrystals had been used to detect binding events taking place at the ligand shell. Thus, Au nanocrystals could colorimetrically determine the successful hybridization of oligonucleotide strands bound to its surface [83, 86]. It has been proposed that colorimetric sensing of heavy metal ions could be obtained by the use of carboxylic acid terminated bifunctional

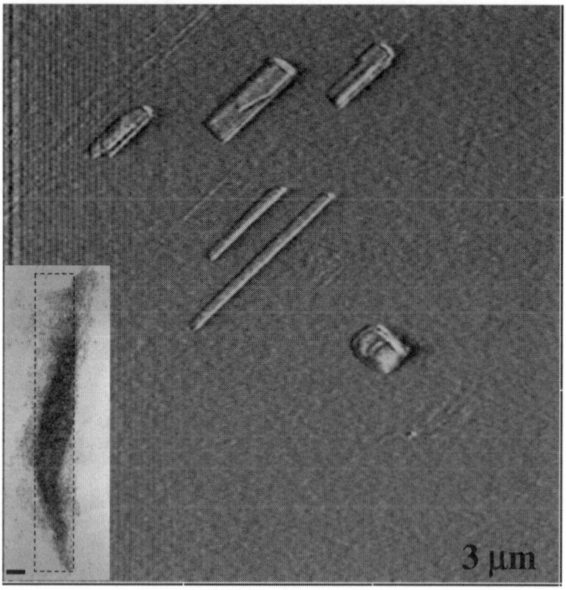

Figure 1.18. Contact AFM scan of a 9-μm² area on mica substrate showing rectangles of various aspect ratios filled with Au nanocrystals. The patterns were obtained by translating a AFM cantilever dipped in a sol across the surface. The inset shows a TEM image of a similar pattern on a holey carbon copper grid, and the dotted line bounds the area sought to be filed. The scale bar in the inset corresponds to 50 nm.

thiols bound to metal nanocrystals [128, 129]. The changes in the electronic absorption spectra of ~5-nm Ag nanocrystals capped with lipoic acid, following the addition of the heavy ions Cu^{2+} and Fe^{2+}, are shown in Figure 1.19. Such a dampening also brings about a change in color. It is apparent that Cu^{2+} ions dampen the plasmon band more effectively than Fe^{2+}. It is hoped that mesoscalar organizations could provide useful substrates consisting of ordered nanocrystals that are required to carry out the above experiments in solid state.

The fact that the physical properties of nanocrystal organizations could be different from that of the isolated particles is being realized. Pellets of monodisperse nanocrystals, obtained by the use of bifunctional ligand that binds to more than one nanocrystal or by applying pressure on dried nanocrystalline matter, have been used for electrical transport measurements [130–133]. Pellets made of small Au and Pd nanocrystals exhibit nonmetallic behavior with specific conductivities in the range of $10^6\,\Omega^{-1}\,cm^{-1}$ [130–132]. The conductivity, however, increases dramatically with an increase in the diameter of the nanocrystals. An insulator metal transition has indeed been reported from pellets made of ~12.5-

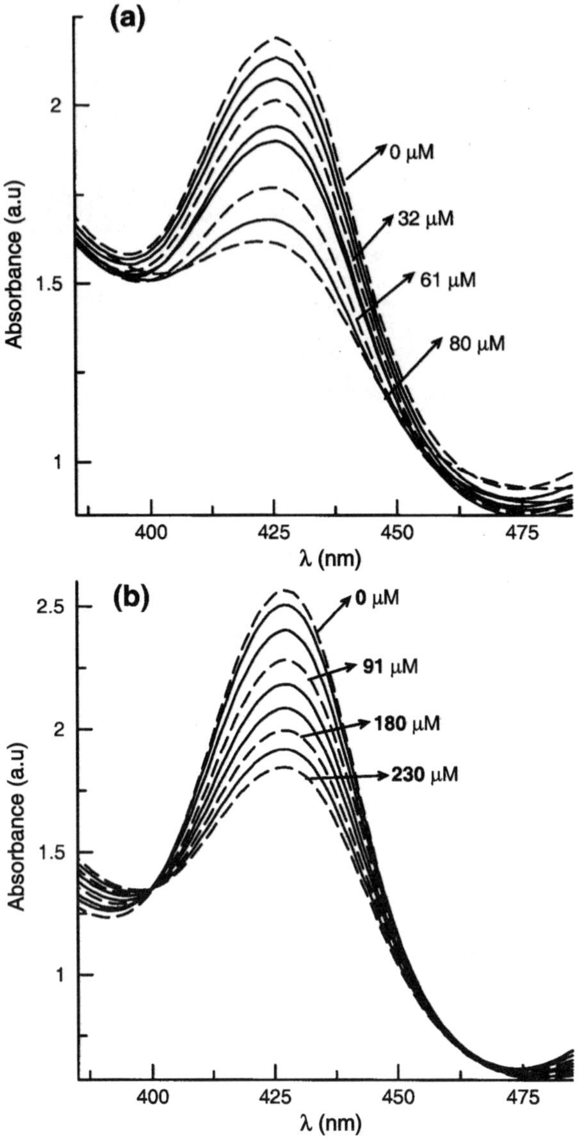

Figure 1.19. Electronic absorption spectra of ~5-nm Ag nanoparticles showing changes accompanying the addition of (a) Cu^{2+} and (b) Fe^{2+} ions. The concentrations of the ions are indicated. (Reproduced with permission from reference 129.)

nm Au and Ag nanocrystals [133]. Electrical transport measurements on layer-by-layer assemblies of nanocrystals on conducting substrates have been carried out by adoption a sandwich configuration [134–136]. Nanocrystalline films with bulk metallic conductivity have been realized with Au nanocrystals of 5-

and 11-nm diameter spaced with ionic and covalent spacers [135, 136]. The conductivity of a monolayered two-dimensional array of metal nanocrystals has been studied with patterned electrodes [137–142]. Structural disorder and inter-particle separation distance are identified as key factors that determine the conductivity of such layers [137–140]. The conductivity of such layers can be enhanced by replacing alkane thiol with an aromatic thiol *in situ* [141, 142]. That the interaction energy of nanocrystals in such organizations can be continually varied by changing the interparticle distance was exploited by Heath and co-workers [143, 144], who prepared a monolayer of Ag (~3 nm) nanocrystals at air–water interface in an LB trough and varied the interparticle distance by applying pressure. A host of measurements including reflectivity and non-linear optical spectroscopic techniques were carried out *in situ*. This study led to the observation of a reversible Mott–Hubbard metal–insulator transition in the nanocrystal ensemble wherein the Coulomb gap closes at a critical distance between the particles. Tunnelling spectroscopic measurements on films of 2.6-nm Ag nanocrystals capped with decanethiol reveal a Coulomb blockade behavior attributable to isolated nanocrystals [144]. On the other hand, nanocrystals capped with hexane and pentane thiol exhibit characteristics of strong interparticle quantum mechanical exchange. Similar behavior was observed in the case of self-assembled two-dimensional arrays of Co nanocrystals and Au nanocrystals [145, 146].

Alternatives to current magnetic storage media are much sought after. It is believed that arrays of interacting superparamagnetic nanocrystals, whose magnetic properties are governed by the interparticle separation, can be effective alternatives to current hard disks. Nanocrystals of Co, when organized into two-dimensional arrays, exhibit a higher superparamagnetic blocking temperature compared to isolated nanocrystals; that is, they display a higher resistance to thermal reversal of their spins than when they are isolated [147]. Sun et al. report a lattice of nanocrystals, each consisting of an Fe core and a Pt shell prepared by heating Fe–Pt alloy nanocrystals [26]. Following phase segregation, the interaction between the nanocrystals increased, leading to ferromagnetic films capable of supporting high-density magnetization reversal transitions. Exchange spring magnets—nanocomposites that consist of magnetically hard and soft phases interacting via magnetic exchange coupling—have been made by carefully annealing the mixed nanocrystal array consisting of Fe–Pt and Fe_3O_4 [148].

6.1. Nanocomputing

Ordered arrays of nanocrystals, in principle, could be thought of as arrays of SET (single-electron transistors), where the electrostatic interaction between neighboring SET acts as wireless communication means. It has been suggested by Korotkov [149] and Lent and co-workers [150] that simple logical operations can be performed on a circuitry consisting of arrays of SET in the form of chains or cells with suitable insulating spacers. An electric field applied in one direction polarizes the strings into either the 0 or the 1 state. Lent's scheme, named quantum cellular automata, instead uses a square cell consisting of five nanocrystals to denote the state of polarization. Preliminary experiments to evaluate the schemes are currently being pursued.

The realization that a self-assembly-driven fabrication process is not capable of producing defect free structures has fueled a search for algorithms that can compute even with defective circuitry. Heath et al. [151] have developed Teramac, a computer that works despite a high concentration of defects in its bank of microprocessors. A more radical solution called *amorphous computing* aims to "engineer pre-specified, coherent behavior from cooperation of large numbers of unreliable parts interconnected in unknown, irregular and time-varying ways" [152–154].

7. CONCLUSIONS

Nanocrystals of metal with diameters in the range 1–50 nm form a class of materials with unusual properties that are size-dependent. Excellent electrical conductivity that primarily characterizes a metallic state becomes a rare entity in small nanocrystals (<2 nm) due to quantum confinement of the electronic states. Similarly, magnetic metals lose much of the coercivity with diminishing size. On the other hand, chemical properties such as reactivity may show up better at smaller sizes due to more number of surface bonding sites and other electronic effects. Considering the importance of nanocrystals in technological applications, a large number of synthesis methods have evolved in recent years which include reverse micelle and sonochemical methods besides laser abalation. Control over size and shape as well as encasing the nanocrystals with ligands of specialized functionalities have become subjects of urgent enquiry. While isolated nanocrystals are interesting by themselves, their organizations, especially those that are capable of self-assembling into well-ordered arrays, have attracted greater attention. Nanocrystals anchored to fragments of DNA or likewise molecules essentially form one-dimensional organizations. When coated with long-chain alkane thiols, nanocrystals exhibit a tendency to assemble into hexagonal arrays on flat substrates. The stability of such a two-dimensional organization depends on the diameter of the nanocrystals and the length of the ligand. Multilayers of nanocrystal arrays can also be made in a programmed way by selecting suitable spacer molecules. However, patterns of nanocrystals can be obtained using scanning probe techniques. Another known mesoscalar aggregation is the giant clusters of nanocrystals with definite nuclearities. It would be ideal to grow crystals of nanocrystals, but such efforts have met with only a limited success to date, giving micron-sized crystals. Nanocrystal organizations may exhibit properties very different from those of the individual. They are amenable to unprecedented control over the lattice, with the size of the nanocrystal and interparticle separation being continuously variable over a range. Exploratory experiments for measuring such collective properties are currently underway in several laboratories around the world.

REFERENCES

1. M. Faraday, *Philos. Trans. R. Soc. London* **147**, 145 (1857).
2. C. R. Berry, *Phys. Rev.* **161**, 848 (1967).
3. C. N. R. Rao, G. U. Kulkarni, P. J. Thomas, and P. P. Edwards, *Chem. Eur. J.* **29**, 27 (2002).
4. P. P. Edwards, R. L. Johnston, and C. N. R. Rao, in *Metal Clusters in Chemistry*, edited by P. Braunstein, G. Oro, and P. R. Raithby, Wiley-VCH, Weinham (1999).
5. R. F. Service, *Science* **274**, 1834 (1996).
6. C. N. R. Rao, G. U. Kulkarni, P. J. Thomas, and P. P. Edwards, *Chem. Soc. Rev.* **29**, 27 (2000).
7. A. N. Shipway, E. Katz, and I. Willner, *CHEMPHYSCHEM* **1**, 18 (2000).
8. M. P. Pileni, *J. Phys. Chem.* **B105**, 3358 (2001).
9. C. B. Murray, C. R. Kagan, and M. G. Bawendi, *Annu. Rev. Mater. Sci.* **30**, 545 (2000).
10. U. Simon, *Adv. Mater.* **10**, 1487 (1998).
11. G. Schmid and L. F. Chi, *Adv. Mater.* **10**, 515 (1998).
12. *Clusters and Colloids: From Theory to Applications*, edited by G. Schmid, VCH, Weinham (1994).
13. J. Turkevich, P. C. Stevenson, and J. Hillier, *J. Discuss. Faraday. Soc.* **11**, 55 (1951).
14. A. I. Kirkland, D. E. Jefferson, D. G. Duff, P. P. Edwards, I. Gameson, B. F. U. Johnson, and D. J. Smith, *Proc. R. Soc. London. A* **440**, 589 (1993).
15. *Physics and Chemistry of Metal Cluster Compounds*, edited by L. J. de Jongh, Kluwer, Dordrecht (1994).
16. T. S. Ahmadi, L. Wang, A. Henglein, and M. A. El-Sayed, *Chem. Mater.* **8**, 428 (1996).
17. M. P. Pileni, *J. Phys. Chem.* **97**, 6961 (1993).
18. I. Moriguchi, F. Shibata, Y. Teraoka, and S. Kagawa, *Chem. Lett.* 761 (1995).
19. M. Platt, R. A. W. Dryfe, and E. P. L. Roberts, *Chem. Commun.* 2324 (2002).
20. C. N. R. Rao, G. U. Kulkarni, P. J. Thomas, V. V. Agrawal, and P. Saravanan, *J. Phys. Chem.* **B107**, 7391 (2003).

21. C. N. R. Rao, G. U. Kulkarni, P. J. Thomas, V. V. Agrawal, U. K. Gautam, and M. Ghosh, *Curr. Sci.* **85**, 1041 (2003).
22. K. Sattler, J. Mhlback, and E. Recknagel, *Phys. Rev. Lett.* **45**, 821 (1980).
23. P. Milani and S. Iannotta, *Cluser Beam Synthesis of Nanostructured Materials*, Springer, Berlin (1999).
24. S. A. Harfenist, Z. L. Wang, R. L. Whetten, I. Vezmar, and M. M. Alvarez, *Adv. Mater.* **9**, 817 (1997).
25. H. N. Vasan and C. N. R. Rao, *J. Mater. Chem.* **5**, 1755 (1995).
26. S. Sun, C. B. Murray, D. Weller, L. Folks, and A. Maser, *Science* **287**, 1989 (2000).
27. N. Sandhyarani, M. R. Reshmi, R. Unnikrishnan, K. Vidyasagar, S. Ma, M. P. Antony, G. P. Selvam, V. Visalakshi, N. Chandrakumar, K. Pandian, Y. T. Tao, and T. Pradeep, *Chem. Mater.* **12**, 104 (2000).
28. S. T. He, S. S. Xie, J. N. Yao, H. J. Gao, and S. J. Pang, *Appl. Phys. Lett.* **81**, 150 (2002).
29. Y.-H. Chen and C.-S. Yeh, *Chem. Commun.* 371 (2001).
30. J.-P. Abid, H. H. Girault, and P. F. Brevet, *Chem. Commun.* 829 (2001).
31. G. Schmid, *Inorg. Synth.* **7**, 214 (1990).
32. M. N. Vargaftik, V. P. Zagorodnikov, I. P. Stolyarov, I. I. Moiseev, V. A. Likholobov, D. I. Kochubey, A. L. Chuvili, V. I. Zaikovsky, K. I. Zamaraev, and G. I. Timofeeva, *Chem. Commun.* 937 (1985).
33. H.-G. Boyen, G. Kastle, F. Weigl, B. Koslowski, C. Dietrich, P. Ziemann, J. P. Spatz, S. Riethmuller, C. Hartmann, M. Moller, G. Schmid, M. G. Garnier, and P. Oelhafen, *Science* **297**, 1533 (2002).
34. T. P. Martin, T. Bergmann, H. Ghlich, and T. Lange, *J. Phys. Chem.* **95**, 6421 (1991).
35. T. Teranishi and M. Miyake, *Chem. Mater.* **10**, 54 (1998), T. Teranishi, H. Hori, and M. Miyake, *J. Phys. Chem.* **B101**, 5774 (1997).
36. S. Link, M. B. Mohamed, and M. A. El-Sayed, *J. Phys. Chem.* **B103**, 3073 (1999).
37. L. Manna, E. C. Scher, and A. P. Alivisatos, *J. Am. Chem. Soc.* **122**, 12700 (2000).
38. P. Zhang and T. K. Sham, *Appl. Phys. Lett.* **81**, 736 (2002).
39. C. M. Niemeyer, *Angew. Chem. Int. Ed.* **40**, 4128 (2001).
40. V. Chechik and R. M. Crooks, *J. Am. Chem. Soc.* **122**, 1243 (2000).
41. P. Mulvaney, L. M. Liz-Marzan, M. Giersig, and T. Long, *J. Mater. Chem.* **10**, 1259 (2002).
42. J. Lin, W. Zhow, A. Kumbhar, J. Wiemann, J. Fang, E. E. Carpenter, and C. J. O'Connor, *J. Solid. State Chem.* **159**, 26 (2001).
43. H. Harai, H. Aizawa, and H. Shiozaki, *Chem. Lett.* **8**, 1527 (1992).
44. K. V. Sarathy, G. Raina, R. T. Yadav, G. U. Kulkarni, and C. N. R. Rao, *J. Phys. Chem.* **B101**, 9876 (1997).
45. K. V. Sarathy, G. U. Kulkarni, and C. N. R. Rao, *Chem. Commun.* 537 (1997).

46. D. I. Gittins and F. Caruso, *Angew. Chem. Int. Ed.* **40**, 3001 (2001).
47. L. O. Brown and J. E. Hutchison, *J. Am. Chem. Soc.* **121**, 882 (1999).
48. (a) D. C. Johnson, R. E. Benfield, P. P. Edwards, W. J. H. Nelson, and M. D. Vargas, *Nature* **314**, 231 (1985); (b) Y. Volokitin, J. Sinzig, L. J. de Jongh, G. Schmid, M. N. Vargaftik, and I. I. Moiseev, *Nature* **384**, 624 (1996).
49. V. Vijayakrishnan, A. Chainani, D. D. Sarma, and C. N. R. Rao, *J. Phys. Chem.* **96**, 8679 (1992).
50. H. N. Aiyer, V. Vijayakrishnan, G. N. Subanna, and C. N. R. Rao, *Surf. Sci.* **313**, 392 (1994).
51. R. Busani, M. Folker, and O. Chesnovsky, *Phys. Rev. Lett.* **81**, 3836 (1998).
52. K. Rademann, O. D. Rademann, M. Schlauf, V. Even, and F. Hensel, *Phys. Rev. Lett.* **69**, 3208 (1992).
53. S. Link and M. A. El-Sayed, *J. Phys. Chem.* **B105**, 1 (2001).
54. S. Link and M. A. El-Sayed, *Int. Rev. Phys. Chem.* **19**, 409 (2001).
55. P. Mulvaney, *Langmuir* **12**, 788 (1996).
56. P. V. Kamat, *J. Phys. Chem.* **B106**, 7729 (2002).
57. G. Mie, *Ann. Phys.* **25**, 377 (1908).
58. G. C. Papavassilliou, *Prog. Solid State Chem.* **12**, 185 (1980).
59. R. Gans, *Ann. Phys.* **31**, 881 (1911).
60. R. Gans, *Ann. Phys.* **47**, 270 (1915).
61. A. C. Templeton, J. J. Pietron, R. W. Murray, and P. Mulvaney, *J. Phys. Chem.* **B104**, 564 (2000).
62. C. P. Vinod, G. U. Kulkarni, and C. N. R. Rao, *Chem. Phys. Lett.* **289**, 329 (1998).
63. M. Rosenblit and J. Jortner, *J. Phys. Chem.* **98**, 9365 (1994).
64. R. A. Perez, A. F. Ramos, and G. L. Malli, *Phys. Rev.* **B39**, 3005 (1989).
65. O. D. Haberlen, S. C. Chung, M. Stener, and N. J. Rosch, *J. Chem. Phys.* **106**, 5189 (1997).
66. S. H. Yang, D. A. Drabold, J. B. Adams, and A. Sachdev, *Phys. Rev.* **B47**, 1567 (1993).
67. C. P. Collier, T. Vossmeyer, and J. R. Heath, *Annu. Rev. Phys. Chem.* **49**, 371 (1998).
68. *Single Charge Tunneling, Coulomb Blockade Phenomena in Nanostructures,* edited by H. Grabert and M. H. Devoret, NATO-ASI Ser. B (1992), p. 294.
69. P. J. Thomas, G. U. Kulkarni, and C. N. R. Rao, *Chem. Phys. Lett.* **321**, 163 (2000).
70. J. Jortner, *Z. Phys. D: Atoms, Molecules and Clusters* **24**, 247 (1992).
71. C. P. Bean and J. D. Livingston, *J. Appl. Phys.* **30**, 1208 (1959).
72. Van de Heer, P. Milani, and A. Chatelain, *Z. Phys. D: Atoms, Molecules and Clusters* **19**, 241 (1991).
73. C. N. R. Rao, V. Vijayakrishnan, A. K. Santra, and M. W. J. Prins, *Angew. Chem. Int. Ed.* **31**, 1062 (1992).

74. A. K. Santra, S. Ghosh, and C. N. R. Rao, *Langmuir* **10**, 3937 (1994).
75. E. Gillet, S. Channakhone, V. Matolin, and M. Gillet, *Surf. Sci.* **152/153**, 603 (1986).
76. (a) D. L. Doering, J. T. Dickinson, and H. Poppa, *J. Catal.* **73**, 91 (1982); (b) D. L. Doering, H. Poppa, and J. T. Dickinson, *J. Catal.* **73**, 104 (1982).
77. U. Heiz, A. Sanchez, S. Abbet, and W.-D. Schneider, *J. Am. Chem. Soc.* **121**, 3214 (1999).
78. M. Valden, X. Lai, and D. W. Goodman, *Science* **281**, 1647 (1998).
79. C. Xu, X. Lai, G. W. Zajac, and D. W. Goodman, *Phys. Rev.* **56**, 13464 (1997).
80. C. P. Vinod, G. U. Kulkarni, and C. N. R. Rao, in *Surface Chemistry and Catalysis*, edited by A. Carley, P. Davies, G. Hutchings, and M. Spencer, Kluwer Academic/Plenum Publishers (2003).
81. A. Sanchez, S. Abbet, U. Heiz, W.-D. Schneider, H. Häkkinen, R. N. Narnett, and U. Landman, *J. Phys. Chem.* **103**, 9573 (1999).
82. A. Terfort, N. Bowden, and G. M. Whitesides, *Nature* **386**, 162 (1997).
83. C. A. Mirkin, R. L. Letsinger, R. C. Mucic, and J. F. Storhoff, *Nature* **382**, 607 (1996).
84. B. A. Korgel and D. Fitzmaurice, *Phys. Rev. Lett.* **80**, 3531 (1998).
85. G. L. Hornayak, M. Krll, R. Pugin, T. Sawitowski, G. Schmid, J. O. Bovin, G. Karrson, H. Hofmeister, and S. Hopfe, *Eur. J. Chem.* **3**, 195 (1997).
86. A. P. Alivisatos, K. P. Johnsson, X. Peng, T. E. Wilson, C. J. Loweth, M. P. Jr. Burchez, and P. G. Schultz, *Nature* **382**, 609 (1996).
87. A. Kumar, M. Pattarkine, M. Bhadbhade, A. B. Mandale, K. N. Ganesh, S. S. Datar, C. V. Dharmadhikari, and M. Sastry, *Adv. Mater.* **13**, 341 (2001).
88. S.-W. Lee, C. Mao, C. E. Flynn, and A. M. Belchar, *Science* **296**, 892 (2002).
89. S. W. Chung, G. Markovich, and J. R. Heath, *J. Phys. Chem.* **B102**, 6685 (1998).
90. E. Dujardin, C. Peet, G. Stubbs, J. N. Culver, and S. Mann, *Nano Lett.* **3**, 413 (2003).
91. M. Brust, M. Walker, D. Bethell, J. D. Schiffrin, and R. Whyman, *Chem. Commun.* 801 (1994).
92. N. Sandhyarani and T. Pradeep, *Chem. Mater.* **12**, 1755 (2000).
93. R. L. Whetten, J. T. Khoury, M. M. Alvarez, S. Murthy, I. Vezmar, Z. Wang, P. W. Stephens, C. L. Clevend, W. D. Luedtke, and U. Landman, *Adv. Mater.* **8**, 428 (1996).
94. B. A. Korgel, S. Fullam, S. Connolly, and D. Fitzmaurice, *J. Phys. Chem.* **B102**, 8379 (1998).
95. C. J. Kiely, J. Fink, M. Brust, D. Bethell, and D. J. Schiffrin, *Nature* **396**, 444 (1998).
96. P. J. Thomas, G. U. Kulkarni, and C. N. R. Rao, *J. Phys. Chem.* **B104**, 8138 (2000).
97. S. Sun and C. B. Murray, *J. Appl. Phys.* **85**, 4325 (1999).

98. C. Petit, A. Taleb, and M. P. Pileni, *J. Phys. Chem.* **B103**, 1805 (1999).
99. M. P. Pileni, *New J. Chem.* 693 (1998).
100. K. Abe, T. Hanada, Y. Yoshida, N. Tanigaki, H. Takiguchi, H. Nagasawa, M. Nakamoto, T. Yamaguchi, and K. Yase, *Thin Solid Films* **327–329**, 524 (1998).
101. G. Schmid, M. Bäumle, and N. Beyer, *Angew. Chem. Int. Ed.* **1**, 39 (2000).
102. B. Kim, S. L. Tripp, and A. Wei, *J. Am. Chem. Soc.* **123**, 7955 (2001).
103. T. Teranishi and M. Miyake, *Chem. Mater.* **11**, 3414 (1999).
104. P. J. Thomas, G. U. Kulkarni, and C. N. R. Rao, *J. Nanosci. Nanotechnol.* **1**, 267 (2001).
105. B. A. Korgel and D. Fitzmaurice, *Adv. Mater.* **10**, 661 (1998).
106. C. J. Kiely, J. Fink, J. G. Zheng, M. Brust, D. Bethell, and D. J. Schiffrin, *Adv. Mater.* **12**, 639 (2000).
107. D. Bargeman and F. V. V. Vader, *J. Electroanal. Chem.* **37**, 45 (1972).
108. K. V. Sarathy, P. J. Thomas, G. U. Kulkarni, and C. N. R. Rao, *J. Phys. Chem.* **B103**, 399 (1999).
109. M. Brust, D. Bethell, C. J. Kiely, and D. J. Schiffrin, *Langmuir* **14**, 5425 (1998).
110. R. Blonder, L. Sheeney, and I. Willner. *Chem. Commun.* 1393 (1998).
111. Y. Liu, Y. Wany, and R. O. Claus, *Chem. Phys. Lett.* **298**, 315 (1998).
112. A. Samokhvalov, M. Berfeld, M. Lahav, R. Naaman, and E. Rabani, *J. Phys. Chem.* **B104**, 8632 (2000).
113. M. Gao, B. Richter, and S. Kirstein, *Adv. Mater.* **9**, 802 (1997).
114. H. G. Fritsche, H. Muller, and B. Fehrensen, *Z. Phy. Chem.* **199**, 87 (1997).
115. H. Feld, A. Leute, D. Rading, A. Benninghoven, and G. Schmid, *J. Am. Chem. Soc.* **112**, 8166 (1990).
116. P. J. Thomas, G. U. Kulkarni, and C. N. R. Rao, *J. Phys. Chem.* **B105**, 2515 (2001).
117. M. Maillard, L. Motte, A. T. Ngo, and M. P. Pileni, *J. Phys. Chem.* **B104**, 11871 (2000).
118. G. Schmid, R. Pugin, T. Sawitowski, U. Simon, and B. Marler, *Chem. Commun.* 1303 (1999).
119. E. Shevchenko, D. Talapin, A. Kornowski, F. Wiekhorst, J. Kötzler, M. Haase, A. Rogach, and H. Weller, *Adv. Mater.* **14**, 287 (2002).
120. H. Ago, T. Komatsu, S. Ohshima, Y. Kuriki, and M. Yumura, *Appl. Phys. Lett.* **77**, 79 (2000).
121. Y. Cui, L. J. Lauhon, M. S. Gudisksen, J. Wang, and C. M. Lieber, *Appl. Phys. Lett.* **78**, 2214 (2001).
122. P. A. Lewis, H. Ahamed, and T. Sato, *J. Vac. Sci. Technol.* **B16**, 2938 (1998).
123. Y.-K. Hong, H. Kim, G. Lee, W. Kim, J. Park, J. Cheon, and J.-Y. Koo, *Appl. Phys. Lett.* **80**, 844 (2002).
124. R. D. Piner, J. Zhu, F. Xu, S. Hong, and C. A. Mirkin, *Science* **283**, 661 (1999).
125. C. A. Mirkin, S. Hong, and L. Demers, *Chem. Phys. Chem.* **2**, 37 (2001).

126. C. R. Martin and D. T. Mitchell, *Anal. Chem.* 322A (1998).
127. D. L. Feldheim and C. D. Keating, *Chem. Soc. Rev.* **27**, 1 (1998).
128. Y. Kim, R. C. Johnson, and J. T. Hupp, *Nano Lett.* **1**, 165 (2001).
129. S. Berchmans, P. J. Thomas, and C. N. R. Rao, *J. Phys. Chem.* **B106**, 4651 (2002).
130. M. Brust, D. Bethell, D. J. Schiffrin, and C. J. Kiely, *Adv. Mater.* **7**, 795 (1995).
131. V. Torma, G. Schmid, and U. Simon, *Chem. Phys. Chem.* **1**, 321 (2001).
132. U. Simon, R. Flesch, H. Wiggers, G. Schn, and G. Schmid, *J. Mater. Chem.* **8**, 517 (1998).
133. M. Aslam, I. S. Mulla, and K. Vijayamohanan, *Appl. Phys. Lett.* **79**, 689 (2001).
134. R. H. Terrill, T. A. Postlewaite, C. Chen, C. D. Poon, A. Terzis, A. Chen, J. E. Hutchinson, M. R. Clark, G. Wignall, J. D. Londono, R. Superfine, M. Falvo, C. S. Johnson, Jr., E. T. Samulski, and R. W. Murray, *J. Am. Chem. Soc.* **117**, 1237 (1995).
135. M. D. Musick, C. D. Keating, M. H. Keefe, and M. J. Natan, *Chem. Mater.* **9**, 1499 (1997).
136. Y. Liu, Y. Wang, and R. O. Clauss, *Chem. Phys. Lett.* **298**, 315 (1998).
137. R. Parthasarathy, X.-M. Lin, and H. A. Jaeger, *Phys. Rev. Lett.* **87**, 186807 (2001).
138. J. Schmelzer, Jr., S. A. Brown, A. Wurl, H. Hyslop, and R. J. Blaikie, *Phys. Rev. Lett.* **88**, 226802 (2002).
139. R. C. Doty, H. Yu, C. K. Shih, and B. A. Korgel, *J. Phys. Chem.* **B105**, 8291 (2001).
140. T. Ogawa, K. Kobayashi, G. Masuda, T. Takase, and S. Maeda, *Thin Solid Films* **393**, 374 (2001).
141. R. G. Osifchin, W. J. Mahoney, J. D. Bielefeld, R. P. Andres, J. I. Henderson, and C. P. Kubiak, *Superlattices and Microstructures* **18**, 283 (1995).
142. R. G. Osifchin, W. J. Mahoney, J. D. Bielefeld, R. P. Andres, J. I. Henderson, and C. P. Kubiak, *Superlattices and Microstructures* **18**, 275 (1995).
143. G. Markovich, C. P. Collier, S. E. Hendricks, F. Ramacle, R. D. Levine, and J. R. Heath, *Acc. Chem. Res.* **32**, 415 (1999).
144. G. Medeiros-Ribeiro, D. A. A. Ohlberg, R. S. Williams, and J. R. Heath, *Phys. Rev.* **B59**, 1633 (1999).
145. A. Taleb, F. Silly, A. O. Gusev, F. Charra, and M.-P. Pileni, *Adv. Mater.* **12**, 633 (2000).
146. T. P. Bigioni, L. E. Harrell, W. G. Cullen, D. K. Guthrie, R. L. Whetten, and P. N. Fist, *Eur. Phys. J.* **D6**, 355 (1999).
147. V. Russier, C. Petit, J. Legrand, and M. P. Pileni, *Phys. Rev.* **B62**, 3910 (2000).
148. H. Zheng, J. Li, J. P. Llu, Z. L. Whang, and S. Sun, *Nature* **420**, 395 (2002).
149. *Molecular Electronics*, IUPAC "Chemistry for the 21st Century" monograph, edited by J. Jortner and M. Ratner, Blackwell Scientific, London (1997).
150. A. O. Orlov, I. Amlani, G. H. Berstein, C. S. Lent, and G. L. Snider, *Science* **277**, 928 (1997).

151. J. R. Heath, P. J. Kuekes, G. S. Snider, and R. S. Williams, *Science* **280**, 1717 (1998).
152. H. Abelson, D. Allen, D. Coore, C. Hanson, G. Homsy, T. F. Knoght, Jr., R. Nagpal, E. Rauch, G. J. Sussman, and R. Weiss, Technical Report A. I. Memo 1665, Massachusetts Institute of Technology, Artificial Intelligence Laboratory, Aug. 1999.
153. D. Coore, R. Nagpal, and R. Weiss, Technical Report A. I. Memo 1614, Massachusetts Institute of Technology, Artificial Intelligence Laboratory, Oct. 1997.
154. H. Abelson, D. Allen, D. Coore, C. Hanson, G. Homsy, T. F. Knoght, Jr., R. Nagpal, E. Rauch, G. J. Sussman, and R. Weiss, *Commun. Assoc. Comp. Mach.* **43**, 5 (2000).

2

METAL-CONTAINING POLYMERS: CRYOCHEMICAL SYNTHESIS, STRUCTURE, AND PHYSICOCHEMICAL PROPERTIES

L. I. Trakhtenberg and G. N. Gerasimov

Karpov Institute of Physical Chemistry (Russian State Scientific Center), Moscow, Russia

1. INTRODUCTION

Metal-containing polymers may be produced by various methods, such as chemical reactions of precursors—in particular, reactions of metal salts in polymer solutions, the treatment of polymers with metal vapors, or the polymerization of various metal–monomer systems [1–4]. Depending on the metal nature and the polymer structure, these processes lead to organometallic units incorporated into polymer chains, metal–polymer complexes, or metal clusters and nanoparticles physically connected with polymer matrix. Of special interest are syntheses with the use of metal vapors. In this case, metal atoms or clusters are not protected by complexones or solvate envelopes and consequently have specific high reactivity. It should be noted that the apparatus and principles of metal vapor synthesis techniques are closely related to many industrial processes with participation of atomic and molecular species [5]—for example, manufacturing devices for microelectronic from different metals and metal containing precursors [6]. Vapor synthesis methods employ varying metals and

Metal-Polymer Nanocomposites, Edited by Luigi Nicolais and Gianfranco Carotenuto
ISBN 0-471-47131-3

organic ligands over a wide range. By such syntheses, one may prepare new organometallic compounds and complexes that are very difficult—and in, many cases, impossible—to synthesize by known methods [7].

Metal vapor techniques provide unique means for cryochemical solid-phase synthesis of metal-containing systems. In this way, metastable compounds, whose existence earlier was only supposed, have been obtained [7]. Besides, cryochemical processes produce stabilized small metal clusters of quantum type, which are the intermediate form of matter between isolated atoms and bulk metal [8, 9]. However, known methods of cryochemical solid-phase synthesis used low-molecular-weight matrices, in which the initial products of such a synthesis can be conserved only at low temperatures, when the matrix is enough rigid to hinder transformation or loss of these products.

In this connection, a cryochemical solid-phase synthesis of metal–polymer systems is of special importance. As a result of such a synthesis, metal clusters and organometallic assemblies formed at low temperatures are buried in a polymer environment, which offers possibilities to stabilize and study these products over a large temperature range. This method was first offered and described in reference 10. The thermal rearrangement of the initial low-temperature system is governed by relaxation processes in polymer matrix. In particular, the aggregation of metal atom clusters to form metal nanocrystals in cryochemically produced metal–polymer systems yields new nanocomposite materials with valuable properties. The study of the mechanism of cluster aggregation, which depends on the characteristics of the polymer matrix, will allow the nanocomposite structure to proceed in the needed direction. Thus, it becomes possible to determine the methods of cryochemical synthesis of metal–polymer materials with predetermined properties.

2. METHODS OF SOLID-PHASE CRYOCHEMICAL SYNTHESIS

Metal-containing polymers can be synthesized by the vapor deposition polymerization of various monomer systems including organometallic compounds and metal–monomer co-condensates. Such co-condensates are produced by simultaneous or layer-by-layer deposition of metal and monomer vapors on substrate plates at low temperatures (usually, 77 K). Polymerization can proceed in different ways. Some metal–monomer systems polymerize during co-condensation (Ge and Sn with acetylene [11], Mg with CN-substituted *p*-xylylene [12]), most probably due to heat released at condensation. In references 13–16, co-condensates of metals (Pd, Ag, Au, etc.) and vinyl monomers prepared at 77 K were thawed, and thus the obtained metal colloids were polymerized in the liquid state under the action of radical initiators.

Such procedures gave composites only with very low metal concentration (0.01–1 wt. %) containing metal particles of rather large size (about 160 nm) [16]. Metal–polymer materials with high metal content can be prepared by polymerization of solid-vapor-deposited metal–monomer co-condensates. The first investigations in this field were carried out in 1959 for vitreous co-condensate of Mg-acrylonitrile, which polymerizes at softening of glass under action of active centers produced by electron transfer from Mg to acrylonitrile [17]. The state of metal in the obtained polymer system has been not characterized. Cryochemical synthesis is based on low-temperature solid-state polymerization in conditions of frozen molecular mobility in the reacting system. Earlier it was shown that a number of monomers, such as acrylonitrile [18], formaldehyde [19–21], or *p*-xylylene (PX) and its derivatives [22–24], polymerize in solid state under γ or UV radiation at 77 K and even close to the liquid helium temperature [18, 20]. It is suggested that such polymerization is caused by specific supramolecular organization of solid-monomer-containing molecular aggregates, which are precursors of polymer chains [25]. In such aggregates, mechanical perturbation arising during the radiation-induced chain generation under certain conditions can lead to the subsequent activationless growth of the polymer chain [26–28].

p-Xylylene compounds containing different substituents are the most suitable monomers for cryochemical synthesis of metal-containing polymers because they polymerize at low temperatures completely without the use of any foreign substances—in particular, initiators or sensitizers. By varying the substituents, one can modify physicochemical properties of the polymer matrix in the rather wide range. PX compounds arise by pyrolysis of corresponding [2,2]-*para*cyclophanes [29]. A scheme of preparation and polymerization of PX compounds is presented in Scheme 2.1.

The solid [2.2]-*p*-cyclophane was sublimated at 383 K, and the vapors were passed through a pyrolitic quartz tube at 873 K. This compound converts into PX without formation of detectable side products under these conditions [29, 30]. As stated in reference 22, solid PX most probably contains polymer chain precursors in the form of monomer stacks transforming into polymer chains as a result of the rather small rotational displacement of molecules without their translation movement. The assumed pattern of the solid-state PX polymerization is presented in Figure 2.1.

Metal-containing poly-*p*-xylylenes were synthesized from PX with organometallic substituents or from PX–metal co-condensates. In the latter case the PX vapors were mixed in the outlet of the pyrolytic tube with metal vapors produced by sublimation of the corresponding metal. Depending on the metal

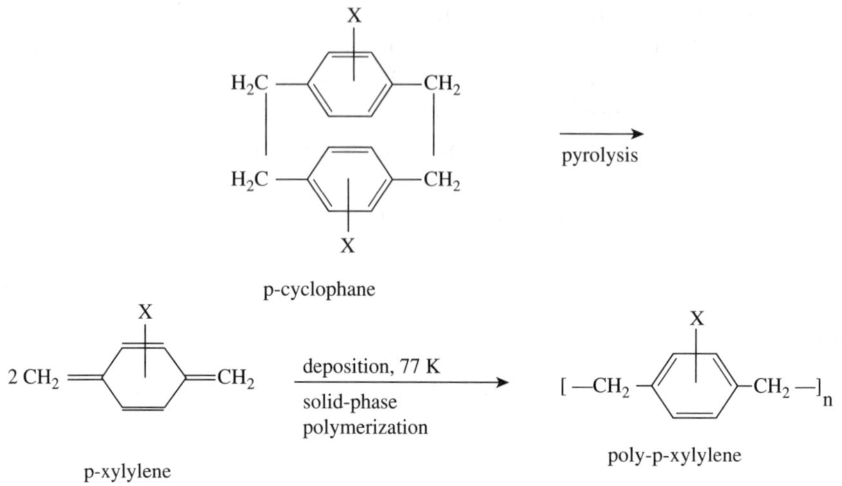

Scheme 2.1. Preparation and polymerization of *p*-xylylene monomers.

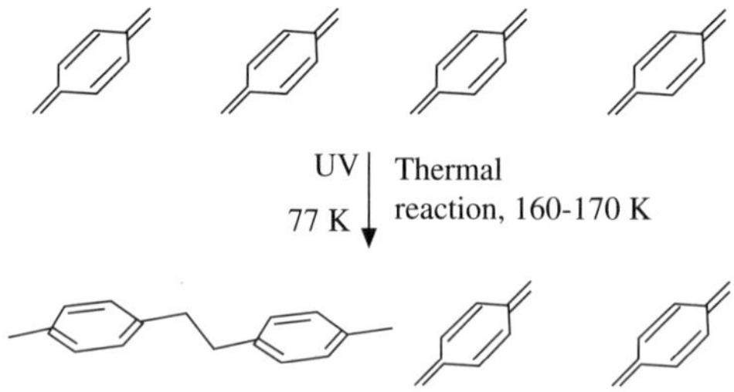

Figure 2.1. The assumed supramolecular structure of solid PX and the PX polymerization pattern.

volatility, such sublimation was performed by heating metal in a special tube supplied with a resistance heater or under the action of an electron (or ion) beam [30, 31]. The PX monomer alone or together with metal vapors was deposited onto a substrate plate, which was cooled to 77 K in an optical cryostat shown schematically in Figure 2.2.

Polymerization was induced by heating an obtained deposit or by its irradiation from a high-pressure mercury lamp. It has been shown that spontaneous thermal polymerization proceeds at temperatures close to 170 K [22], but under UV radiation PX polymerize completely even close to 77 K [21–23].

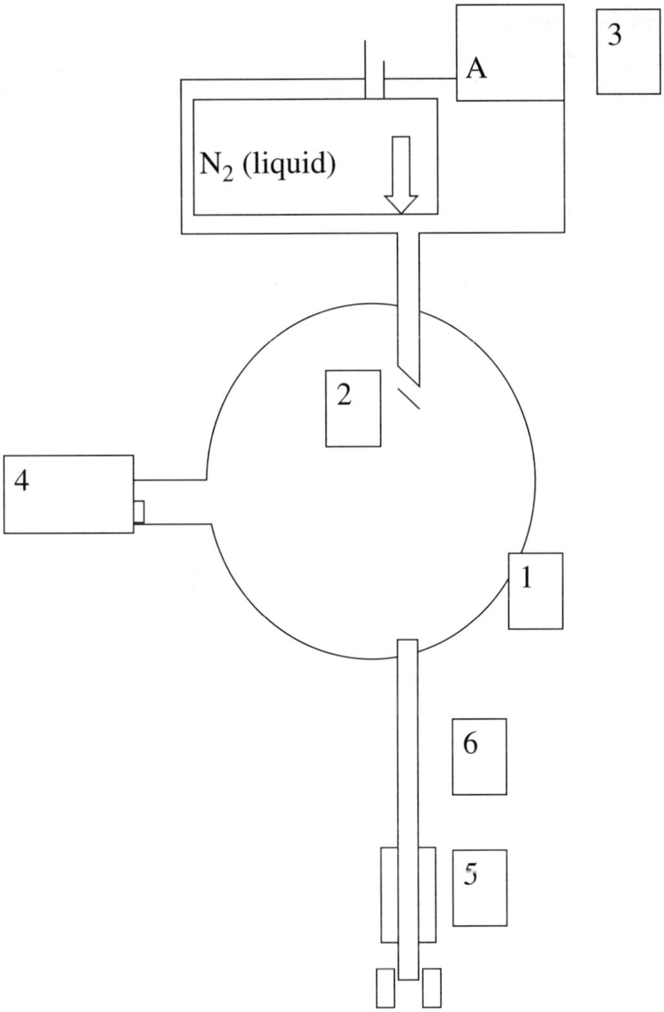

Figure 2.2. Schematic of apparatus for synthesis of a metal/SC-polymer nanocomposite: 1, vacuum chamber; 2, substrate holder; 3, resistance measurements (for metal control in the film); 4, heater (metal evaporation); 5, heater (cyclophane evaporation); 6, heater (cyclophane pyrolysis).

3. STRUCTURE OF CRYOCHEMICALLY SYNTHESIZED METAL–POLYMER COMPLEXES

Depending on the metal nature and monomer structure, the vapor deposition cryopolymerization of metal–monomer systems yields a possibility producing metal-containing polymers of different structure. Polymer containing organometallic groups or metal–polymer complexes can be prepared using preliminary synthesized organometallic monomer or monomer and metal chemically interacting to one another during vapor co-deposition. In the later case, *p*-xylylene monomers are particularly attractive because of their high reactivity and ability to form complexes with metals due to low-lying π-orbitals. In the absence of chemical or coordination bonds between the metal and the monomer, the vapor deposition cryochemical synthesis gives metal clusters physically incorporated in a polymer matrix. The stability of low-temperature products of such a synthesis, as well as their transformation as a result of secondary thermal processes in matrix, depends on the structure and physicochemical properties of matrix.

3.1. Polymers Containing Organometallic Units

Organometallic derivatives of poly-*p*-xylylene (PPX) with Ge and Sn atoms covalently bonded to polymer chains have been synthesized by a vapor deposition technique using bridged [2,2]-*para*cyclophanes with corresponding organometallic groups [32, 33]. Pyrolysis of these cyclophanes, along with polymerization of the *p*-xylylene monomer mixture, is shown in Scheme 2.2.

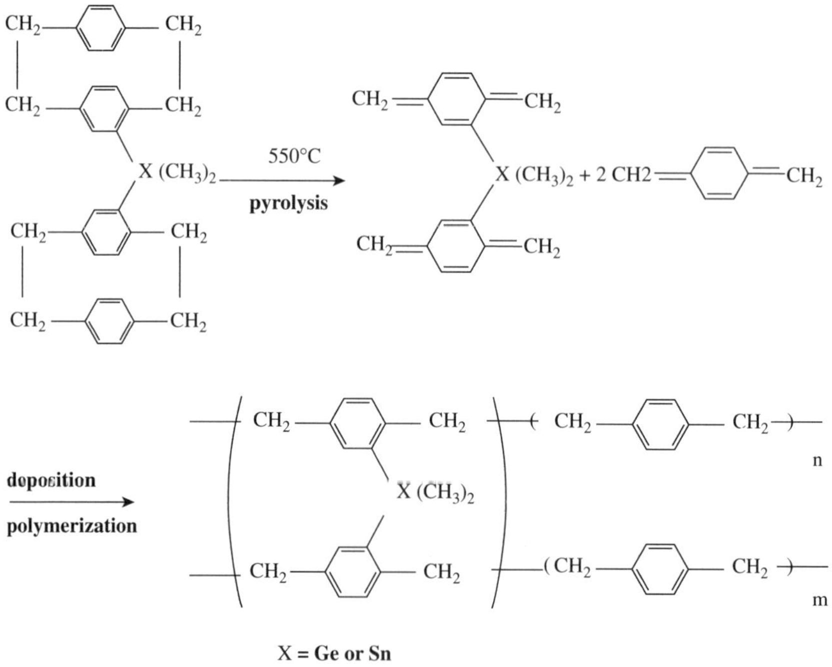

Scheme 2.2. Vapor-deposition preparation of poly-*p*-xylylenes containing organogermanium and organotin units.

It has been shown that the temperature of the deposition and polymerization of the monomer system strongly influences the structure of obtained polymers. Deposition of monomers at 77 K, followed by thermal low-temperature solid-state polymerization of the formed monomer composite, leads to crystalline block copolymers, whose structure, most probably, corresponds to that of microcrystalline monomer deposits. At the same time, deposition and polymerization of monomers at 283 K in conditions of high molecular mobility results in amorphous block copolymers or liquid-like oligomers [33]. The thermal decomposition of organotin units of corresponding crystalline copolymer in air yields SnO_2 microcrystals. Organogermanium blocks of the crystalline block copolymer decompose in an inert atmosphere, giving Ge microcrystals in a PPX matrix. Thermal treatment of amorphous block copolymers in the same conditions also leads to a total loss of Ge-organic structures, however, without formation of Ge crystals [33]. So, one can conclude that the supramolecular organization of the polymer produced by the cryochemical synthesis determines the direction of thermal decomposition of organometallic polymer fragments.

Polymers with organometallic units incorporated in a polymer structure can be also synthesized by the low-temperature solid-state polymerization of vapor deposited metal–monomer co-condensates, in which metal atoms or clusters form intermediate chemical or coordination bonds with monomer molecules and active centers of polymerization. As has been shown in reference 11, Ge and Sn atoms react with C_2H_2 during the deposition of metals together with a monomer at 77 K, producing acetylene–metal copolymers with composition $(C_2H_{2.7}Ge_{0.72})_x$ and $(C_2H_{2.6}Sn_{0.70})_x$. These metals (M) incorporated in polyacetylene structure are probably in the M(II) and M(IV) oxidation state. According to reference 11, the most probable structure of prepared copolymers can be presented by Scheme 2.3. These copolymers are weakly paramagnetic but lose

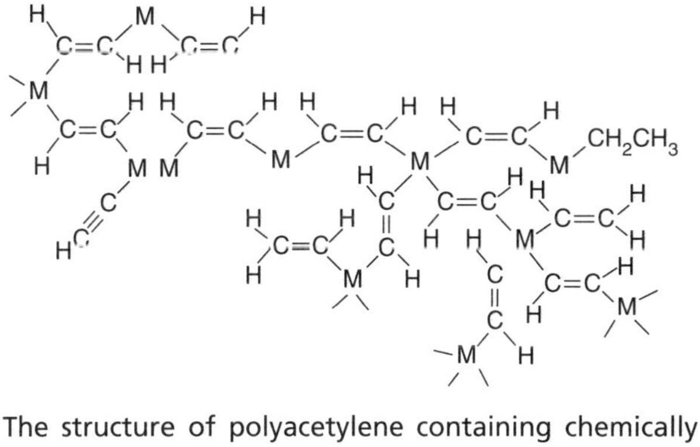

Scheme 2.3. The structure of polyacetylene containing chemically bounded Ge and Sn.

the paramagnetism as a result of their irreversible oxidation in O_2 with the formation of metal oxides [11].

The structure and polymerization of the vapor-deposited PX–Mg monomer system were described in references 23 and 34–37. Taking into consideration IR and UV–Vis spectral data, it was concluded that the co-condensation of Mg with PX at 77 does not lead to the PX polymerization but results in formation of charge transfer complexes where PX is in the benzenoid form [35, 36] (Scheme 2.4).

In the UV–Vis spectrum of this co-condensate, there is a very intense band with maximum close to 340 nm (Figure 2.3) [36]. This band lies between absorption bands of ionic organometallic compounds in the range 420–470 nm and those of organomagnesium derivatives with predominantly covalent Mg–C bonds in the range 280–300 nm [38, 39]. Thus, the bonds in complexes of PX with a magnesium atom (or cluster Mg_n) are neither purely ionic not covalent, but have intermediate structure. This was confirmed by quantum chemical calculations [36]. Polymerization under UV radiation at 77 K is accompanied by the shift of band from 340 to 310 nm (Figure 2.3), which means the transformation of initial complexes into organomagnesium units of PPX backbone chains [$–CH_2–C_6H_4–CH_2–Mg_n–$] with Mg–C bonds [35, 36] close to σ bonds of organomagnesium compounds. The fact that the absorbance maximum of Mg–C bonds in Mg–PPX is at 310 nm and not in the range 270–300 nm indicates some deformation of these bonds as compared to those of usual stable organomagnesium compounds. Initial monomer complexes are located in defects of monomer crystal between stacks shown in Figure 2.1 and do not sufficiently disturb the stack structure. The formation of organomagnesium units of PPX at low temperatures in conditions of frozen molecular mobility can be explained by the fact that the Mg–C bonds around Mg_n are rather weak and are deformed in crystalline lattice to fit to polymer chain. The assumed pattern of the process is shown in Figure 2.4.

The stability of synthesized organomagnesium structures depends on atomic metal content X_{at}—that is, $N_M/N_M + N_P$, where N_M and N_P are the amount of metal atoms and polymer chain units, respectively. At high X_{at} of Mg (>40

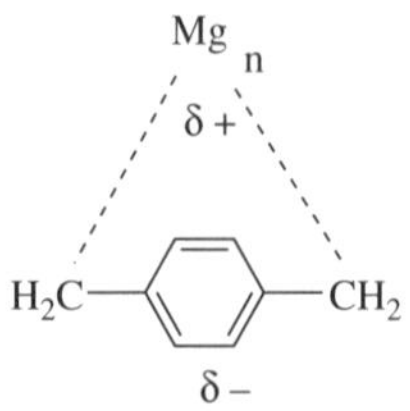

Scheme 2.4. The assumed structure of Mg–*p*-xylylene complex.

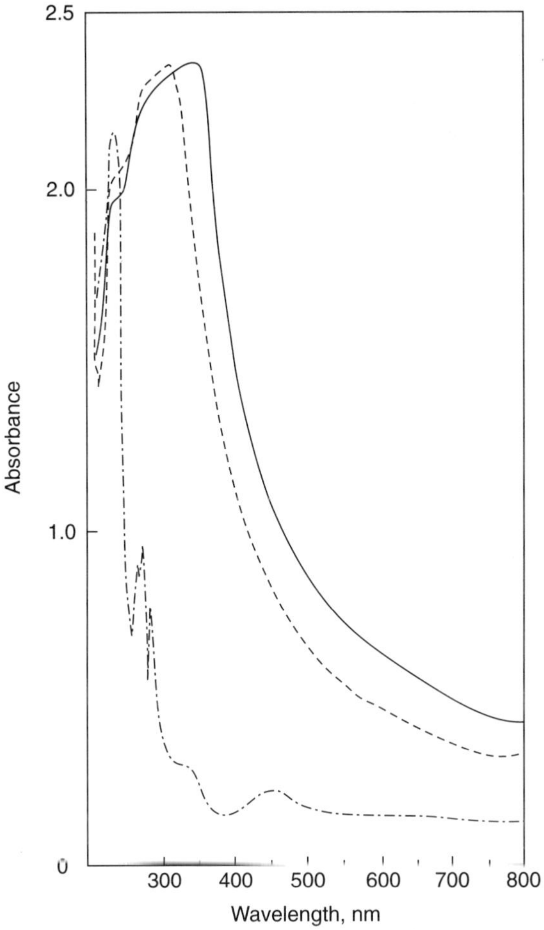

Figure 2.3. The UV–Vis spectra of the PX–Mg films with an Mg concentration of about 30 mol. %. Solid line, original co-condensate at 77 K; dashed line, after UV irradiation at 77 K (complete polymerization); dot–dashed line, after 36 hr of storage at room temperature in a vacuum.

at. %) these structures are stable in vacuum or inert atmosphere at room temperature and may be of interest as organomagnesium catalysts. Stability of obtained structures in this case is probably provided by the additional Mg–C bonds arising as a result of interaction between organomagnesium centers of PPX chains. This interaction leads to the formation of a tetrahedral configuration of Mg–C bonds that takes place in crystalline organomagnesium compounds [40, 41]. At X_{at} of Mg $\leq$ 20 at. %, organomagnesium PPX units are destroyed at room temperature even in vacuum. In the spectrum of Mg–PPX

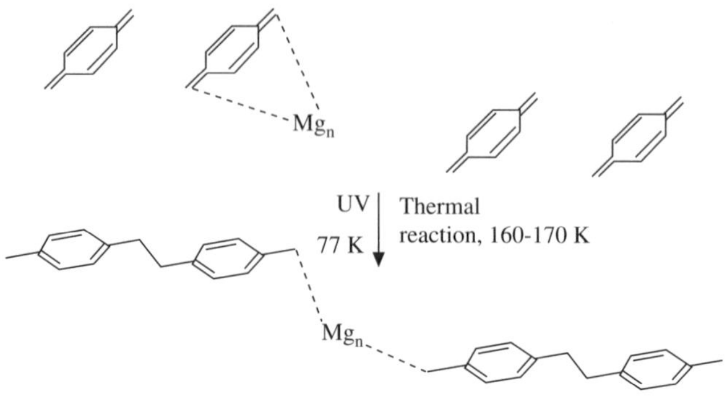

Figure 2.4. The assumed schematic pattern of producing organomagnesium PPX units during polymerization of a PX–Mg co-condensate.

annealed in vacuum at room temperature along with polymeric band, there are two new weak bands at 345 and 440 nm (Figure 2.3) that suggest the formation of Mg_3 clusters. Other Mg clusters, which probably form from organomagnesium structures, could not be observed on a background of PPX absorption [36]. In air at room temperature, organomagnesium structures and Mg clusters oxidize with the formation of MgO nanocrystals of mean size about 35 Å immobilized in PPX matrix [36].

The vapor deposition of PX with a C≡N substituent in the quinoid ring (CN-PX), together with Mg at 77 K, leads to polymerization of this monomer during deposition with the resulting formation of the corresponding substituted PPX (CNPPX) containing organomagnesium structures analogous to those in Mg-PPX [12]. But vapor co-deposition of Mg and CN–PX at 290 K accompanied by polymerization yields only CNPPX containing π complexes between Mg and nitrile groups of polymer [12]. The influence of the deposition temperature on polymer structure in this case was explained by the fact that the monomer polymerization and formation of intermediate Mg–monomer complexes differently depend on temperature. During low-temperature co-deposition the polymerization of the solid monomer phase proceeds rather slowly, and Mg incorporates into this phase before the monomer converts to polymer. At 290 K the rate of polymerization of deposited CN–PX is so high that Mg has no time to associate with a monomer and forms complexes only with C≡N substituents of polymer chains [12].

The co-condensation of Mn with PX at 77 K yields complexes of two types: charge transfer complexes analogous to those of Mg–PX and *d*-π complexes of quinoid PX involving *d* orbitals of Mn and π orbitals of PX [36, 42]. Polymerization at 77 K under irradiation is accompanied by the destruction of *d*–π complexes and transformation of charge transfer complexes to organomanganese PPX units similar to organomagnesium ones [36, 42]. These units decompose at room temperature even in vacuum yielding clusters of Mn atoms. Clusters produced by the decomposition of organomanganese units and complexes form Mn nanocrystals in the interior of the PPX matrix. According to the data of TEM, the main crystal size is about 50 Å [36].

3.2. Polymers Containing Metal Clusters and Nanocrystals Physically Incorporated in Polymer Matrix

As was shown in the previous section, inclusions of zero-valent metal or inorganic metal compounds in polymer matrices (semiconductor compounds; in particular, metal oxides are of most interest, because the electron structure of semiconductor particles strongly depends on their size) can be prepared by the destruction of cryochemically synthesized organometallic polymer structures. However, in most cases metal does not interact chemically with a monomer, so that metal clusters and nanocrystals physically incorporated in polymer matrix are formed directly at vapor deposited cryochemical synthesis [4, 24, 43, 44]. The primary products of cryochemical syntheses involving such metals or semiconductors are small atomic or molecular clusters. When such a composite is warmed up, these particles combine into metal or semiconductor nanocrystals. The temperature range of this process depends on the structure of the polymeric matrix. The stabilizing properties of the matrix can be enhanced to the extent that small atomic and molecular clusters remain intact in this matrix for a long time even at room temperature. This can be done by varying substituents in PPX polymer chains.

The metal particles, both clusters and nanocrystals, in the composite films are stabilized without any specific coordination bonds between the particulate surface and the polymer environment or without any special stabilizing compounds. The particles have only weak (Coulomb and van der Waals) interaction with the environment, and their size is due mainly to the stiffness of the matrix, in the interior of which the particles are formed. This is the primary advantage of the cryochemical solid-state technique over other conventional methods. The technique allows growth of particles over a wide size range, including small clusters, in nonpolar as well as hydrophobic polymer matrices. The concentration of nanoparticles can also vary widely up to high values, providing useful properties resulting from their interaction. The structure and properties of nanocomposites are determined by the fraction of volume occupied by metal particles that is volume content of metal X_V. Therefore, this parameter is mainly used further at the description of composites. It relates to above-defined atomic content X_{at} as $X_V = \rho_P X_{at}/[\rho_P X_{at} + \rho_M(1 - X_{at})M_P/M_M]$, where ρ_P and ρ_M are the density of polymer and metal, respectively, and M_P is the molecular weight of the polymer chain unit and M_M is the atomic weight of the metal. The vapor-deposited low-temperature synthesis allows preparing composites with X_V up to 50%.

Co-condensation of gold vapor and the diacetylene monomer at 77 K followed by warming up the condensate gave a polymer with conjugate π bonds, which contained gold nanoclusters and had unusual optical properties [45]. Metal–polymer composites prepared by thermal polymerization of low-temperature co-condensates of acrylic acid or methyl acrylate with some metals were described in reference 46. Polymer composites contained metal particles of size in the range 5–15 nm.

Cryochemical synthesis of Ag–PPX systems and their structures were studied in references 24, 34, 36, 37, and 44. The simultaneous vapor deposition of PX-, CN–PX-, and Cl-substituted PX (Cl–PX) with Ag at 77 K does not lead to complexation or the formation of any organometallic compounds [24, 36, 44]. In the case of PX and Cl–PX, such deposition proceeds without polymerization. The co-deposition of CN–PX with Ag is accompanied by the partial polymerization of monomer. The initial condensates at 77 K contain a small amount of Ag nanocrystals that can be revealed and characterized using UV–Vis spectroscopy because such crystals have the specific absorption band of surface electron plasmons of about 430 nm [3] (Figure 2.5 and 2.6). UV irradiation of these condensates at 77 K leads to the total conversion of monomers to the corresponding polymers (PPX, ClPPX, and CNPPX). However, intensity (D_{cr}), maximum position (λ_{max}), and half-width ($\Delta_{1/2}$) of the nanocrystals plasmon band do not practically change in the course of cryopolymerization (Figures 2.5 and 2.6, Table 2.1).

Because D_{cr} is proportional to the content of Ag nanocrystals and $\Delta_{1/2}$ is in inverse proportion to their mean size ($\bar{\mathbf{d}}$) [3, 47], one may conclude that the state and amount of Ag nanocrystals were not affected by the cryopolymerization [24, 36, 44]. Sharp growth of D_{cr} at heating of obtained metal–polymer films specifies that the main part of Ag at 77 K is in a form of small noncrystalline Ag_n clusters, which aggregate with formation of nanocrystals under the action of thermal relaxation processes in the polymer matrix. In UV–Vis spectra of PPX films on a background of the PPX absorption, it is possible to observe only absorption bands of Ag_n with $n > 15$, which, according to reference 9, should be in an "open" range of PPX spectrum at $\lambda > 320$ nm. Because in this spectral range there are no absorption bands except for the absorption band of Ag nanocrystals, it may be concluded that clusters stabilized in obtained films contain less than 15 atoms [24, 44].

A solid polymer matrix hinders the aggregation of cryochemically prepared Ag clusters so that substantial part of Ag in the prepared Ag–PPX systems remains in noncrystalline form at ambient and even higher temperatures. As was stated in references 24, 36, and 44, all of the Ag introduced into the investigated PPX matrices transforms to nanocrystals after annealing at 373 K.

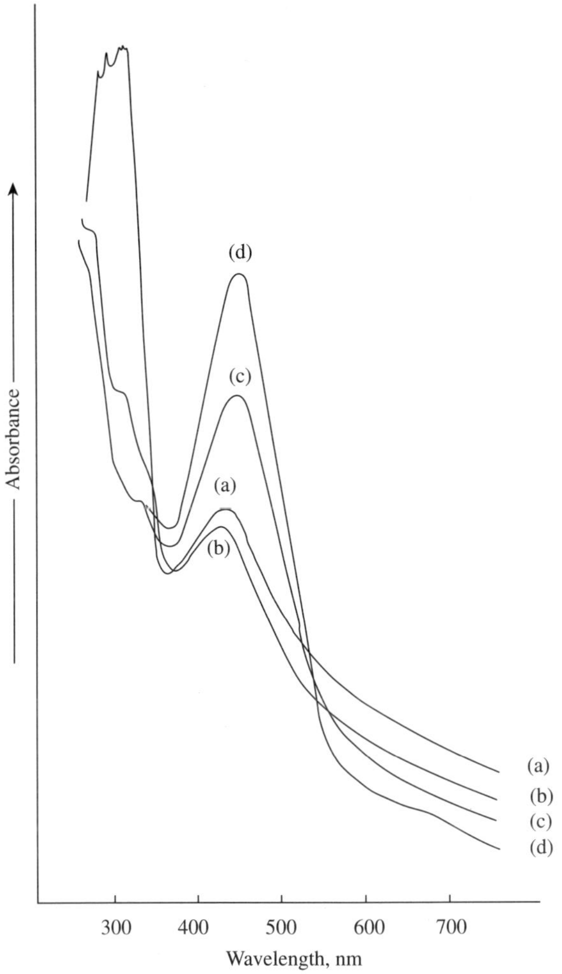

Figure 2.5. The UV–Vis spectra of solid films at 80 K: (**a**) Films obtained by co-condensation of Ag and ClPX at 80 K; (**b**) the same films polymerized under UV irradiation at 80 K; polymerized films after annealing at (**c**) 293 K and (**d**) 373 K.

Therefore the ratio of D_{cr} in a spectrum of an examined film to $(D_{cr})_0$ in a spectrum of the same film annealed at 373 K characterizes a degree of the cluster crystallization (Table 2.1). Spectral data show that the position of plasmon band for nanocrystals in Ag–PPX films heated from 77 to 298 K shifts to the long-wavelength range and then does not change at heating from 298 to 373 K (Table 1). The same effect was observed for Ag–ClPPX films [24].

It should be noted that, unlike Ag nanocrystals of Ag–PPX nanocomposites with λ_{max} of a plasmon band in the range 430–445 nm, nanocrystals

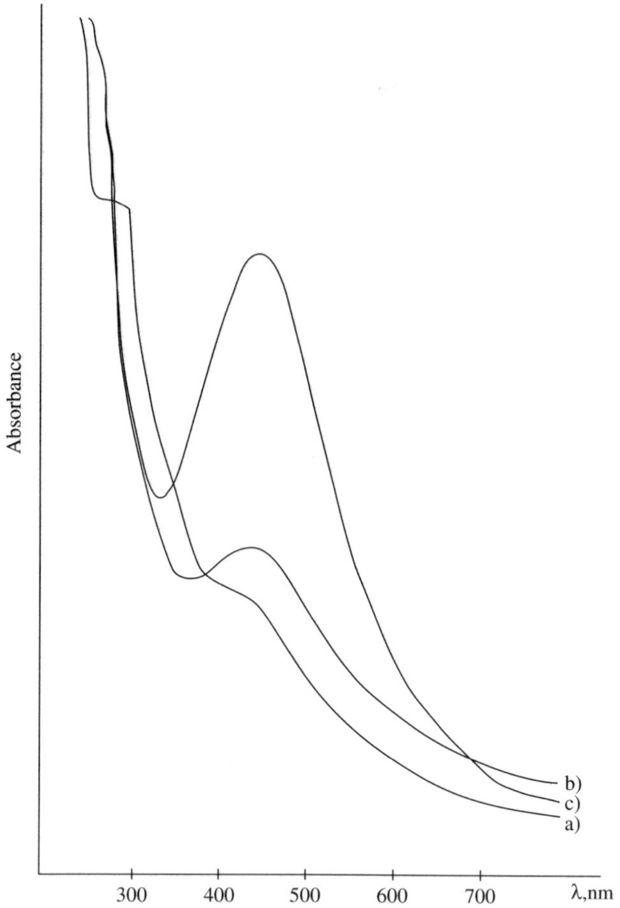

Figure 2.6. The UV–Vis spectra of solid films at 80 K: (**a**) Films obtained by co-condensation of Ag and CNPX at 80 K; (**b**) the same films polymerized under UV irradiation at 80 K and annealed at 293 K; (**c**) polymerized films after annealing at 373 K.

TABLE 2.1. Spectral Characteristics and Content of Ag Nanocrystals in the Solid System During the Synthesis of an Ag–PPX Nanocomposite

System	λ_{max} (nm)[a]	$\Delta_{1/2}$ (nm)[a]	D_{cr} (at λ_{max})[a]	Content of Ag Nanocrystals (at. %)
PX–Ag at 77 K	430	93	0.24	4.2
PPX–Ag at 77 K	430	93	0.24	4.2
PPX–Ag at 298 K	435	96	0.28	5.0
PPX–Ag at 320 K	443	98	1.10	16.3
PPX–Ag at 373 K	443	98	1.22	18.0

[a] Designations are supplied in the text (see above).

prepared by reduction of Ag^+ ions in solution of poly (*N*-vinylpyrrolidone) and protected by molecules of this polymer have a plasmon band with λ_{max} at 410 nm [48]. As is specified in reference 48, the UV–Vis spectrum of nanocrystals depends on their size and form as well as on the embedding matrix. The plasmon band of Ag nanocrystals in reference 48 coincides with that of modeling spherical nanoparticles with a smooth ideal surface, which were theoretically treated from different points of view in references 49 and 50. The nanocrystals of this type form in various liquid media, such as organic solution [3, 48] or the softened quasi-liquid glass [49, 50], where there are no steric hindrances for the growth of equilibrium crystals without surface defects. At the same time, barriers for aggregation of clusters or atoms to metal nanocrystals in the solid system that arises during the cryochemical solid-state synthesis favor the formation of crystals with structural defects, especially on the crystal surface. This increases the surface diffuseness of such nanocrystals resulting in the red wavelength shift of the surface plasmon resonance [51]. The defreezing of the polymer matrix and corresponding alteration of the metal–dielectric interface can explain the further increase of λ_{max} of the plasmon band.

The cluster aggregation to nanocrystals depends on the metal content. In the obtained Ag–ClPPX films the relative part of metal, stabilized in cluster form at 293 K, decreases from 90% to 20% while increasing the total metal content from 0.25 to 2 vol. % [44]. The stability of clusters is greatly influenced also by the structure of the polymer matrix [36, 44]. Comparison of spectra in Figures 2.5 and 2.6 shows that in cryochemically synthesized films, the Ag–CNPPX stability of Ag clusters is much higher than that in similar films Ag–ClPPX. The data on $D_{cr}/(D_{cr})_0$ (see above) in the spectra of various films show that in cryochemically synthesized Ag–CNPPX films heated to ambient temperature, up to 70% from total incorporated Ag (in amount of about 2 vol. %) is stabilized as small noncrystalline clusters [44]. A similar result has been reported in reference 36 for Ag–PPX (Table 2.1). But in Ag–Cl–PPX containing the same total Ag, the fraction of Ag clusters at 295 K is only about 20% [24, 44].

The rather high stability of Ag clusters in Ag-containing PPX can be explained by features of the PPX amorphous phase where clusters are mainly localized. Due to the absence of substituents in PPX chains, this phase differs probably from a similar phase of substituted PPX in higher density and consequently hinders the cluster aggregation in the greater degree. In CNPPX (as compared with ClPPX) there is probably the specific interaction between electron-donating Ag clusters and electrophilic CN groups of the polymer environment that provides a high stability of clusters.

The $\bar{\mathbf{d}}$ value estimated from $\Delta_{1/2}$ of the nanocrystal surface plasmon band in spectra of Ag–ClPPX is about 45 Å [44]. Spectral data for cryochemically prepared Ag–PPX [36] testify to the fact that $\bar{\mathbf{d}}$ is much the same. The calculation based on $\Delta_{1/2}$ of the 110-reflection peak of Ag nanocrystals in an X-ray diffractogram of Ag–ClPPX film gave a $\bar{\mathbf{d}}$ value close to 50 Å [24] (Figure 2.7). The diffractogram of a model system obtained by layer-by-layer deposition of a ClPX monomer and Ag at 77 K, followed by photoinduced polymerization at the same temperature and annealing at 293 K, is also shown in Figure 2.7. Although the interfacial energy on the boundary Ag–polymer matrix in the layered system and the polymerized co-condensate is the same, the value of $\bar{\mathbf{d}}$ in the layered system is approximately equal to 120 Å [24]. So $\bar{\mathbf{d}}$ of Ag nanocrystals in a co-condensate is determined mainly by steric restrictions on the crystal growth in the interior of solid matrix.

The Ag-containing PPX was investigated by TEM. The TEM pattern demonstrates globular particles of sizes between 20 and 120 Å quite homogeneously dispersed within the polymeric matrix [36]. Histogram of the particle size distribution (Figure 2.8) shows that the main particle size is between 40 and 60 Å, which is in good agreement with the results of X-ray and spectral measurements for Ag–ClPPX. Similar histograms were determined by TEM for Pb- (Figure 2.9), Zn-, and Cd-containing nanocomposite PPX films prepared

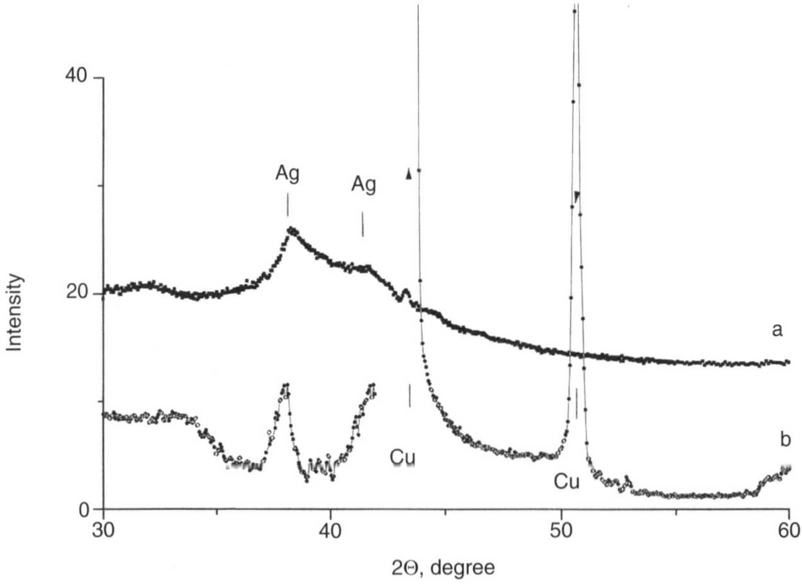

Figure 2.7. X-ray diffractograms of Ag–poly(chloro-*p*-xylylene) systems: (**a**) co-condensate Ag–C1PX polymerized at 80 K and annealed at 293 K; (**b**) layer-by-layer deposition of C1PX and Ag followed by polymerization and annealing at 293 K. X-ray diffractogram peaks of Cu foil are used as a standard.

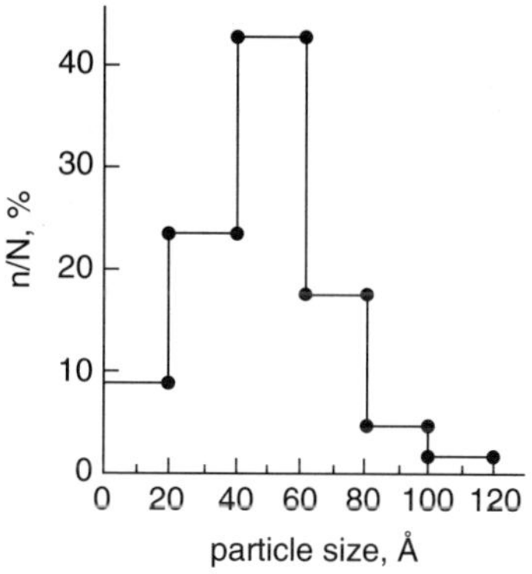

Figure 2.8. Histogram of the size distribution for PPX–Ag films.

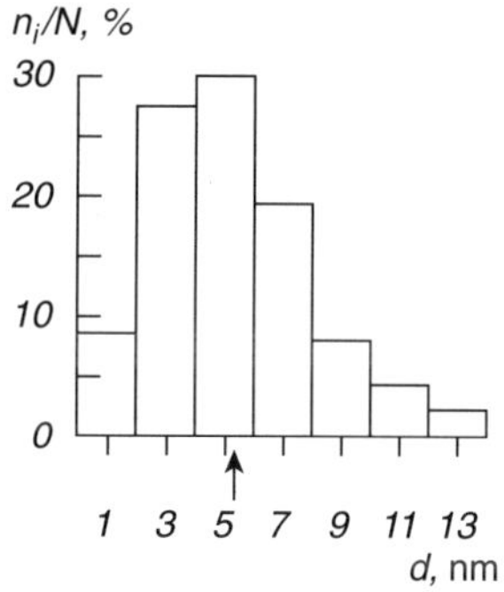

Figure 2.9. Histogram of the size distribution for PPX–Pb films. Arrow shows the mean size of Pb nanocrystals.

by vapor deposition cryochemical synthesis [52]. The value $\bar{\mathbf{d}}$ of metal nanocrystals in these films is about 50 Å. Approximately the same size $\bar{\mathbf{d}}$ (about 45 Å) has been evaluated from $\Delta_{1/2}$ of X-ray diffraction peak for semiconductor PbS nanocrystals in PbS–PPX nanocomposite prepared by the vapor deposition of PX together with PbS at 77 K followed by heating of obtained monomer system to 293 K [53].

So, the mean size of metal or semiconductor crystals in cryochemically prepared PPX nanocomposites depends almost not at all on the nature of such crystals but is determined by the polymer matrix. In particular, data on the width of the nanocrystals plasmon band [44] (cf. Figures 2.5 and 2.6) specify that $\bar{\mathbf{d}}$ of nanocrystals formed from Ag clusters in CNPPX is, approximately, in one and a halftime less than that of analogous nanocrystals in PPX and ClPPX.

Based on spectral data, it has been concluded that the change of Ag content in the cryochemically synthesized films Ag–PPX and Ag ClPPX from 0.3 to 2 vol. % (that is, 4–20 at. %) does not influence $\bar{\mathbf{d}}$ of Ag nanocrystals formed as a result of the cluster aggregation [36, 44]. The same result has been obtained for Pb–PPX films with Pb content varied from 0.01 to 1 vol. % [52]. Unlike other similar systems, PbS–PPX nanocomposites were studied at high PbS loading in the range from 4.7 to 10.2 vol. % of PbS. In this case according to X-ray data, $\bar{\mathbf{d}}$ of PbS nanocrystals also does not depend on PbS content [53]. Thus, it may be concluded that the independence of the nanocrystal mean size from the metal or semiconductor content is a general feature of nanocomposites resulted from solid-state synthesis.

The growth of nanoparticles in a polymerized system has been considered in reference 54. This process was modeled by assuming instantaneous nucleation followed by deposition of metal atoms on the surface of growing nucleus due to diffusion of atoms incorporated in polymer to the nucleus. At modeling it was taken into account that in a polymerized system the diffusion coefficient of metal atoms strongly decreases with an increase in conversion of monomer to polymer in relation to a decrease of free volume of the system during polymerization [54]. The model allows establishing the dimensionless parameters that determine the formation of nanocomposite films and their structure during classical polymerization processes [55]. But at the solid state, cryochemical synthesis processes of polymerization and formation of nanocrystals occur separately: During low-temperature polymerization the aggregation of metal atoms and small metal clusters is frozen and proceeds after polymerization at heating of obtained polymer system.

The polymer structure resulting from the low-temperature solid-state polymerization corresponds to the structure of solid microheterogeneous monomer deposit and contains deep and shallow traps. It is possible to assume that growing crystal nuclei are formed from metal clusters immobilized in deep traps. Metal atoms and clusters situated in shallow traps are released as a result of thermal relaxation processes and join with growing nuclei. The relationship between deep and shallow traps is governed by polymer composite morphology and properties. In this case the concentration of nuclei is proportional to metal content and thus the mean size of nanocrystals; that is, the mean amount of metal per nucleus does not depend on metal content. From this point of view, the fact that crystallization in CNPPX produces Ag nanocrystals smaller than those formed in other PPX matrices can be explained by an increase of the relative part of specific deep traps for Ag clusters in a polymer matrix with electron-accepting CN groups. This leads to an increase of the relative part of crystalline nuclei in the total amount of Ag clusters and, as a consequence, leads to reduction of the mean size of nanocrystals.

Properties of a polymer composite containing metal or semiconductor nanocrystals depend not only on the mean size of nanocrystals, but also on their size distribution. The dependence of the crystallite size distribution on PbS content was investigated for cryochemically synthesized nanocomposite PbS–PPX films in the range from 4.7 to 10.2 vol. % of PbS. The distribution curves obtained from analysis of wide-angle X-ray scattering (WAXS) data is presented in Figure 2.10 [53, 56, 57]. As can be seen, the distribution is not very broad, extending from 20 to 150 Å; the main population of PbS crystallites has a size much less than 100 Å. The curve for nanocomposites with 4.7 vol. % of PbS is similar to histograms of the crystallite size distribution determined by TEM for nanocomposite films Ag–PPX and Pb–PPX with low metal content (Figures 2.8 and 2.9). It means that, as well as the average size of crystals, distribution of the sizes does not depend on the nature of the metal or semiconductor incorporated in a polymer as a result of cryochemical synthesis. The few exceptions are the systems distinguished by specific interaction between metal clusters, which are precursors of nanocrystals, and polymers, as, for example, the system Ag–CNPPX (see above).

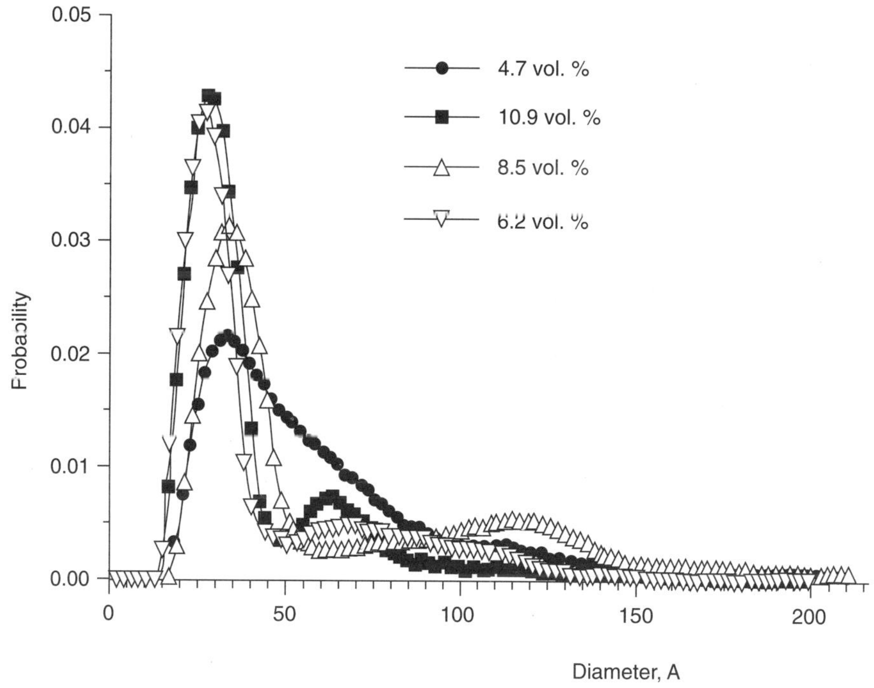

Figure 2.10. Size distribution of PbS crystallites versus PbS concentration in the PPX–PbS composite.

A PbS–PPX nanocomposite with 4.7 vol. % of PbS was investigated also by the small-angle X-ray scattering (SAXS) method, which characterizes different PbS inclusions in PPX matrix regardless of their internal structure. The size distribution of PbS inclusions calculated from SAXS data is similar to the WAXS crystallite size distribution on Figure 2.10. This result signifies that PbS nanoparticles do not aggregate in matrix but are distributed in PPX mostly as small crystallites [53].

It has been shown that the arrangement of nanocrystals in composite films is determined by the PPX supramolecular structure. The SAXS curve at relatively small PbS content of 4.7 vol. % has a broad maximum (Figure 2.11), which specifies some order in an arrangement of PbS nanocrystals. In this case, the PPX matrix retains its semicrystalline structure [53]. In such a matrix the nucleation and subsequent growth of inorganic nanocrystals takes place mostly in amorphous regions where there is enough free volume. It is possible that the observed SAXS maximum corresponds to an average periodicity between PbS nanocrystal assemblies in different amorphous regions separated by PPX lamellar crystallites. The characteristic parameter of this supramolecular periodicity is about 250 Å, which approximately coincides with thickness of PPX lamellas [53].

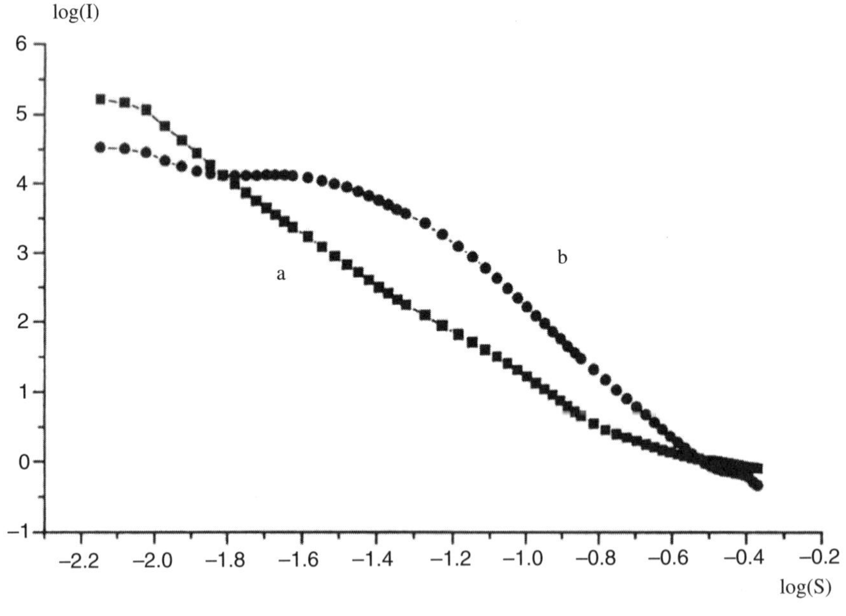

Figure 2.11. Small-angle X-ray scattering of PPX–PbS composite film with 4.7 vol. % of PbS (b) and of pure PPX film (a).

An increase in the concentration of PbS leads to the change of the crystal size distribution, which acquires a bimodal form. Recently, it was shown that this type of bimodal crystal size distribution takes place not only in PPX–PbS samples, but also in PPX-containing Ag nanocrystals [58]. Unlike other analogous systems, the composites obtained by the cryochemical solid-state synthesis demonstrate a marked increase in the percentage of small crystals with increasing content of inorganic component (metal or semiconductor) incorporated in a polymer. Figure 2.10 shows that the contribution of PbS crystallites of approximately 30-Å size increases from 20% to 80% with an increase in the overall PbS content from 4.7 to 10 vol. %. At the same time, the high PbS loading favors the formation of rather large crystals in the 70-Å size range. For this reason the crystal mean size determined from the half-width of the X-ray diffraction line does not change significantly and remains in the range 40–50 Å.

The reasons for such an unusual change of the crystal size distribution curve with increasing metal content are not precise investigated now. Preliminary data show [56] that an increase in the content of metal or semiconductor inclusions decreases the polymer matrix crystallinity and polymer crystal size. It is believed that this process is accompanied by the formation of a large number of local defect sites where the introduced inorganic component nucleates in the polymer matrix. The nucleation at high inorganic loading increases the rigidity of a system, which retards the crystal growth associated with the reorganization of system and redistribution of free volume. Therefore, the main part of crystals has a small size. At the same, time an increase in metal (or semiconductor) content can lead to formation of spatial defects with high free volume where nanocrystals have the ability to coagulate, giving rather large crystalline clusters. It may be supposed that such defects arise under strains accompanying nucleation at high loading.

4. PHYSICOCHEMICAL PROPERTIES OF SYNTHESIZED METAL–POLYMER FILM MATERIALS

Vapor-deposition solid-state synthesis can produce composites with high (close to percolation threshold) content of stable metal particles of varying size, trapped inside different polymer matrices. Due to the high metal content obtained, nanocomposites possess important valuable properties. First, the effects caused by the local behavior of isolated particles are sharply amplified (for example, the density of magnetic data recording in composite containing ferromagnetic monodomain nanoparticles). Furthermore, in nanocomposites with high metal content, one should take into account the interaction between nanoparticles, which determines cooperative behavior of the system of metal nanoparticles immobilized in polymer matrix. This interaction manifests itself mainly as charge transfer processes between nanoparticles [59]. In particular, these processes substantially influence electrophysical and dielectric properties as well as catalytic activity of metal–polymer materials synthesized by the vapor deposition technique.

4.1. Conductivity and Photoconductivity

Metal–polymer systems obtained as a result of vapor deposition cryochemical synthesis contain stabilized small nonmetallic clusters of metal atoms and metal nanocrystals, but only metal nanocrystals participate in conductivity processes. The data on Ag-containing PPX composites [44] testify to the fact that the relative part of metal stabilized in cluster form at ambient temperature sharply decreases with increasing total metal content even in the range 0–2 vol. %. Therefore, at measurements of conductivity, which were carried out at considerably higher metal concentrations, it was accepted that as a first approximation, all metal is as nanocrystals.

When metal particles are isolated in a polymer so that the interaction between them can be neglected, the conductivity of a composite is determined by that of a polymer matrix. In such composites, metal nanoparticles can only inject carriers into a polymer but do not influence substantially the conductivity process [59]. This is the case of metal–PPX films prepared by the cryochemical vapor deposition technique if such films contain metal or semiconductor nanocrystals in amounts less than 4–5 vol. % [30]. The conductivity of such composites follows classical ohmic current–voltage relationship [30] and Arrhenius dependence on temperature [57]. At higher metal contents the mechanism of conductivity in synthesized composite films changes under influence of the interparticle interaction. In PPX films containing Pb nanoparticles in amounts from 5 to 10 vol. %, the dependence of current I on voltage U looks like $\ln I \sim U^{1/2}$ [30]. This dependence is characteristic for a current caused by the tunnel electron transfer between nanoparticles [60]. As is stated in reference 59, the tunnel current arises because at the certain metal content the nanoparticles participating in processes of tunnel electron transfer form infinite cluster penetrating all sample of a composite (percolation threshold). The barrier for the interparticle electron transfer decreases with increase in number of particles in infinite cluster and corresponding reduction of distances between them. As a consequence, probably, the I–U relationship of the conductivity in synthesized Pb–PPX nanocomposites returns to the ohmic one at the rise of Pb content above 10 vol. % [30]. The percolation transition to tunnel conductivity precedes the transition to metalic conductivity resulted from direct contacts between metal nanoparticles. In PPX films containing Ag nanocrystals the conductivity is of metal type at metal content about 7 vol. %: conductivity of composite increases with lowering temperature proportional to

$(1 + \alpha T)^{-1}$ as that of block metals, but coefficient α is approximately in 2.5 times less than value α_0, characteristic for block silver [57].

PPX composite films containing semiconductor nanocrystals of PbO (formed by oxidation of Pb nanocrystals) and PbS show photoconductivity [30, 53]. In the PbO–PPX films the photocurrent value (I_{ph}) in the wavelength range 250–350 nm does not depend on wavelength, and the ratio between I_{ph} and dark current is close to 10^4. Then I_{ph} is gradually reduced up to zero at an increase of wavelength to 450 nm [30].

In the PPX films containing semiconductor PbS nanocrystals of mean size about 45 Å, the photoconductivity has been found even at 630 nm. This wavelength is close to the long-wave edge of the electron absorption of nanocrystals determined from UV–Vis of the films [53]. I_{ph} in PbS–PPX is proportional to the 0.8–1 power of the light intensity, and the activation energy of photoconductivity is 10^{-1}–10^{-2} eV. The energy of light quantum at wavelength 630 nm is about 2 eV, whereas the electronic work function of PbS nanocrystal is higher than 4 eV. This circumstance has led to a conclusion that observed photoconductivity is caused by the photoinduced tunnel electron transfer [53]. The conclusion proves to be true by the low value of the activation energy of photoconductivity and by the nonlinear voltage dependence of photocurrent, which follows the relation $\ln(I_{ph}) \sim U^{1/2}$ characterizing tunnel current (see above). The photocurrent response time is 0.2–0.5 sec [53]. Such a high (for the electron photo detachment) value of the response time suggests that photocurrent proceeds with participation of intermediate long-lived electron–hole pairs generated by light in semiconductor nanocrystals and localized presumably on their defects [cf. reference 61].

4.2. Sensor Properties

Cryochemically synthesized PPX films containing metal or semiconductor nanocrystals have a fine porous structure caused by features of solid-state polymerization. Due to this structure, molecules of gaseous substances readily penetrate into polymer film from the environment. Synthesized composite films demonstrate valuable strong sensor effects resulting from a marked influence of some low-molecular-weight molecules, diffusing into the polymer and adsorbed onto nanocrystals, on the film conductivity [4, 30, 62–64]. Such effects are characteristic of films whose conductivity is governed by electron transfer between nanoparticles.

PPX films containing PbO nanocrystals are sensitive sensors on humidity. The conductivity of the PPX film with about 10 vol. % of PbO nanocrystals versus the air humidity is presented in Figure 2.12. PbO nanocrystals in a PPX matrix are formed by oxidation of Pb nanocrystals in Pb–PPX nanocomposite films produced by cryochemical synthesis (see above).

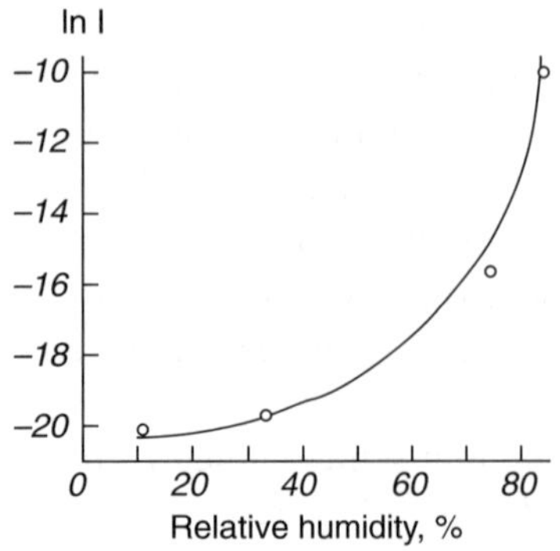

Figure 2.12. Current *I* (in amperes) in PPX–PbO films versus air humidity.

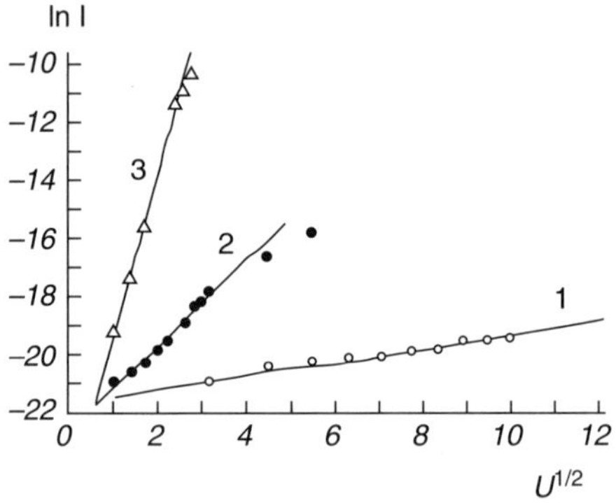

Figure 2.13. Current *I* (in amperes)–voltage *U* (in volts) relationship for PPX–PbO films at various air humidity levels: 1, 11%; 2, 75%; 3, 85%.

It is believed that the adsorption of water molecules on the boundary between the semiconductor PbO and the PPX matrix gives localized ionic states that lead to the lowering of the barrier for the interparticle electron transfer [65]. The current–voltage relationship at various values of air humidity (Figure 2.13) testifies to the tunneling mechanism of the conductivity [30]. The complex character of the conductivity dependence on humidity in Figure 2.12 can be explained by the nonlinear dependence of the tunneling barrier on the amount of water adsorbed on nanoparticles. The influence of air humidity on the film conductivity is reversible: After the replacement of humid air with dry air, the conductivity comes back quickly to an initial value for dry air (Figure 2.14). As seen from Figure 2.14, direct and reverse response times are about 10–15 sec.

It should be noted that the essential rise of conductivity under water vapors has been observed in reference 59 for the composite CuS–polyvinyl alcohol

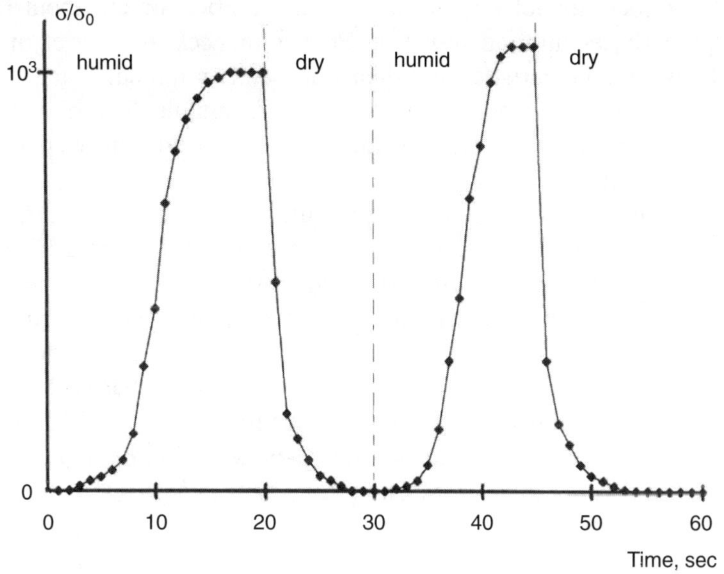

Figure 2.14. Film conductivity σ variations in response to repeated replacement of dry air by humid air, and vice versa. σ_o is film conductivity in dry air.

film obtained by the classical liquid-phase method and contained semiconductor CuS nanocrystals. The effect has been explained by the fact that the absorption of water vapors increases the polymer permittivity, resulting in an increase of the electron tunnelling probability between nanoparticles [59]. In this case the model is in rather good accordance with experimental data because polyvinyl alcohol is a hydrophilic polymer and absorbed water is dissolved and distributed evenly in polymeric matrix. Therefore, macroscopic characteristics of a polymer environment can be used for the description of the effect of water. But this model is not applicable to nanocomposite films based on hydrophobic PPX, where molecules of water "are pushed out" from the polymer and located on semiconductor particles. Such films are much more influenced by atmospheric water than are similar composite films based on the hydrophilic polymer in reference 9.

PPX films containing PbO nanocrystals show also a strong increase in their conductivity under the action of small quantities of ammonia and ethanol vapors in gaseous environment [30, 62, 63]. The effect from ammonia was observed in films with 5 vol. % of PbO. It is believed that the adsorption of ammonia lowers barriers for the electron transfer to a greater degree than that of water, which is why the sensitivity to ammonia appears at a lower concentration of semiconductor particles. The increase of conductivity of PbO–PPX nanocomposite films under ethanol vapors was revealed at PbO content about 15 vol. % [63]. Apparently, ethanol adsorbed on PbO influences the barrier of electron transfer between nanoparticles less than water and ammonia. The responses of conductivity to ammonia and ethanol are also reversible: The film conductivity returns to its initial value after the removal of these substances from the surrounding atmosphere.

Sensor response to low hydrogen pressures at ambient temperature for the PPX film containing Pd nanocrystals is presented in Figure 2.15 [64]. The electrical resistance of the film increases under the influence of hydrogen, presumably because of dissociative adsorption of hydrogen molecules on Pd nanocrystals that leads to the rise of the electron work function of these nanocrystals [66] and correspondingly to the increase of barrier to the tunnel electron transfers between nanocrystals. After removal of hydrogen, the film resistance returns to its initial value within 30–60 sec (Figure 2.15) so such films can be used as a reversible sensor on hydrogen.

It should be noted that the influence of hydrogen on the resistance of Pd–PPX films depends on voltage (Figure 2.16). The tunnel conductivity of initial film in the absence of hydrogen has the linear current–voltage relationship; but under the action of hydrogen, this relationship gets strongly nonlinear character and follows the equation $\ln I \sim U^{1/2}$ (see above). Hydrogen penetrating in the film disrupts the ways of tunnel current of initial film. At the same time, it is known that the absorption of hydrogen by Pd particles leads to a distortion of the Pd lattice and can give essential surface roughness of particles. As a result, the electric field between nanoparticles can rise very strongly, exceeding by several orders of magnitude the average intensity of a field in the film. This leads to an increase in the field effect on the tunnel conductivity and to the appearance of new ways for tunnel current. Therefore, as can be seen in Figure 2.16, at rather low voltages the decrease in conductivity following the action of hydrogen is replaced by an increase in conductivity.

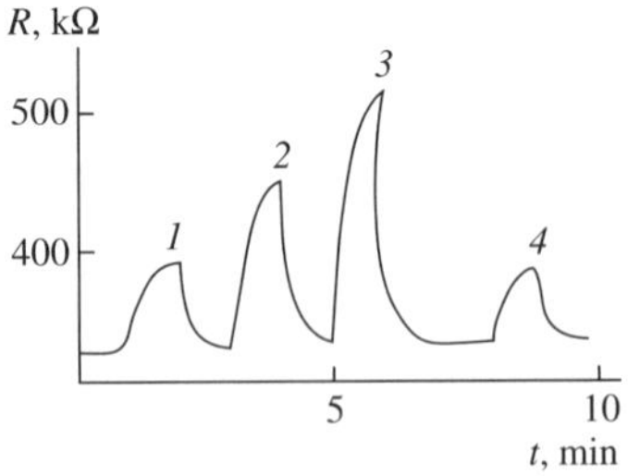

Figure 2.15. Time evolution of the resistance (*R*) of the PPX–Pd films brought into contact with hydrogen at various pressures (mm Hg): 1, 0.1; 2, 0.25; 3, 10; 4, 0.1.

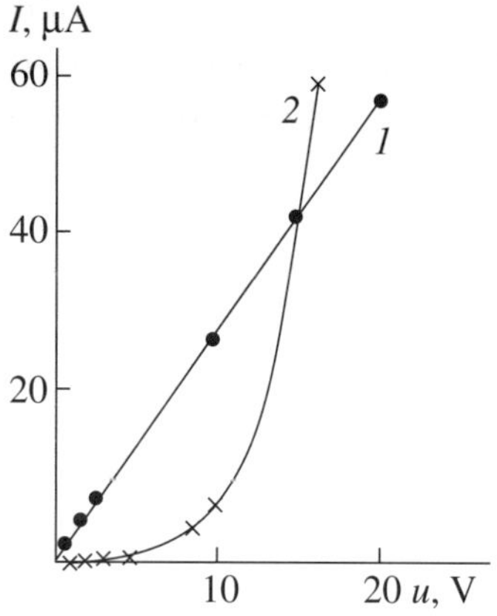

Figure 2.16. Current (I) versus voltage (U) at ambient temperature for PPX with Pd nanocrystals: 1, in pure air; 2, in hydrogen at 1 atmosphere.

4.3. Ferromagnetic Properties. Data Storage

The films with a high concentration of ferromagnetic nanoparticles have an application in new high-density magnetic data processing systems. Features of ferromagnetic materials consisting of domains determine magnetic data recording, each characterized by a certain direction of magnetic momentum. In the absence of an external magnetic field the magnetic moments of all domains are chaotic and the resultant moment of the ferromagnetic material is zero. With externally applied local magnetic field, the magnetic moments of domains in this area become ordered and aligned along the field, thus magnetizing the material. Such a system response is indeed a form of information recording since the direction of domain momentum is retained after the magnetic field has been removed. In this case the limiting recording density is determined, to a first approximation, by the size of domains. However, in conventional ferromagnetic materials, which are employed in magnetic hard disks, the domains interact with each other, thus limiting the density attainable for a given material size. Specialists from Seagate (a high-technology firm) have indicated that conventional magnetic disks will reach the limiting density of 2×10^{10} bit/in.^{2} in the near future.

In polymer composite materials, ferromagnetic nanoparticles constitute monodomains isolated in the nonmagnetic polymer matrix. In such materials the magnetic moments of nanoparticles do not interact with each other, because electron exchange between ferromagnetic particles is insignificant at interparticle distances larger than 1 nm. Therefore, each nanoparticle can be considered as one information bit, which is in the 0 or 1 logic state depending on the direction of magnetic moment relative to the direction of the external magnetic field. In metal–polymer films prepared by vapor-deposition solid-state synthesis with high composition of ferromagnetic nanoparticles, the density of the data recording has the potential to increase significantly. It has been estimated that the density of films with nanoparticles of about 5-nm size and approximately 5-nm spacing can reach about 10^{13} bit/in.^{2} [4]. Such films can be used for writing, reading, and storage of the information.

The recorded information can be read by invoking the effect of negative magnetoresistance, which is the diminution of electrical resistance of the material in a magnetic field. This effect has been observed in various nanocomposite materials containing magnetic nanocrystals in a matrix of nonmagnetic metal or dielectric [67, 68]. Magnet moments of nanocrystals were oriented by a strong magnetic field. Negative magnetoresistance of composite films with ferromagnetic nanoparticles incorporated in dielectric matrix is caused by the fact that the probability of the tunnel electron transfer between neighboring nanoparticles increases if these nanoparticles have parallel orientation of spin magnet moments [69].

The phenomenon of the negative magnetoresistance has been observed recently in PPX composite films containing partially oxidized ferromagnetic Fe nanoparticles [4]. The relative change of conductivity in a magnetic field with $H = 11$ kOe is 34% at voltage around 30 V [4]. After removal of the magnetic field, the film conductivity returns to its original value; that is, the effect of magnetic field on conductivity is reversible. Total oxidation of Fe nanoparticles destroys the effect. Specific negative magnetoresistance in synthesized films containing heterogeneous nanoparticles (oxide phase around core of Fe) is not explained now. It is believed that this effect is due to interaction of magnet and dipole moments of such nanoparticles.

4.4. Dielectric Properties

Films of pure PPX and of PPX composites resulting from cryochemical solid-state synthesis were studied by the dielectric spectroscopy method [70]. Dielectric spectroscopy has proven very useful for studying the structure and dynamics of polymer materials, as well as the transport mechanism of charge carriers. In order to study features of the polymer structure, dielectric test methods were used because of their high sensitivity to morphological changes.

The typical three-dimensional plot of the imaginary (ε'') part of the complex dielectric function, $\varepsilon^*(\omega) = \varepsilon' - i\varepsilon''$, versus frequency and temperature is displayed in Figure 2.17 for the PPX composite film with 8 vol. % of Cu. The complex dielectric behavior of the dielectric losses can be described in terms of several distributed relaxation processes with some common features for all the samples. The $\varepsilon''(f, T)$ cuts at a constant frequency plane (1 kHz) for pure

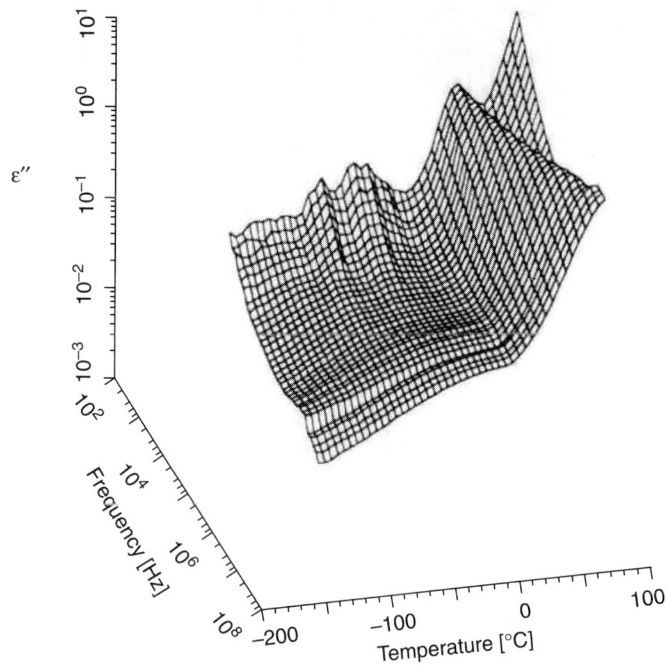

Figure 2.17. Three-dimensional plot of the frequency and temperature dependence of the dielectric losses ε″ for the sample PPX + Cu, 8 vol. %. The accuracy of the measured dielectric losses is estimated to be better than 3%.

matrix PPX and polymer matrix PPX with the addition of Cu (8 vol. %) and Zn (8 vol. %) represent the temperature dependence of the dielectric losses (Figure 2.18). A constant frequency of 1 kHz is chosen because this frequency shows all relaxation transitions.

In the temperature interval of −70 to 0°C and in the low-frequency range, an unexpected dielectric relaxation process for polymers is detected. This process is observed clearly in the sample PPX + Cu. In the sample PPX + Zn, only traces of this process can be observed, and in PPX + PbS as well as in a pure PPX matrix the process completely vanishes. The amplitude of this process essentially decreases when the frequency increases, and the maximum of dielectric losses have almost no temperature dependence. This is a typical dielectric response for percolation behavior [71, 72]. This process may relate to electron transfer between the metal nanoparticles through the polymer matrix. Data on electrical conductivity of metal-containing PPX films (see above) show that at metal concentrations higher than 5 vol. % there is an essential probability for electron transfer from one particle to another and thus such particles become involved in the percolation process.

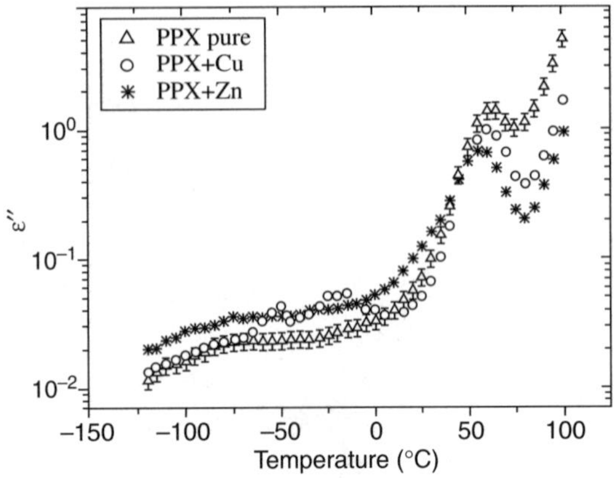

Figure 2.18. Temperature dependence of the 1-kHz frequency behavior of the dielectric losses ε″ of samples of pure PPX (Δ), PPX + Cu, 8 vol. % (○), and PPX + Zn, 8 vol. % (*).

Analysis of dielectric relaxation parameters of this process allowed us to determine the fractal properties of the percolation cluster [70]. The dielectric response for this process in the time domain can be described by the Kohlrausch–Williams–Watts (KWW) expression

$$\Psi(t/\tau) \sim \exp\left[(t/\tau)^{\nu}\right]$$

where Ψ is the dipole correlation function, τ is the average relaxation time, and ν is the stretched parameter, $0 < \nu \leq 1$. It was shown [73] that in complex fractal systems, for the relaxation caused by the charge transfer along the ramified path, the stretched parameter ν could be related to the fractal dimension D_f of the percolation cluster,

$$\nu = D_f/3$$

For the sample PPX + Cu the calculated fractal dimension D_f is equal to 2.609 [70]. It should be noted that the above-mentioned size distribution of metal nanoparticles leads to the mutual charging of such particles in the percolation cluster. This effect is discussed in the following section in connection with catalysis by nanoparticles. As stated in reference 70, the specific low-temperature peak of dielectric losses in the synthesized composite samples PPX + Cu is probably due to the interaction of electromagnetic field with mutually charged Cu nanoparticles immobilized in the PPX matrix. The minor appearance of this peak in PPX + Zn can be explained by oxidation of Zn nanoparticles.

The next relaxation process is typical for amorphous polymers and can be assigned to the α-relaxation that appears in the whole frequency range and in the temperature interval from 50°C to 100°C (Figures 2.17 and 2.18). This process is well-observed for all samples. It corresponds to the glass–rubber transition of the amorphous phase.

The peak of the dielectric loss of this process reflects its viscoelastic nature by obeying the time–temperature superposition principle, wherein the peak is shifted to higher temperatures for shorter times (higher frequencies) and vice versa. This process has been described by the Havriliak–Negami empirical formula [74]

$$\varepsilon*(\omega)\sim\Delta\varepsilon\Big/\left[1+(i\omega\tau')^{\alpha}\right]^{\beta}$$

Here Δε is the dielectric strength and τ′ the mean relaxation time. The parameters α and β describe the symmetric and asymmetric broadening of the relaxation process. The temperature dependencies of the relaxation times of the observed α-relaxation process for pure PPX, PPX + Cu, and PPX + Zn samples demonstrate an Arrhenius behavior with the energies of activation 196 kJ/mol, 187 kJ/mol, and 201 kJ/mol, respectively, and correlate with the activation energies of the α-process in most known polymer materials [75].

4.5. Catalytic Activity

The adsorption of chemical compounds on nanoparticles of synthesized composites not only leads to electrical sensor effects, but also can result in new catalytic processes caused by the surface properties of such nanoparticles [61]. The important catalytic properties of PPX–metal composites prepared via cryochemical vapor deposition synthesis were discovered while studying the isomerization of 3,4-dichlorobutene into *trans*- and *cis*-1,4-dichlorobutene catalyzed by Pd nanoparticles of Pd–PPX [64] and the reaction C–Cl bonds metathesis in the mixture of *n*-decane with CCl_4 catalyzed by Cu nanoparticles of Cu–PPX [76].

Catalytic isomerization of 3,4-dichlorobutene was studied at 100°C in toluene. The ratio of *trans*- to *cis*-1,4-dichlorobutene for the reaction with the usual palladium catalyst is about 10. The same result was obtained by catalytic isomerization on Pd–PPX with low concentration of Pd nanoparticles. But the selectivity of the reaction decreases with increasing Pd concentration: The yield of *trans*-1,4-dichlorobutene decreases while the yield of *cis*-1,4-dichlorobutene remains constant. This result shows that the change in the catalytic properties of the composite is determined by interactions between nanoparticles rather than by the size effects. During catalytic reaction under the influence of Pd–PPX films, where the volume content of Pd nanoparticles is close to the percolation threshold, the *trans*-to-*cis* ratio for produced isomers of 1,4-dichlorobutene is about 2.9—that is, close to equilibrium value of this ratio.

Specific catalytic properties of synthesized Pd–PPX nanocomposites have been explained by charge transfer between nanoparticles at metal concentrations close to percolation threshold, when distances between nanoparticles decrease up to several nanometers and electrons can tunnel from one nanoparticle to another [59]. The effect of distances between nanoparticles on their catalytic properties has been observed at studies of oxidation of CO by Pd particles (2.8–13 nm) immobilized on the surface of MgO [77]. Nanoparticles immobilized in polymer matrix have rather wide size distribution (see above). At the same time, the energy of the Fermi level of a small metal particle depends strongly on its size and form [8, 78]. Electron transfer between particles of different size results in their mutual charging that leads to equalization of their electrochemical potentials [78]. The charging of metal nanoparticles should greatly affect catalytic reaction. In particular, the isomerization of 3,4-dichlorobutene on positively charged Pd nanoparticles proceeds, most probably, via intermediate carbonium ions to give the mixture of 1,4-dichlorobutene isomers close to the equilibrium mixture [64].

Catalytic reaction of the C–Cl bonds' metathesis under the influence of a Cu–PPX nanocomposite film was studied in the mixture of *n*-decane with CCl_4 at molar ratio CCl_4/n-decane = 4 : 1:

$$CCl_4 + C_{10}H_{22} = CHCl_3 + C_{10}H_{21}Cl \tag{1}$$

The reaction was conducted in the absence of oxygen at 180°C for 4 hr. The data on the yield of chlorodecanes are presented in Table 2.2 along with values of specific catalytic activity, which is defined as the number of molecules of chlorodecanes produced per Cu atom of the nanocomposite film during 1 hr of

TABLE 2.2. Dependence of Conductivity and Catalytic Activity of Nanocomposites Cu–PPX on Cu Content

Content of Cu (vol. %)	Electrical Resistance at 77 K (Ω)	Electrical Resistance at 298 K (Ω)	Yield of Chlorodecanes (mol. %)	Specific Catalytic Activity
1.3	∞	∞	Traces	Traces
3.5	7.0×10^8	3.5×10^8	4	130
5.0	10.2×10^3	8.3×10^3	15	500
7.0	3.3×10^3	645	20	650
10.3	66	29	35	1150
14	16.2×10^3	3.1×10^3	14	450

the reaction. The data on the reaction are related to the data on the electrical resistance (*R*) of the composite films.

At very small concentration of Cu (X_{cu}) around 1 vol. %, the *R* value of the composite film is close to that of pure PPX. In this case, Cu nanoparticles fully isolated in matrix do not interact with one another and do not influence the composite conductivity. But is seen in the table, the increase of X_{Cu} even to 3.5 vol. % leads to the sharp fall of *R* as a result of charge transfer between Cu nanoparticles. The probability of such transfer rises exponentially with the decrease of distance between particles—that is, with X_{Cu} increasing.

At the same time, the lowering of *R* with temperature in the range from 77 to 298 K indicates that even at the maximal Cu concentration investigated, the composite film is the system of metal "islands" in dielectric matrix. The true percolation threshold [59], at which the film shows the metallic behavior characterized by the increase of *R* with temperature, is not achieved. At the same time, the data on *R* (Table 2.2) suggest that at X_{Cu} in the range 5–10 vol. % the obtained Cu-nanoparticle ensemble in the PPX matrix is close to the percolation threshold. But the further increase of X_{Cu} up to 14% gives rise to explosive aggregation of nanoparticles with formation of isolated large-scale metal inclusions: This process is accompanied by the substantial decrease in the number of metal nanoparticles and the corresponding rise of *R* (Table 2.2).

At very low X_{Cu}, the catalytic activity of fully isolated and noninteracting Cu nanoparticles in PPX (see Table 2.2) is very small. The increase of X_{Cu} leads to the strong gain in catalytic activity, which achieves the maximum near the percolation threshold under the conditions of the most effective interaction between metal nanoparticles isolated in the polymer matrix. The metal aggregation of metal nanoparticles results in the fall of the catalytic activity. The maximal specific activity is 1150, which is much greater than the activity of all known catalysts in this reaction. For comparison, the same reaction of C–Cl bond metathesis was investigated on the special prepared catalyst containing 1 mass % of high dispersed metallic Cu deposited on silica. In conditions analogous to those of the reaction on the nanocomposite Cu–PPX film, the specific activity of this catalyst is 4. Moreover, it has low selectivity: In this case the formation of by-products from condensation processes accompanies the reaction, whereas the Cu–PPX catalyst gives monochlorosubstituted decanes only.

The initiation of the reaction (1) is carried out as a result of dissociation C–Cl bonds with formation of initiating radicals CCl_3:

$$CCl_4 \rightarrow CCl_3 + Cl \tag{2}$$

The energy expenditure of this reaction in gaseous phase (E_i^v) differs from that of the same reaction in the adsorption state on Cu (E_i^{ad}) by the value $E_i^{ad} - E_i^v = Q^{ad}(Cl) + Q^{ad}(CCl_3) - Q^{ad}(CCl_4)$, where $Q^{ad}(X)$ is the heat of adsorption of corresponding molecule. Most likely the heat of adsorption of CCl_4 on Cu is less than that of adsorption of Cl or radical CCl_3, so that $E_i^{ad} - E_i^{v} > 0$, and the adsorption of CCl_4 on Cu particles should favor its dissociation and initiation of reaction. But obtained data (Table 2.2) lead to the conclusion that this effect is rather insignificant at very low metal content, when isolated Cu particles in PPX are located far apart so the charge transfer between particles is practically impossible and they remain neutral. The increase of reactivity with the rise of metal content can be explained by the mutual charging of Cu nanoparticles. Presumably, negatively charged particles formed in this case, among positively charged ones, facilitate the process of initiation. Because CCl_4, CCl_3, and Cl have the significant electron affinity (E_A is about 2 eV for CCl_4, 1.5 eV for CCl_3, and 3.6 eV for Cl), Q^{ad} should increase due to the very great E_A of Cl atom. Consequently, one would expect the increase of the rate of CCl_4 dissociation.

The effect of charge on the dissociation of CCl_4 should be especially strong on roughness of the nanoparticle surface where the electron work function decreases because of the sharp increase of the charge density on such roughness. In this connection, it should be noted that nanoparticles have the high percentage of atoms on or near the surface (for example, a 5-nm CdS-particle has about 15% of the atoms on its surface) [61]. The structure and properties of such vast interface between the particle and surrounding medium depend, to a large measure, on this medium. Unlike nanoparticles formed in the gaseous phase and characterized by a quasi-equilibrium surface, which is nearly the same as that of bulk metal, nanoparticles produced and immobilized in a solid or highly viscous matrix can have the defect rough surface, because of hindering formation of equilibrium surface structure from matrix. As a result, catalytic properties of such nanoparticles can be greatly influenced by a polymer matrix (cf. reference 79).

5. CONCLUSIONS

The cryochemical vapor deposition synthesis of metal–polymer films (from the gaseous state to the solid polymer one, bypassing the liquid phase) allows the production of both new organometallic structures and new valuable composites with high concentrations of nano-sized metal or semiconductor particles.

Solid-state synthesis of this type draws the special attention in connection with an opportunity to prepare and stabilize small clusters of metal or semiconductor atoms in a polymeric matrix at ambient and even higher temperatures. Size-dependent quantum states in such clusters, which are the intermediate form between metal atoms and bulk metal, can give new electronic, optical, and magnetic effects [80]. The structure and properties of clusters in synthesized polymer systems are still almost unknown. The study of clusters in the rather wide (up to 300–320 K) temperature range and in various polymer matrices can give insight into important basic and applied problems of cluster science.

The cluster aggregation at heating of cryochemically synthesized systems gives polymer composites containing metal or semiconductor nanocrystals. The composites prepared in this manner, unlike other analogous materials, demonstrate a marked increase in the percentage of small crystals with increasing content of inorganic component (metal or semiconductor) incorporated in a polymer, although the mean size of nanocrystals thus remains practically constant. It should be noted that the effect of the metal particle size on the energy of Fermi level and work function of a particle becomes appreciable for nanoparticles of size less than 5 nm [8]. Correspondingly, the mutual charging of metal inclusions in a polymer composite can significantly affect physical and chemical properties of composite if the concentration of such nanoparticles is great enough to provide tunnel electron transfer between them in all volume of a composite, because it takes place in the synthesized composites, where they are the main part of total number of particles even at high metal content.

The technique of the low-temperature solid-state polymerization of the vapor-deposited metal (or semiconductor)–monomer co-condensates provides a possibility to vary the structure of a polymer matrix in a rather wide range. In this connection, it should be noted that not only *p*-xylylene and its derivatives containing various substituents including strong polar ones [24, 44], but also other monomers (e.g., reference 18), can be used as initial compounds. Although the study of the cryochemical solid-state synthesis, as well as structure and properties of obtained materials, is only in an initial stage, now it is clear that this synthesis gives much promise for producing new intriguing metal-containing polymers that can be of great importance in various fields of science and technique.

REFERENCES

1. D. Woehrle, in *Macromolecule–Metal Complexes*, edited by F. Ciardelli, E. Tsushida, D. Woehrle, Springer-Verlag, Berlin, (1996), p. 100.
2. A. D. Pomogailo, A. S. Rozenberg, I. Ye. Uflyand, *Nanoparticles of Metals in Polymers*, Chemistry Moscow (2000), pp. 1–673.
3. M. P. Andrews and G. A. Ozin, *Chem. Mater.* **1**, 174 (1989).
4. L. I. Trakhtenberg, G. N. Gerasimov, and E. I. Grigoriev, *Russian J. Phys. Chem.* **73**, 209 (1999).
5. K. J. Klabunde, *Free Atoms, Clusters and Nanoscale Particles*, Academic Press, San Diego (1994), p. 311.

6. M. Ritala, K. Kukli, A. Rahtu, P. I. Raisanen, M. Leskela, T. Sajavaara, and J. Keinonen, *Science* **288**, 319 (2000).
7. M. Moskovits, in *Cryochemistry*, edited by G. A. Ozin, John Wiley & Sons, New York, (1976), pp. 1–594.
8. E. Shumacher, *Chimia* **42**, 357 (1988).
9. M. P. Andrews and G. A. Ozin, *J. Phys. Chem.* **90**, 2922 (1986).
10. M. A. Petrukhina, L. N. Alexandrova, V. V. Zagorskii, G. B. Sergeev, I. E. Kardash, and G. N. Gerasimov, *Proceedings of Fifth Soviet Union Conference on Low Temperature Chemistry*, December 10–13, 1991, Moscow University, Moscow (1991), p. 130.
11. R. W. Zoelner and K. J. Klabunde, *Inorg. Chem.*, **23**, 3241 (1984).
12. V. A. Sochilin, I. E. Kardash, and G. N. Gerasimov, *Polym. Sci. (Russia)* **37**, 1938 (1995).
13. G. Gardenas T., K. J. Klabunde, and H. Habdas, *Chem. Mater.* **1**, 481 (1989).
14. G. Gárdenas T., C. Retamal C., and K. J. Klabunde, *J. Appl. Polym. Sci.* **49**, 15 (1991).
15. G. Gárdenas T., C. Retamal C., and K. J. Klabunde, *Polym. Bull. (Berlin)* **25**, 315 (1991).
16. G. Gárdenas T. and C. Muňoz D., *Makromol. Chem.* **194**, 3377 (1993).
17. V. A. Kabanov, G. B. Sergeev, V. P. Zubov, and V. A. Kargin, *Vysokomol. Soed. (Polym. Sci.)* **1**, 859 (1959).
18. G. N. Gerasimov, S. M. Dolotov, and A. D. Abkin, *Radiat. Phys. Chem.* **15**, 405 (1980).
19. C. Chachaty and M. Magat, *J. Polymer Sci.* **48**, 139 (1960).
20. G. N. Gerasimov and A. D. Abkin, *Khim. Fiz.* (*Chem. Phys. Rep.*) **3**, 170 (1984).
21. E. S. Mansueto and Ch. A. Wight, *J. Am. Chem. Soc.* **111**, 1900 (1989).
22. L. M. Alexandrova, L. V. Shundina, G. N. Gerasimov, and I. Ye. Kardash, *Polym. Sci. (Russia)* **35**, 361 (1993).
23. L. N. Alexandrova, V. A. Sochilin, G. N. Gerasimov, and I. E. Kardash, *POLIMEX-93 International Symposium on Polymers*, November, 1993, Cancun, Mexico, Preprints, p. 150.
24. G. N. Gerasimov, V. A. Sochilin, S. N. Chvalun, L. V. Volkova, and I. Ye. Kardash, *Macromol. Chem. Phys.* **197**, 1387 (1996).
25. N. N. Semenov, *J. Polym. Sci.* **55**, 563 (1961).
26. V. I. Goldanskii, *Nature* **279**, 109 (1979).
27. S. F. Timashov and L. I. Trakhtenberg, *Russ. J. Phys. Chem.* **67**, 209 (1993).
28. M. V. Basilevsky, G. N. Gerasimov, S. I. Petrochenko, and V. A. Tikhomirov, *Chem. Phys.* **55**, 259 (1981).
29. L. Alexandrova and R. Vera-Graziano, in J. C. Salamone, ed., *Polymeric Materials Encyclopedia*, Vol. 9, CRC Press, Boca Raton, FL (1996), p. 7180.
30. G. N. Gerasimov, E. I. Grigoriev, A. E. Grigoriev, P. S. Vorontsov, S. A. Zavialov, and L. I. Trakhtenberg, *Chem. Phys. Rep.* **17**, 1247 (1998).

31. E. Kay, *Z. Phys. D—Atoms, Molecules and Clusters* **3**, 251 (1986).
32. H. Hopf, G. N. Gerasimov, S. N. Chvalun, V. I. Rosenberg, E. L. Popova, E. V. Nikolaena, E. I. Grigoriev, L. I. Trakhtenberg, and S. A. Zavyalov, *Adv. Mater. Chem. Vap. Depos.* **3**, 197, (1997).
33. G. N. Gerasimov, E. L. Popova, E. V. Nikolaeva, S. N. Chvalun, E. I. Grigoriev, L. I. Trakhtenberg, V. I. Rosenberg, and H. Hopf, *Macromol. Chem. Phys.* **199**, 2179 (1998).
34. G. Sergeev, V. Zagorsky, and M. Petrukhina, *J. Mater. Chem.* **5**, 31 (1995).
35. L. N. Alexandrova, V. A. Sochilin, G. N. Gerasimov, and I. E. Kardash, *Polymer* **38**, 721 (1997).
36. L. Alexandrova, E. Sansores, E. Martinez, E. Rodrigez, and G. Gerasimov, *Polymer* **42**, 273 (2001).
37. L. I. Trakhtenberg, G. N. Gerasimov, L. N. Aleksandrova, and V. K. Potapov, *Radiat. Phys. Chem.* **65**, 479 (2002).
38. A. A. Arest-Yakubovich, R. V. Basova, and E. A. Birman, *Vysokomol. Soed. (Polym. Sci.)* **B21**, 226 (1979).
39. H. F. Ebel and B. O. Wagner, *Chem. Ber.* **104**, 307 (1971).
40. E. Weiss, *J. Organomet. Chem.* **2**, 314 (1964).
41. E. Weiss, *J. Organomet. Chem.* **4**, 101 (1965).
42. L. Alexandrova, D. Likhachev, S. Muhl, R. Salcedo, G. Gerasimov, and I. Kardash, *J. Inorg. Organomet. Polym.* **8**, 157 (1998).
43. L. I. Trakhtenberg, G. N. Gerasimov, E. I. Grigoriev, A. E. Grigoriev, I. E. Kardash, S. V. Radzig, P. S. Vorontsov, and S. A. Zavialov, *Proceedings of Second International Conference on Low Temperature Chemistry*, edited by J. R. Durig and K. J. Klabunde, Bk Mk Press, Kansas (1996), p. 211.
44. G. N. Gerasimov, E. V. Nikolaeva, E. I. Smirnova, V. A. Sochilin, and L. I. Trakhtenberg, *Doklady Phys. Chem.* **380**, 184 (2001).
45. A. W. Olsen and Z. H. Kafafi, *J. Am. Chem. Soc.* **113**, 7758 (1993).
46. G. B. Sergeev, *Nanochemistry*, Moscow University, Moscow, 2003, p. 85.
47. K. P. Charle, F. Frank, and W. Schulze, *Ber. Bunsenges. Phys. Chem.* **88**, 350 (1984).
48. C. Carotenuto, G. P. Pepe, and L. Nicolais, *Eur. Phys. J.* **B16**, 11 (2000).
49. U. Kreibig and C. von Fragstein, *Z. Phys.* **224**, 307 (1969).
50. L. Genzel, T. P. Martin, and U. Kreibig, *Z. Phys.* **B21**, 339 (1975).
51. P. Apell and Å. Ljungbert, *Solid State Commun.* **44**, 1367 (1982).
52. G. B. Bushueva, V. V. Zagorskii, G. M. Zinenkova, M. A. Petrukhina, O. V. Revokatov, and G. B. Sergeev, *Russ. Phys. Bull.* **61**, 1871 (1997).
53. E. V. Nikolaeva, S. A. Ozerin, A. E. Grigor'ev, E. I. Grigor'ev, S. N. Chvalun, G. N. Gerasimov, and L. I. Trakhtenberg, *Mater. Sci. Eng.* **C8–9**, 217 (1999).
54. L. I. Trakhtenberg, A. B. Rabinovich, V. A. Kaminskii, and G. N. Gerasimov, *Chem. Phys. Rep.* **21**, 69 (2002).

55. O. J. Ilegbusi, M. Iguchi, and W. Wahnsiedler, *Mathematical and Physical Modeling of Materials Processing Operations*, Chapman & Hall/CRC, New York (1999).
56. E. V. Nikolaeva, Ph.D. Thesis, Karpov Institute of Physical Chemistry, Moscow, 1999.
57. S. A. Ozerin, M.D. Thesis, Moscow Physical Technical Institute, Moscow, 1999.
58. L. I. Trakhtenberg, Yu. F. Krupyanskii, and G. N. Gerasimov, *Chem. Phys. Rep.* (in press).
59. D. Yu. Godovski, *Adv. Polym. Sci.* **119**, 58 (1995).
60. J. G. Simmons, *J. Appl. Phys.* **35**, 2472 (1964).
61. Y. Wang and N. Herron, *J. Phys. Chem.* **95**, 525 (1991).
62. G. Sergeev, V. Zagorsky, M. Petrukhina, S. Zav'yalov, E. Grigor'ev, and L. Trakhtenberg, *Anal. Commun.* **34**, 113 (1997).
63. L. I. Trakhtenberg, E. Axelrod, G. N. Gerasimov, A. E. Grigoriev, E. I. Grigoriev, S. A. Zav'yalov, and Yu. Feldman, *Sci. Isr.—Technol. Adv.* **1**, 572 (1999).
64. P. S. Vorontsov, E. N. Golubeva, E. I. Grigor'ev, S. A. Zavyalov, L. M. Zavyalova, and L. I. Trakhtenberg, *Russ. J. Phys. Chem.* **72**, 1912 (1998).
65. R. M. Hill, *ERA Research Report*, No. 5232 (1967), p. 1.
66. P. D. Cobden, B. E. Nieuwenhuyas, and V. V. Gorodetskii, *Appl. Catal.* **A188**, 69 (1999).
67. T. Furubayasci and I. Nakatani, *J. Appl. Phys.* **79**, 6250 (1996).
68. L. R. Schlep, A. Fert, P. Fettar, P. Holady, S. F. Lee, J. L. Maurice, F. Petroff and A. Vaurěs, *Phys. Rev.* **B56**, 5747 (1997).
69. B. Fugimori, S. Mitani, S. Ohnuma, *J. Appl. Phys.* **79**, 4733 (1996).
70. L. I. Trakhtenberg, E. Axelrod, G. N. Gerasimov, E. V. Nikolaeva, and E. I. Smirnova, *J. Non-Cryst. Solids* **305**, 190 (2002).
71. Yu. Feldman, N. Kozlovich, I. Nir, and N. Garti, *Phys. Rev.* **E51**, 478 (1995).
72. A. Gutina, E. Axelrod, A. Puzenko, E. Rysiakiewicz-Pasek, N. Kozlovich, and Yu. Feldman, *J. Non-Cryst. Solids* **302**, 235 (1998).
73. Yu. Feldman, N. Kozlovich, Yu. Alexandrov, R. Nigmatullin, and Ya. Ryabov, *Phys. Rev.* **E54**, 5420 (1996).
74. A. Schynhals, in *Dielectric Spectroscopy of Polymeric Materials: Fundamentals and Applications*, edited by James P. Runt, John J. Fitzgerald, American Chemical Society, Washington, DC (1997), pp. 81–106.
75. S. Havriliak and S. Negami, *Polymer* **8**, 161 (1967).
76. L. I. Trakhtenberg, G. N. Gerasimov, E. I. Grigoriev, S. A. Zav'yalov, O. V. Zagorskaya, V. Yu. Zufman, and V. V. Smirnov, in *Studies in Surface Science and Catalysis,* Vol. 130, 12th ICC, Part B, edited by A. Corma, F. V. Melo, S. Mendioroz and J. L. G. Fierro, Elsevier, Amsterdam (2000), p. 941.
77. C. Becker and C. R. Henry, *Surf. Sci.* **352–354**, 457 (1996).
78. E. L. Nagaev, *Phys. Usp.* **162**, 49 (1992).
79. M. C. M. Alves, and G. Tourillon, *J. Phys. Chem.* **100**, 7566 (1996).
80. J. I. Brauman, *Science* **271**, 889 (1996).

3

CONTROLLED PYROLYSIS OF METAL-CONTAINING PRECURSORS AS A WAY FOR SYNTHESIS OF METALLOPOLYMER NANOCOMPOSITES

A. D. Pomogailo, A. S. Rozenberg, and G. I. Dzhardimalieva

Institute of Problems of Chemical Physics, Russian Academy of Sciences, Chernogolovka Moscow Region, Russia

The processes of thermal decomposition play an important role in various practical fields including metallurgy, mineral conversion processes, pyrogenic chemical technology, pyrolysis and thermal treatment of wood and coal, combustion of natural and synthetical fuels, coal carbonization, the prevention of grain spontaneous ignition, and so on. There is still a need for scientifically well-founded approaches to control the thermal decomposition of materials. Many studies focus attention on the chemical transformations and reactions in the pyrolized systems and on the creation of mathematical models of such processes as the major parts of the theory of chemical decomposition of complex substances.

Thermolysis is widely used in the synthesis of nano-sized metallopolymer composites. This is a result of the specifics of the preparation of such materials. In spite of a variety of the methods for the preparation of metallopolymer nanocomposites, there are two principal routes: "bottom-up" and "top-down."

Metal-Polymer Nanocomposites, Edited by Luigi Nicolais and Gianfranco Carotenuto
ISBN 0-471-47131-3

Widely spread bottom-up processes consist of nanoparticle assembling from the separate atoms or ions by means of their reduction in the presence of a polymer matrix (or its precursor). Such a process can be represented by the following stages [1]:

A chain generation (the stage of the M_1 active particle formation)

$$M \xrightarrow{k_0} M_1 + M_j \tag{1}$$

A chain (cluster) growth

$$M_j + M \xrightarrow{k_j} M_{j+1} \qquad (1 < j < \infty) \tag{2}$$

Formation of the products (P, P*)

$$M + M_j = M_1 + P^* \tag{3}$$

From the viewpoint of kinetics, all these reactions can be considered as a chain process. In the general case, the bottom-up processes can be carried out by "wetting" and "drying" means. For the former, at least, one of the components (or, at least, on some stage of the multistage process) should be used as a solution or a dispersion in solvent. The "drying" method allows us to exclude totally the solvent from the technology. Due to a low solubility of components and, in some cases, their chemical instability in solution as well as the difficulties in removing a solvent (including the ecological reasons), the drying processes are more preferable for synthesis of metallopolymer nanocomposites. However, the number of such methods is limited, and technologically final adjustment methods are quite rare.

One such potential method can be pyrolysis—a new and rapidly developing field in the nanocomposites science. This route consists of a controlled pyrolysis of metallopolymers, organic and nonorganic acid salts, metal complexes, and metal-containing compounds in the presence of polymers. Such processes can be carried out into a gaseous phase (the different variants of the CVD, CVD synthesis), liquids (in high-temperature solvents, in the suspension decomposition products from the high-molecular-weight liquids or polymer melts [2]). Of great interest is the thermal decomposition in the solid state (in vacuum or in a self-generating atmosphere of the gaseous transformation products). The processes temperature can be kept constant (isothermal conditions) or be program-controlled (nonisothermal conditions, thermal analysis mode). The processes rate can be controlled by the evolving gaseous product amounts, the change of the initial or final products, or the spectral and magnetic properties of the reacting system as a whole. Below we will consider the thermolysis of the most common precursors, namely, metal-containing polymers, the high-volatile metal-forming compounds such as metal carbonyls ($M_a(CO)_b$) and carbonitrosyls ($M(CO)_x(NO)_y$), formates, acetates, and some π-allyl complexes. Naturally the metal-containing monomers are of special interest, as is the computer modeling of such processes.

1. METAL-CONTAINING POLYMERS AND THEIR THERMAL DECOMPOSITION

There are three main methods for synthesis of metal-containing polymers. Their spreading degree can be divided as follows: interaction of metal compounds (MX_n) with linear functionalized polymers when a backbone has not been affected (polymer-analogous transformations, method A); preparation of metallopolymers by a condensation of appropriated precursors; in this case the metal ion is embedded into a main polymer chain, and its removal is accompanied by a total rupture of the chain (method B); and, finally, a recently developed method, namely, polymerization and copolymerization of metal-containing monomers (method C). The method of synthesis of metal-containing polymers seems to significantly affect their properties. For example, in method A not all functional groups of polymer participate in the complexation, and remaining vacant groups alter the properties of the metallopolymer formed. In method B each unit of polymer contains an equivalent of chelate-bonded metal. Such polymers are known to be stable up to 520–720 K, and some of them even up to 900 K [3]. The thermal transformations of metallopolymers of the condensation type are well-studied because there was a need to obtain thermostable polymers. At the same time, the investigations of thermolysis of carbon-chain polymers (A and C) are limited, although the considerable efforts in this area have been undertaken [4–9].

Kinetics of thermal decomposition of such polymers can be described by the equation for one-stage chemical reaction [10]:

$$dw/dt = k(w - w_r)^n, \qquad k = k_0 e^{-E/RT} \tag{4}$$

where k_0 is the preexponential coefficient; E is the energy of activation; n is the reaction order; $w = M/M_0$, $w_r = M_r/M_0$; M_r, is the mass of the residue nondecomposed at the temperature of experiment, and M_0 and M are the initial and current mass of the sample, respectively; w_r is the member taking into account the nondecomposed solid residue of polymer.

As a rule, the complexation of the initial precursors with functional groups of polymer precedes the formation of metallocomposites [11]. Essentially, the macromolecular complex formed undergoes thermolysis. For a long time the Pt and Pd colloids were known to be stable to heating [12]. The parameters of thermal decomposition depend on the thermostability of the polymer itself, on the stability of the macromolecular complex formed, on the intermolecular interaction, which diminishes with temperature, and on the heating temperature. It should be noted that the formation of macrocomplex itself can lead to a partial destruction of polymer. For example [13], this is possible due to a rupture of polymethylmethacrylate chains under the influence of the mechanical forces, namely, from the tensile stresses (σ) occurring at the formation of the complex:

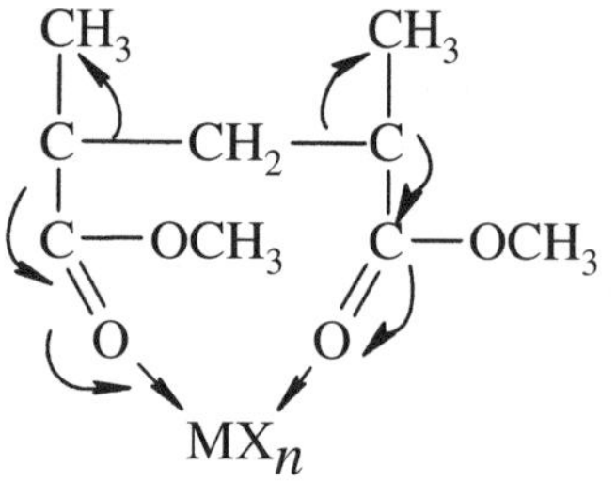

Scheme 3.1.

The tension leads to a decrease of the C–C bond energy to the value $E = E_0 - \gamma\sigma$, where γ is the structure coefficient determining the distribution of tensile stresses in the bonds (see reference 14). Not only the tensile stress, but also the energy of the M–O bond and the conformation of the cyclic complex, is important because a local stress of C–C bonds diminishes the potential barrier of their rupture. A mutual effect of the tensile stress and thermal motion have been analyzed in reference 15.

In principal, to describe the thermal destruction of polymer chains the mathematical models [16], which take into account the differences between the rates of decomposition of "complexed" (N_z) and free (N_f) chain units, can be used:

$$dc/d\tau = k_1 N_z + k_2 N_f \tag{5}$$

where k_1 and k_2 are the constants of the reaction rate of decomposition of the chain units bonded with MX_n and free ones, respectively. However, there is no still such analysis. The transformations are considerably more complicated for the thermolyzed polymer composites containing metal halogenides, such as the systems MX_n – PMMA (M = Cr, Mn, Zn; X = Cl, Br) [17]. For example, the $CrCl_3$–PMMA composite lost under pyrolysis at 523–773 K is about 62% in mass. The volatile fraction contains monomer, carbon oxides, HCl, methane traces, and nonidentified organic products (in the condensed phase along with nonsaturated oligomers, the highly dispersed solid chromium anhydride and oxides are present [18–22]. A quantitatively similar picture of PMMA destruction was observed (at 373–873 K) for $MnCl_2$ [23] when one of the final solid pyrolysis products is manganese oxide formed during the PMMA manganese ionomer thermolysis. The compound $CuBr_2$ (in concentrations of 5–10%) has a thermostabilizing effect for PEO [24].

The metallopolymeric systems pyrolysis permits us to produce some other important and interesting products. For example, the carbon films, containing cobalt nanoparticles, can be produced by a pyrolysis of macrocomplex ($CoCl_2$ with PAN) at 1570 K [25]. The thermotreated PAN complexes with copper halogenides has an electrical conductivity [26], and the macrocomplexes containing two metals (e.g., Co, Ni, or Rh) demonstrates a synergy in thermostabilization [27]. The Fe, Co, and Ni nanoparticles can be used as catalysts for a low-temperature graphitization of amorphous carbon [28]. Such examples are very numerous and demonstrate that metal acetylacetonates in polymer matrices can perform a double function; that is, along with their decay that leads to the formation of metal-containing nanoparticles, they interact with the matrix, initiating its depolymerization and destruction [29].

Such investigations are numerous including those concerning macromolecular π-complexes, but most of them provide only qualitative information. Here let's only note that the products of thermal decomposition of metal salts in polymer matrices are usually highly dispersed metal oxide particles. Their size is determined by the competition between thermolysis and sintering. A decrease in the probability of sintering can be attained by a fast decomposition of salts at the high temperatures followed by cooling of the products formed [30].

2. THE PECULIARITIES OF THERMOLYSIS OF TRANSITION METAL POLYACRYLATES

The pyrolysis of transition metal carboxylates (in the example of Ni^{2+}) has an essential place due to both (a) the quantitative characteristics of the main stages and (b) the properties of the products obtained. Such metallopolymers can be prepared by two principally different routes: by the above-mentioned polymer-analogous reactions, namely, by the interaction of polyacrylic acid with metal salts Ni^{2+}, (PAA-Ni^{2+}), Co^{2+}, Fe^{3+}, and so on (method A), or by radical polymerization of the appropriated monomers—for example, nickel acrylate, $[Ni(CH_2CHCOO)_2]_n$, to give nickel polyacrylate (NiPAcr) (method C). The kinetic peculiarities of metallopolymers thermolysis were compared with the behavior of their low-molecular-weight analogues: nickel propionate,

$Ni(CH_3CH_2COO)_2 \cdot 2H_2O$ $Ni(Prop)_2$, and acrylate, $Ni(CH_2CHCOO)_2 \cdot 2H_2O$ $Ni(Acr)_2$. The synthesis and characterization of such products are given in reference 4.

Let us consider the kinetic behavior of PAA–Ni^{2+} and NiPAcr at their thermolysis. It is known that carboxyl groups of PAA, as well as their low-molecular-weight analogues, are able to chemically bond transition metal compounds. The polymer containing a number of functional groups is a polydentate ligand. This defines its behavior in the complexation reaction. Thus, using various physicochemical methods, it was shown [31] that the interaction of $Ni(CH_3COO)_2$ with PAA is an equilibrium and reversible reaction, it being known the equilibrium is attained immediately after mixing of the components. It was found there are both a "single-point" binding of Ni(II) (the C_1 form) and the Ni(II) binding with two units of the same (the C_2 form) or two different (C_3) polymer chains. The general scheme of such a reaction can be represented as follows:

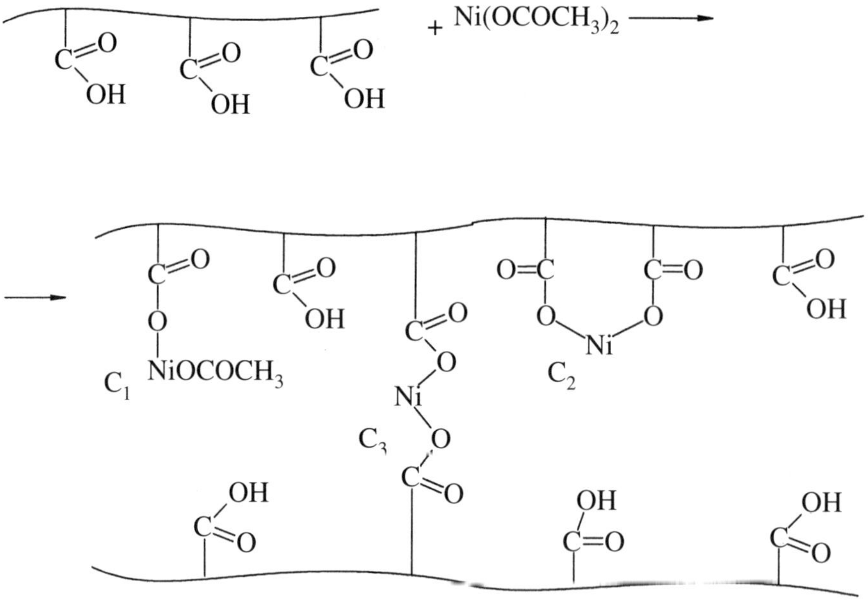

Scheme 3.2.

As a rule, the content of the C_1 form is an order of magnitude less than that for C_2; in the polymer the intramolecular cyclization usually takes place. The portion of Ni(II) ions that binds two chains (the form C_3) is quite small. The complexation reactions with macromolecular ligands are appreciably influenced by a cooperative character of the process—that is, the dependence of reactivity of macroligand functional groups on the state of another units [32]. In particular, for this reason (so called "the effect of neighbor") many carboxylic groups of macroligand do not bind Ni(II) ions. Their portion in the system considered is about 0.3. At the same time for nickel polyacrylate, each unit of the chain contains an equivalent of Ni^{2+}; that is, there is no the vacant carboxylic groups. Such a polymer chain is strongly strained. Thermal transformations of the compound considered are accompanied by gas evolution. The kinetics of gas evolution was recorded using a membrane zero-manometer (Figure 3.1). The kinetic parameters of the process are given in Table 3.1. The kinetics of gas evolution depends on T_{exp}. The change of the m_0/V ratio (at constant T_{exp}), where m_0 is the mass of sample and V is the volume of a reaction vessel, practically does not affect the rate of gas evolution. The character of conversion $\eta(t)$ depends on the nature of metallopolymers and is satisfactorily approximated by the following types of integral equations (here $\eta = (\alpha_{\Sigma,t} - \alpha_{\Sigma,0})/(\alpha_{\Sigma,f} - \alpha_{\Sigma,0})$ is the degree of conversion; $\alpha_{\Sigma,0}$, $\alpha_{\Sigma,t}$, and $\alpha_{\Sigma,f}$ are the mole amounts of

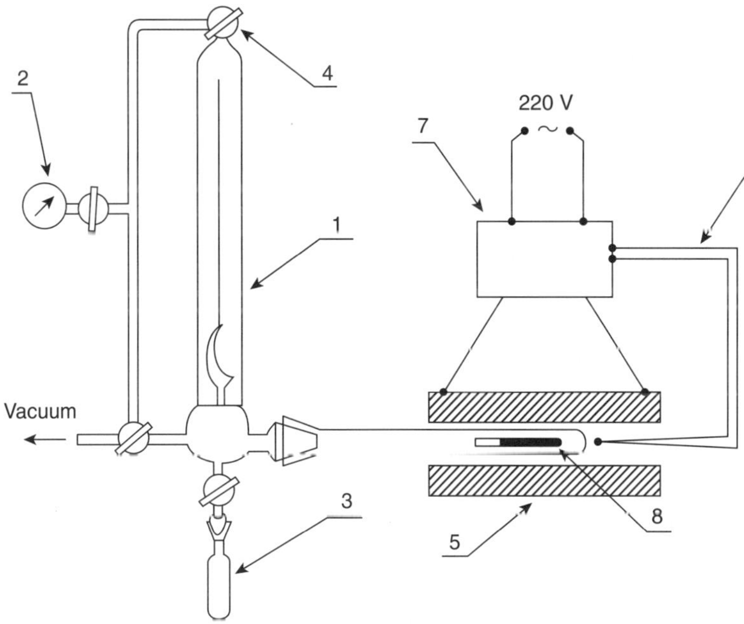

Figure 3.1. A schema of the apparatus for the kinetic investigation of the gas evolution. 1, the reaction vessel; 2, vacuum gage; 3, gutters for sampling of the gaseous products; 4, vacuum cocks; 5, the heating furnace with rectangular profiles of temperatures, high-temperature thermocouple; 7, thermoregulator; 8, ampule with sample.

TABLE 3.1. The Dependence of Kinetic Parameters on the Temperature of Thermolysis of Ni-Containing Precursors

			η_∞, $\alpha_{\Sigma,f} = A\exp[-\Delta H/(RT)]$			k, $\xi_0 = A\exp[-E_a/(RT)]$	
Compound	T_{exp} (K)	η_∞, $\alpha_{\Sigma,f}$	A, sec^{-1}	ΔH (kJ/mole)	k_i, ξ_0	A, sec^{-1}	E_a (kJ/mole)
$NiAcr_2$[a]	573–633	η_∞	2.6	4.6	k_1	1.7×10^{17}	242.5
					k_2	7.5×10^{8}	156.8
NiPAcr	563–598	$1 - \eta_\infty$	1.45×10^{-6}	−61.9	k_1	4.5×10^{8}	143.0
					ξ_{01}	7.1×10^{11}	151.7
					k_2	5.9×10^{5}	107.0
					ξ_{02}	$(3.2 \pm 0.3) \times 10^{-2}$ ≈ const	
PAA-Ni^{2+}	573–643	$\alpha_{\Sigma,f}$	5.38×10^{3}	39.3	k_1	2.1×10^{5}	87.8
					ξ_0	8.1×10^{5}	81.5
Fe_2NiAcr_9[b]	603–643	η_∞	4.4×10^{7}	96.1	k_1	6.1×10^{6}	129.6
					k_2	0.6×10^{2}	79.4
$Ni(Prop)_2$	603–643	$\alpha_{\Sigma,f}$	0.32×10^{2}	16.7	k_1	3.1×10^{8}	117.0

[a] η_∞ are in the range of 0.29 (573 K) to 0.69 (633 K).
[b] η_∞ are in the range of 0.34 (603 K) to 1.0 (643 K).

gaseous products evolved per mole of the initial compound in the beginning, at the moment of t, and in the end of gas evolution, respectively). Thermolysis of the polymer PAA–Ni^{2+} (obtained by method A) is described by the equation for the rate of an autocatalytic reaction of the first order:

$$\eta(t) = \xi_0\left(e^{k1t} - 1\right)\left(1 + \xi_0 e^{k1t}\right)^{-1} \tag{6}$$

The kinetics of thermolysis of NiPAcr (obtained by radical polymerization of nickel acrylate (method B) reveal the more complex picture (Figure 3.2).

At a small degree of conversion the $\eta(t)$ dependence is autocatalytic and is satisfactorily described by equation (6). In the region of low T_{exp} (at the constant m_0/V) the $\eta(t)$ dependence reaches the maximum value of $\eta_{max}(t)$ and then in the end of gas evolution falls to η_∞.[1] This equation is true up to $t \approx \tau$ when the experimental curves of $\eta(t)$ begin to deviate from equation (6). At $t > \tau$ the $\eta(t)$ is transformed into an expression approximated by the equation

$$\eta(t) = \xi_{01}\left(e^{k1t} - 1\right)\left(1 + \xi_{01} e^{k1t}\right)^{-1} - (1 - \eta_\infty)\xi_{02}\left(e^{k2t} - 1\right)\left(1 + \xi_{02} e^{k2t}\right)^{-1} \tag{7}$$

[1] Here $\eta(t) = 0.5(\alpha_{\Sigma,t} - \alpha_{\Sigma,0})/(\alpha_{\Sigma,f} - \alpha_{\Sigma,0})$.

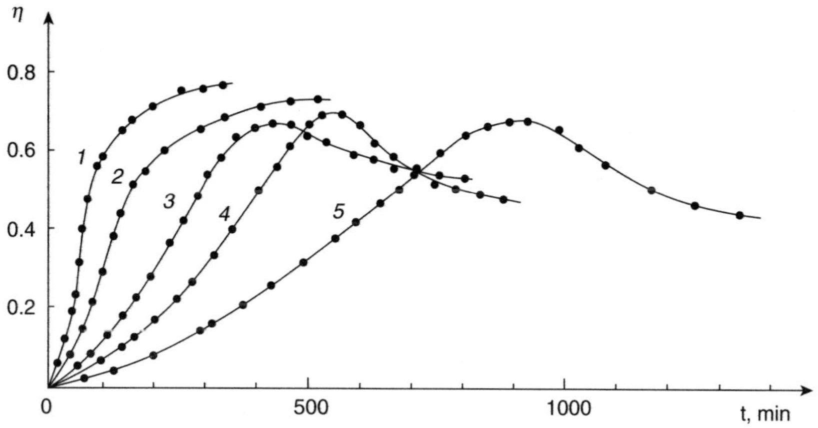

Figure 3.2. The influence of T_{exp} on the $\eta(t)$ dependence during thermal transformation of NiPAcr at the constant ratio $(m_0/V) \times 10^3 = 2.09\,g\cdot cm^{-3}$: 1, 598; 2, 593; 3, 583; 4, 573; 5, 563 K. Solid line, calculated data; dots, the experimental data.

Taking into account that $\tau \approx 1/k_1$ and $\alpha_\Sigma(\tau) \approx 1{,}1 \pm 0{,}1$ = constant, the conversion regime that corresponds to equation (7) is realized at the same level of gas evolution. Increasing T_{exp} causes the shifting of $\eta_{max}(t)$ to the lesser t, a decrease in $\Delta\eta = \eta_{max}(t) - \eta_\infty$, and finally a degeneration of $\eta_{max}(t)$: The dependence of $\eta(t)$ becomes a monotonically accelerating relationship (Figure 3.2). The same picture is also observed at the change of the ratio m_0/V (at the constant T_{exp}). Thus, decreasing of the m_0/V ratio results in both a decrease of the value of $\eta_{max}(t)$ in the line of smaller time of conversion due to a decrease of τ (Figure 3.3, curves 1 and 2) and in the disappearance of η_{max} at higher T_{exp} (Figure 3.3, curves 3 and 4). A decrease of $\alpha_\Sigma(\tau)$ is also observed. It should be noted that the similar adsorption of evolved gaseous products during the conversion has been observed earlier during the investigation of thermal transformation of nickel formate [33] and other compounds [34].

For comparison of the thermal stability of the compounds being studied, it is convenient to consider the initial rates of gas evolution: $W_0 = \xi_0 k_1(1 + \xi_0)^{-2}$. In the range of 573–633 K the metallopolymer being studied can be placed into a series PAA–Ni^{2+} < NiPAcr in which their stability is increased. Figure 3.4 illustrates the yield of the gaseous products during thermolysis of NiPAcr. In an earlier stage of the conversion, the rate of gas evolution is quite high. CO_2 is essentially a single product. The major amount of the product is evolved and then the rate of conversion becomes appreciably lower (Figure 3.4, curve 2). Only when $\alpha_{\Sigma,t} \approx 0.8$ do CO, H_2, and CH_4 appear. The character of their accumulation is different. If the yield of H_2 and CH_4 is monotonically increased with

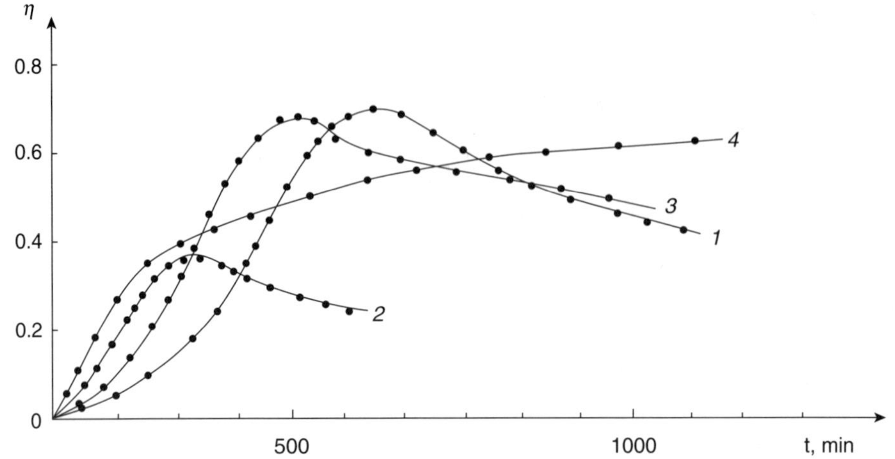

Figure 3.3. The effect of $(m_0/V) \times 10^3$ on the $\eta(t)$ dependence during the NiPAcr conversion: 1, $(m_0/V) \times 10^3 = 2.09\,g{\cdot}cm^3$, $T_{exp} = 573\,K$; 2, $1.46\,g{\cdot}cm^3$, 573 K; 3, $2.09\,g{\cdot}cm^3$, $T_{exp} = 583\,K$; 4, $1.46\,g{\cdot}cm^3$, 583 K.

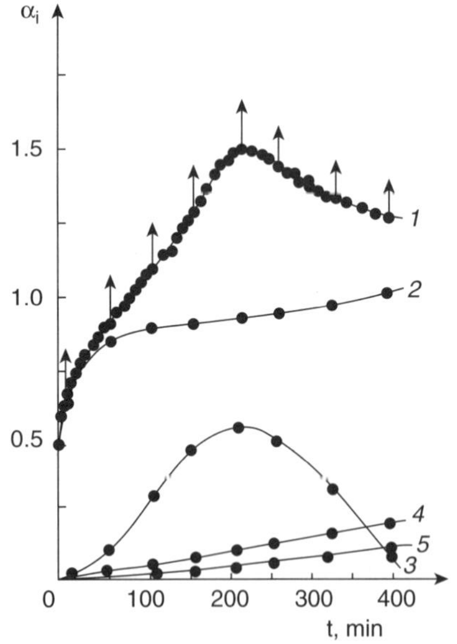

Figure 3.4. The dependence of yield of gaseous products on time during thermolysis of NiPAcr (573 K): 1, $\alpha_{\Sigma,t}$; 2, $\alpha_{CO_2,t}$; 3, $\alpha_{CO,t}$; 4, $\alpha_{CH_4,t}$; 5, $\alpha_{H_2,t}$. The arrow indicates the points of sampling for mass-spectrometry analysis.

the time of conversion, the accumulation of CO passes through a maximum and then falls to the low values (Figure 3.4, curve 3). With T_{ex} the maximum of CO_2 evolution is shifted to the small time range and at $T_{ex} > 583\,K$ is degenerated, as is observed for $\eta(t)$. It is essential that with the falling of the CO yield, some increasing of the rate of CO_2 evolution takes place. Thus, the falling of the total gas evolution in the end of thermolysis of NiPAcr is most probably caused by a consumption of CO at the end of conversion.

The data of IR studies in the example of NiPacr and $Ni(acr)_2$ allowed us to observe the order degree of macromolecular structure during thermolysis by monitoring the character of an appearance and disappearance of the band of regularity (BR) [35]. The bands of regularity are the adsorption bands of the longitudinal vibrations of the lengthy regular parts of a polymer chain. Their intensity is monotonically increased with an increase of a chain length.[2] In the case of NiPAcr, δ_{C-H} at $836\,cm^{-1}$ can belong to BR and ν_{C-H} at $2935 \pm 5\,cm^{-1}$ can be used as the inner standard. Then the $\Psi = D_{836}/D_{2935}$ ratio will characterize the average (the intermediate value between the number-average and number-weight ones [35]) length of the regular part of the polymer chain. For the initial NiPAcr we obtain $\Psi = 0.66$, which can correspond to the average length of the regular part containing 6–10 units. Thermolysis of NiPAcr (T_{exp} = 573 K) at the loss of mass $\Delta m/m_0 = 0.11$ results in the increasing of Ψ to 0.77, but already at $\Delta m/m_0 \approx 0.28$ the BR has disappeared ($\Psi = 0$), which indicates the destruction of the regular structure in NiPAcr.

The solid products of the thermolysis of PAA–Ni^{2+} and NiPAcr (Table 3.2) are the mixture of two phases in the partially X-ray-amorphous matrix. One of them is a well-crystallized phase of metallic Ni, and another is the particularly crystallized phases of nickel oxide (for NiPAcr) and nickel carbide (for PAA–Ni^{2+}). For all the polymers, decarboxylation of metal-containing groups is a source of the largest portion of the total gaseous products. The major product is CO_2:

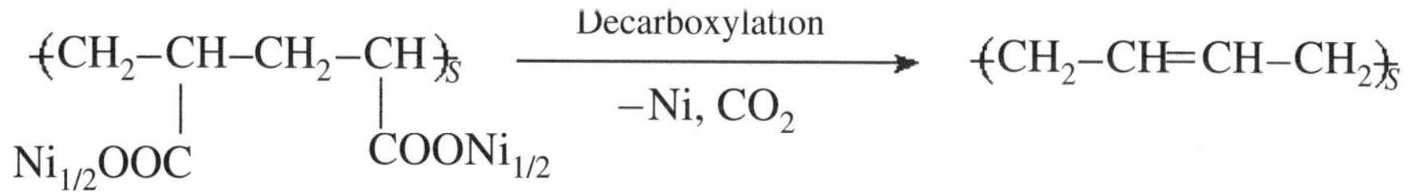

Scheme 3.3.

[2] In the IR spectra of carbon-chain polymers the BR are the bands of C–H deformation vibrations δ_{C-H} ($720–950\,cm^{-1}$) of the backbone, and the bands of stretching vibrations ν_{C-H} ($2800–3000\,cm^{-1}$) are least sensible to the structure of polymer and can be used as an inner standard.

TABLE 3.2. The Composition of Solid-Phase Products of Thermal Decomposition of Ni-Containing Precursors ($T_{exp} \approx 643$ K)

Compounds	Composition
$NiAcr_2$	Ni (~43 wt. %), NiO (~wt. 35%), Ni_3C (~22 wt. %)
NiPAcr	Ni (~wt. 75%), NiO (~25 wt. %)[a]
PAA-Ni^{2+}	Ni(~84 wt. %), NiC[a](~16 wt. %)
Fe_2NiAcr_9	Ni,[b] NiO,[b] Ni_3C[b]
$Ni(Prop)_2$	Ni,[b] NiO,[b] Ni_3C[a]

[a] Partially crystalized phase.
[b] X-ray amorphous phase.

TABLE 3.3. The Yield of Gaseous Products at the End of Thermolysis of Ni-Containing Precursors with Temperature

Compounds	T_{exp} (K)	$\alpha_{\Sigma,f}$	α_{CO_2}	α_{CH_4}	α_{H_2}	α_{CO}
NiAcr	573–643	0.62–2.05	0.55–1.62	0.04–0.38	0.04–0.01	~0.01
NiPacr	563–598	0.98–1.8	0.72–1.50	0.11–0.23	0.10–0.01	0.07–0.14
PAA–Ni^{2+}	573–643	2.06–3.25	1.88–2.74	0.07–0.40	0.09–0.04	0.02–0.07
Fe_2NiAcr_9	603–643	3.34–3.72	2.45–2.52	0.11–0.25	0.75–0.53	0.45–0.60
$Ni(Prop)_2$	603–643	0.87–1.38	0.63–1.02 (0.16–0.30)[a]	0.02–0.04	0.04–0.01	≤0.01

[a] $\alpha_{C_2H_4}$.

H_2 is likely to appear during the course of a catalytic dehydrogenation of decarboxylated polymer matrix under the influence of the Ni nanoparticles formed. However, the contribution of such a process in the total balance of gas evolution is obviously quite small; this is also confirmed by the low values of α_{H2} (Table 3.3).

CO can be formed in the oxidation of metallic Ni formed:

$$Ni_{(s)} + CO_{2(g)} = NiO_{(s)} + CO_{(g)} - 38.5\ \mathrm{kJ/mol} \tag{8}$$

A decrease in the CO yield in the range of high degree of gas evolution is apparently caused by its consumption in the Ni carbide formation:

$$3Ni_{(s)} + 2CO_{(g)} = Ni_3C_{(s)} + CO_{2(gr)} + \sim 122\ \mathrm{kJ/mol} \tag{9}$$

$$4Ni_{(s)} + CO_{(g)} = Ni_3C_{(g)} + NiO_{2(s)} + \sim 84\ \mathrm{kJ/mol} \tag{10}$$

The appearance of CH_4 in the end stages of thermolysis (Figure 3.4) can be determined by the possibility of its formation using the several routes with the participation of the Ni_n phase when the Ni particles reach certain sizes. This can be a direct interaction of CO_2 and/or Co with metallic Ni to yield NiO [8]:

$$2Ni_{(s)} + CO_{2(g)} + 2H_{2(g)} = CH_{4(g)} + 2NiO_{(s)} + 170.1\ \text{kJ/mole} \quad (11)$$

$$Ni_{(s)} + CO_{(g)} + 2H_{2(g)} = CH_{4(g)} + NiO_{(s)} + 208.6\ \text{kJ/mole} \quad (12)$$

Alternatively, it can be their catalytic (Ni_n) hydrogenation by H_2:

$$CO_{2(g)} + 4H_{2(g)} \rightarrow CH_{4(g)} + 2H_2O_{(g)} + 165.1\ \text{kJ/mole} \quad (13)$$

$$CO_{(g)} + 3H_{2(g)} \rightarrow CH_{4(g)} + 2H_2O_{(g)} + 206.1\ \text{kJ/mole} \quad (14)$$

Another way can consist of a catalytic destruction of decarboxylated matrix as follows:

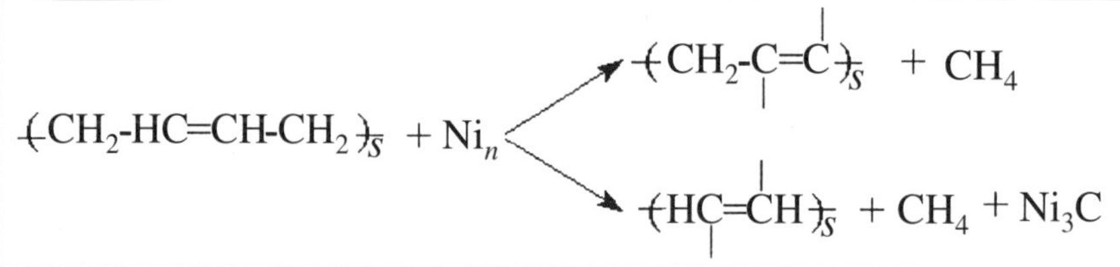
(15)
(16)

The calculations of the composition of solid-phase products with consideration of the balance $\alpha_{Ni} + \alpha_{NiO} + \alpha_{Ni3C} = 1$ and their comparison with data on α_{CH4} have showed that the most probable routes for the formation of CH_4 are the reactions (15) and (16).

Thus, the kinetic peculiarities of solid-phase thermal transformations of metallopolymers is the result of thermal stability of metal carboxylate groups. Their reactivity is determined first of all by a spatial organization of the nearest surroundings—that is, the microstructure of polymer chains. This is confirmed by the presence of bands of regularity, as well as by their appearance on earlier stages of thermolysis followed by their evolution during thermolysis. From the viewpoint of solid-phase topography, metallopolymers are characterized by heterogeneity of the particle reactivity. The qualitative and quantitative composition of gaseous and condensed products essentially depend on the secondary processes with participation of the metal-containing phases formed.

Let us consider the magnetic properties of solid-phase products of the metallopolymer thermolysis. Such products are ferromagnetic. The magnetization of the initial $Ni(Acr)_2$, $\sigma_s(300\,K)$, equals $0.209\,Gs \cdot cm^3 \cdot g^{-1}$; and the specific magnetic susceptibility, $\chi_\sigma(300\,K)$, equals $0.255 \times 10^{-4}\,cm^3 \cdot g^{-1}$. Respectively we estimate the magnetic moment μ_{eff} of 3.40 mB; this value is close to the spin moment $\mu_s = 2.83$ mB, which is characteristic for octahedron symmetry of Ni^{2+}. During thermolysis the σ_s and χ_σ change (Table 3.4).

TABLE 3.4. The Change of Magnetic Properties of the Products During Thermal Transformation of $NiAcr_2$ ($T_{exp} = 643\,K$)

$\Delta m/m_0$	σ_s ($Gs \cdot cm^3 \cdot g^{-1}$)		σ_F ($Gs \cdot cm^3 \cdot g^{-1}$)		$\chi_\sigma \times 10^5$ ($cm^3 \cdot g^{-1}$)		η_F	
(wt. %)	300 K	77 K	300 K	77 K	300 K	77 K	300 K	77 K
0	0.209	—	0	—	2.55	—	0	0
19.1	0.235	0.675	0.024	0.041	2.19	6.70	4.4×10^{-4}	7.5×10^{-4}
27.1	0.323	1.447	0.084	0.155	2.53	11.8	1.54×10^{-3}	2.84×10^{-3}
35.4	—	2.177	—	1.240	—	9.9	—	2.28×10^{-2}
46.0	0.93	2.184	0.624	1.949	3.24	11.6	1.14×10^{-2}	1.74×10^{-2}
51.2	14.36	14.95	12.80	13.78	16.7	12/3	0.235	0.235

Notes: $\sigma_s = \sigma(9446\,Oe)$ represents the magnetization in the field of 9446 Oe; $\sigma_F \Rightarrow \sigma(0)$ represents extrapolation of the magnetization to zero field; $\chi_\sigma \cdot = \chi(9446\,Oe)$ represents the specific magnetic susceptibility in the field of 9446 Oe; $\eta_F = \sigma_F/\sigma_s(Ni)$ represents the mass portion of metallic Ni assuming that all ferromagnetism of the sample comes from Ni, σ_{s},(Ni) $= 54.5\,Gs \cdot cm^3 \cdot g^{-1}$.

The observed nonmonotonic dependence of the σ_s and χ_σ on the time of thermolysis is caused by a drastic increase of ferromagnetic phase formation rate at the end of conversion. At the end of gas evolution, $\eta_F \approx 0.23$–0.25; that is, only 25% of Ni atoms are in a ferromagnetic phase, and the remaining Ni atoms are in a low-magnetic phase. The yield product has the following magnetic characteristics: $\sigma_s(300\,K) = 14.36$ (300 K) and 14.95 (77 K) $Gs \cdot cm^3 \cdot g^{-1}$; $\chi_\sigma \times 10^5 = 16.7$ (300 K) and 12.3 (77 K) $cm^3 \cdot g^{-1}$; $H_c = 8.94$ (300 K) and 53.6 (77 K); $j_r = 0.02$ (300 K) and 0.133 (77 K), where H_c and j_r are the coercive force and the coefficient of rectangle, respectively.

Thus, thermolysis of even relatively simple metallopolymer systems takes place on the multichannel routes, and the properties of the yield products are chiefly determined by the thermolysis conditions.

3. THE PECULIARITIES OF THERMOLYSIS OF METAL-CONTAINING MONOMERS

This perspective method of metal–polymer nanocomposites formation involves the synthesis of both a nanoparticle and its stabilizing polymer matrix in one place (essentially in one stage). Such an approach is conceptually unique, and the systems under consideration are chemically self-regulating ones and thus they have embodied the best solution of the given problem (i.e., the nanoparticles formation and stabilization in polymer systems). Although at present the method is realized on transition metal acrylates and maleinates only [36], there are no principal limitations for its use with the other types of metallomonomers.

The kinetic studies of the thermal transformations were carried out in the self-generating atmosphere of many simple and clustered acrylates [4–9]: $Co(CH_2{=}CHCOO)_2 \cdot H_2O(CoAcr_2)$, $Ni(CH_2{=}CHCOO)_2 \cdot H_2O(NiAcr_2)$, $Cu_2(CH_2{=}CHCOO)_4(CuAcr_2)$, $Fe_3O(OH)(CH_2{=}CHCOO)_6 \cdot 3H_2O(FeAcr_3)$, their co-crystallizates $[Fe_3O(OH)(CH_2{=}CHCOO)_6] \cdot [Co(CH_2{=}CHCOO)_2]_{2.4}(FeCoAcr_{10})$ and $[Fe_3O(OH)(CH_2{=}CHCOO)_6] \cdot [Co(CH_2{=}CHCOO)_2]_{1,5} \cdot 3H_2O$ (Fe_2CoAcr_9), $NiAcr_2$ and $FeAcr_3$ co-crystallite with atomic ratio of $[Fe] : [Ni] \approx 2$ (Fe_2NiAcr_9), maleinate $Co(OCOCH{=}CHCOO) \cdot 2H_2O$ (CoMal), and acid maleinate $Fe_3O(OH)(OCOCH{=}CHCOOH)_6 \cdot 3H_2O$ $(FeMal_6)$. It should be noted that the IR spectra of Fe_2NiAcr co-crystallite cannot be considered as a superposition of the adsorption spectra of two individual compounds. In particular, for $Ni(Acr)_2$ in the region of 920–1020 cm^{-1}, the shifting of the band from 967 to 973 cm^{-1} is observed. Also a new sharp adsorption band at 1001 cm^{-1} appears. The appearance of new sharp bands in the IR spectra of the Fe_2NiAcr_9 co-crystallite is interpreted as a formation of a co-crystallite with a perfect crystal structure. Some common features of thermal behavior of the monomers being studied can be presented as follows [37–45].

The thermal decomposition is accompanied by a gas evolution and mass loss of the samples. A commonness of the processes lies in the fact that the transformations go through three main macrostages with different temperatures:

1. The initial monomers dehydration (desolvation) at 303–473 K
2. The stage of a solid-state homo- and copolymerization of the dehydrated monomer (at 473–573 K)
3. The produced polymer decarboxylation to a metal-containing phase and oxygen-free polymer matrix at $T_{ex} > 523$ K (for copper acrylate, $T > 453$ K) with an intense gas emission

The gas evolution kinetics (Table 3.5) (with reference to a value $\eta(t)$) for all compounds being studied in a general way can be approximated adequately by the expression

$$\eta(t) = \eta_{1f}[1 - \exp(-k_1\tau)] + (1 - \eta_{1f})[1 - \exp(-k_2\tau)] \qquad (17)$$

where k_1 and k_2 are effective rate constants; $\eta_{1f} = \eta(t)$ at $k_2 t \rightarrow 0$ and $k_1 t \rightarrow \infty$; $\tau = t - t_{\Pi}$; t_{Π} is the sample warming-up time.

Thermolysis of the low-molecular-weight analogue $Ni(Prop)_2$ follows the law of first-order reaction:

$$\eta(t) = 1 - e^{-k1t}(1) \qquad (18)$$

TABLE 3.5. Thermolysis Kinetic Parameters of Transition Metal Unsaturated Carboxylates

			η_{1f}, $\Delta\alpha_{\Sigma,f} = A\exp[-\Delta H/(RT)]$			$k = A\exp[-E_a/(RT)]$	
MR_n	T_{ex} (K)	η_{1f}, $\Delta\alpha_{\Sigma,f}$	A	ΔH (kJ/mole)	k	A	E_a (kJ/mole)
$CuAcr_2$	463–513	η_{1f}	1.8×10^4	48.1	k_1	9.5×10^{11}	154.7
		$\Delta\alpha_{\Sigma,f}$	3.6	12.5	k_2	9.2×10^{11}	163.0
$CoAcr_2$	623–663	η_{1f}	1.0	0	k_1	3.0×10^{14}	238.3
		$\Delta\alpha_{\Sigma,f}$	1.55	0	k_2	0	0
	473–573	η_{1f}	1.0	0	k_1	4.2×10^{21}	246.6
$FeAcr_3$		$\Delta\alpha_{\Sigma,f}$	1.6×10^2	25.5	k_2	0	0
	573–643	η_{1f}	1.0	0	k_1	1.3×10^6	127.5
		$\Delta\alpha_{\Sigma,f}$	1.7×10^2	26.3	k_2	0	0
$NiAcr_2$	573–633	η_{1f}	2.6	1.1	k_1	1.7×10^{17}	242.4
		$\Delta\alpha_{\Sigma,f}$	1.4×10^{11}	125.4 (<613 K)			
		$\Delta\alpha_{\Sigma,f}$	1.2	10.5 (>613 K)	k_2	7.5×10^8	156.8
$FeCoAcr_{10}$	613–663	η_{1f} = 0.45(663 K) – 0.65(613 K)			k_1	2.3×10^{12}	206.9
		$\Delta\alpha_{\Sigma,f}$	5.25×10^2	7.5	k_2	6.0×10^8	137.9
Fe_2CoAcr_9	613–663	η_{1f} = 0.35(663 K) – 0.50(613 K)			k_1	2.6×10^{12}	204.8
		$\Delta\alpha_{\Sigma,f}$	1.9×10^2	6.0	k_2	6.6×10^5	125.4
CoMal	613–643	η_{1f}	1.0	0	k_1	1.6×10^6	125.4
		$\Delta\alpha_{\Sigma,f}$	1.3×10^2	23.4	k_2	0	0
$FeMal_6$	573–643	η_{1f}	0.59×10^7	23.4	k_1	3.3×10^7	133.8
		$\Delta\alpha_{\Sigma,f}$ = 4.78(573 K) – 7.40(643 K)			k_2	1.0×10^7	110.8

and the monomer precursors $Ni(acr)_2$ and $Fe_2Ni(Acr)_9$ decompose according to the equation of the first-order rate of two parallel reactions:

$$\eta(t) = \eta_\infty\left(1 - e^{-k1t}\right) + \left(1 - \eta_\infty\right)\left(1 - e^{-k2t}\right) \tag{19}$$

It is interesting that the kinetics of thermolysis for the $Fe_2Co(Acr)_9$ co-crystallite is also approximated by equation (7). The effective energy of activation for the reaction rate k_1 and k_2 are higher (204.8 and 125.4 kJ/mole) than that for $Fe_2Ni(acr)_9$ (129.6 and 79.4 kJ/mole, respectively). The same is true for the values of the preexponential factors (2.6×10^{12} and $6.6 \times 10^5\ sec^{-1}$). At the same time, the values of k_1 for $Fe_2Ni(Acr)_9$ and k_2 for Fe_2CoAcr_9 are close to each other (Table 3.5).

The studied $MAcr_n$ compounds can be ordered according to their transformation initial rates (i.e., by gas emission ability decrease) in the following manner: Cu ≥ Fe > Co > Ni.

The analysis of the gaseous products thermal transformation and the solid product composition (decarboxylated polymer, including metal or its oxide) allows us to determine a general scheme of the metal acrylate thermal transformations:

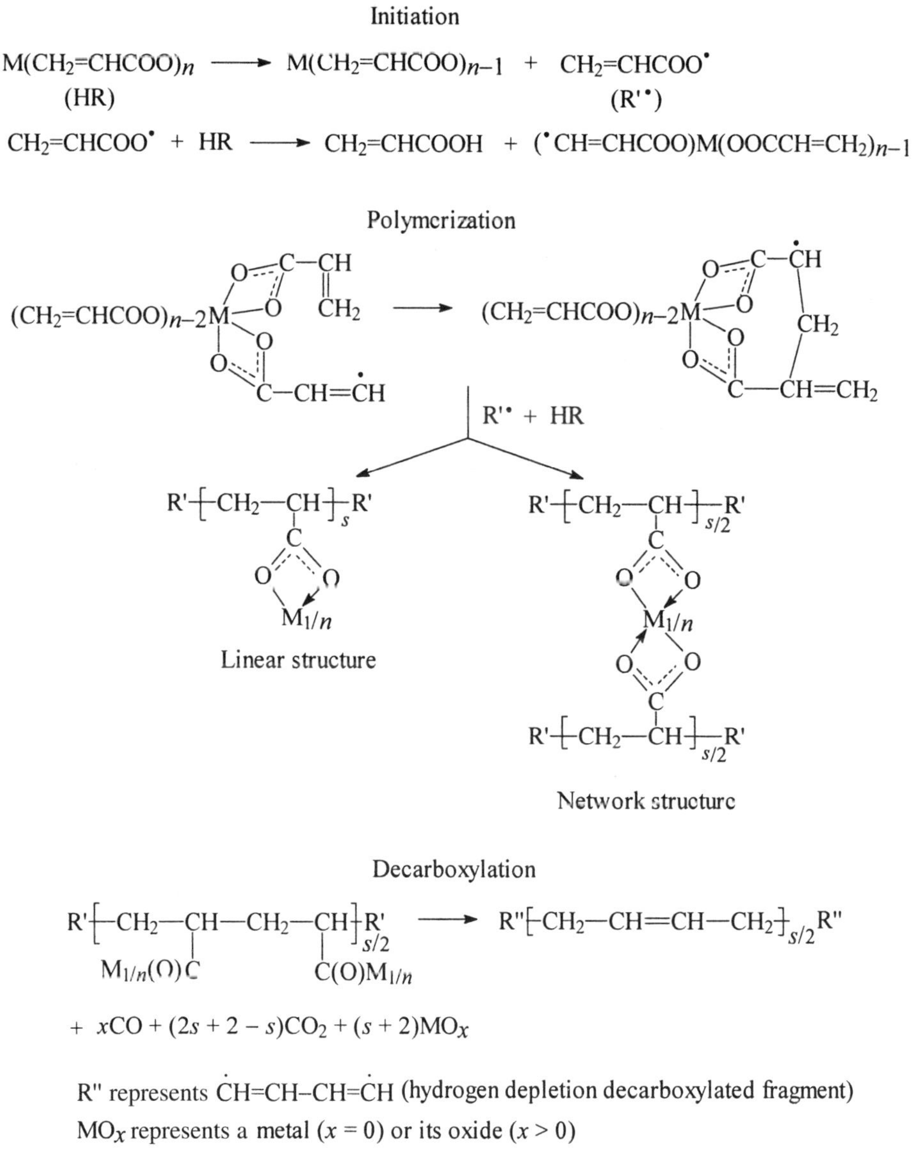

Scheme 3.4.

One of the main transformations is an origin of acrylic CH_2=CHCOO radical in the primary decomposition act that initiates a metal-containing monomer polymerization with a subsequent decarboxylation of metal-containing units. The process temperature has a pronounced effect on the products' yield and their composition. Using EXAFS data [41, 46], it was shown that even at the dehydration stage the metal ligand's surroundings become changed, and during the process such reconstruction is growing. At high temperatures the decarboxylation is accompanied by virtually complete removal of the oxygen-containing units from the polymer matrix.

The studies of nanocomposites' specific surface and its topography [47–49] (Table 3.6) show that they are powders without crystallinity at the distances correlating with transmitted light wavelength. The produced samples have a high value of specific surface (15–30 $m^2 g^{-1}$) and a corresponding dispersity. The mean size of the metal-containing particles (using data on S^f_{sp} and assuming the complete polymer decarboxylation) is ~20–30 nm. In some special cases ($CuAcr_2$, $CoAcr_2$, and partially $NiAcr_2$) a dispersion of the big aggregates is observed that results in a decrease of particles mean size and a growth of S_{sp}. At deep stages of the metal carboxylate pyrolysis, the small (<1 µm) opaque particles are observed. Sometimes the fractal-type chain structures of 50–70 µm in length consisting of the agglomerates from the 6–7 primary particles are generated. At the $Ni(Acr)_2$ (T_{exp} = 593 K), thermolysis on earlier stages of conversion the BR with Ψ = 1.27 (δ_{C-H} 830 cm^{-1}, ν_{C-H} 2930 cm^{-1}) appeared which disappeared with conversion and were not observed at T_{exp} – 573 K. The appearance of the BR also indicates that polymerization precedes the major gas evolution. The yield product of thermolysis of Fe_2NiAcr_9 is quite X-ray amorphous.

TABLE 3.6. Dispersity of Starting Metal Carboxylate Samples and the Thermolysis Products

Sample	$S_{0,sp}$ (m^2/g)	$S_{f,sp}$ (m^2/g)	$L_{OM,av}$ (µm)
$CuAcr_2$	14.7	48.0 (463 K)–53.8 (473 K)–43.8 (503 K)	5–50
$CoAcr_2$	20.2	24.1 (623 K)–42.1 (663 K)	100–150
$FeAcr_6$	15.0	15.0	1–5
$NiAcr_2$	16.0	55.0–60.5	60–100
$FeCoAcr_{10}$	9.0	13.6	5–10
Fe_2CoAcr_9	8.1	11.3	10–15
Fe_2NiAcr_9	8.5	13.5	100–200
CoMal	30.0	30.0	5–70
$FeMal_6$	24.0	26.0	30–50

In the case of $Ni(Prop)_2$ a partially crystallized phase (this means that X-ray diffraction spectra has a set of most intensive peaks characteristic for the given phase) of nickel carbide, Ni_3C, is observed. As follows from the X-ray diffraction data, the product of $Ni(Acr)_2$ thermolysis is a mixture of three well-crystallized phases $Ni–NiO–Ni_3C$ with the ratio of $Ni:NiO:Ni_3C \approx 0.51:0.81:1.0$ [Ni, a face-centered cubic, $a = 0.3524 \pm 0.002$ nm (0.3538 nm); NiO, a face-centered cubic, $a = 0.420 \pm 0.02$ nm (0.4177 nm); Ni_3C, $a = 0.265$ nm, $c = 0.433$ nm (in the parentheses the reference data are given).] Decreasing the T_{exp} to 613 K leads to an amorphization of the product. According to the X-ray data, the particle sizes are ~6.0 nm (Ni, NiO) and ~13.0 nm (Ni_3C); that is, the highly dispersed metal-containing phases are formed.

The topography and composition of the solid-phase products were studied using the electron microscopy and electron diffraction methods, and the results show a morphologically similar picture. There are the electron-dense metal-containing particles of oxides with a near-spheric form. They are presented as the individual particles and aggregates (3–10 particles), and they are uniformly distributed through the matrix space with a lower electronic density. The particles have a narrow size distribution with mean diameter $\bar{d}$ = 4.0–9.0 nm (Figure 3.5), and the distance between them in matrix is 8.0–10.0 nm. At the same time, there are some big aggregates in the form of cubic crystals 10.0–20.0 nm in size. The uniformity of the metal-containing particles' space distribution and the narrow size distribution suggest that the decarboxylation and new phase formation processes are largely homogeneous. The estimates show [9] that the mean distance between forming nanoparticle centers is in the range of 7.5–13.5 nm—that is, near the above-mentioned value of 8.0–10.0 nm.

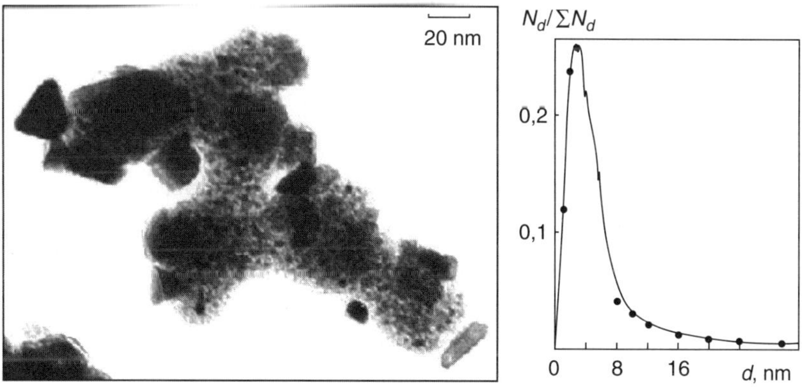

Figure 3.5. Transmission electron micrographs of the product of thermolysis of $Co(OOCCH{=}CHCOO)_2H_2O$ (623 K) and the particle's size distribution.

In the range of 573–633 K the compounds under study can be placed in a series in which their stability is increased: $Ni(Prop)_2$ < PAA-Ni^{2+} < $Ni(acr)_2$ ≤ $Fe_2Ni(acr)_9$ ≤ NiPacr.

It should be noted that the complication of the spatial organization of metal carboxylated groups leads to an enhancement of their thermal stability. The increasing of the portion of metal carboxylated groups in a polymer chain reveals the same effect. It was found that the end and intrachain metal carboxylic groups of metallopolymer and copolymer formed during thermal transformations also differ in their thermal stability.

In such a manner in the system of transition metal unsaturated carboxylates the nanoparticles synthesis was combined with their synchronous stabilization by the forming decarboxylated matrix [47, 48]. It is safe to assume that further investigations of metal-containing precursor thermolysis in high-molecular-weight media will extend our notions of the interactions and processes in the metal–matrix systems.

Thus, thermolysis of salts of unsaturated mono- and dicarboxylic acid and their polymers usually results in the formation of metal oxides in polymer matrix. At the same time, there exist systems where only metal nanoparticles are formed upon decomposition of some salts. For example, the salts of *m*-carborandicarboxylic acid decompose (at 500–650 K, 2–4 hr) as follows [50]:

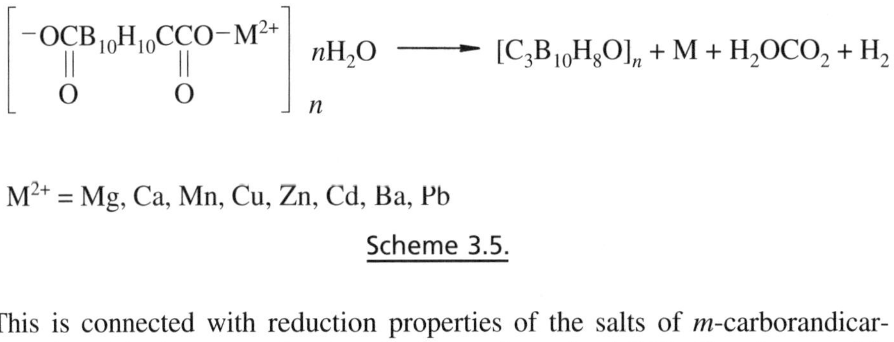

M^{2+} = Mg, Ca, Mn, Cu, Zn, Cd, Ba, Pb

Scheme 3.5.

This is connected with reduction properties of the salts of *m*-carborandicarboxylic acid when, upon heating, hydrogen is evolved [51]. Carrying out this reaction in polymers (for example, polyiminoimides, phenol-formaldehyde resins, etc.) yields the materials containing metal nanoparticles (5–30 nm, depending on temperature) and nonvolatile boron-containing derivatives.

4. THE FORMATION OF METAL SOLS IN POLYMERS BY THERMAL DECOMPOSITION OF THEIR CARBONYLS

Such reactions in the presence of polymers are the most used way to obtain the polymer compositions having a great fraction (up to 90% in mass) of the colloidal metal particles. Moreover, at present it is the main method to produce the ferromagnetic nanoparticles being incorporated into polymers due to a macromolecule chemisorption on the nanoparticles directly at their formation moment (*in situ*) including the processes without solvents.

The metal carbonyls are thermolyzed according to the following scheme:

$$Fe(CO)_5 \rightarrow Fe_2(CO)_9 \rightarrow Fe_3(CO)_{12} \rightarrow Fe_n^0$$

Scheme 3.6.

The mechanism of this process includes a chain ion–radical clusterization which in the case of iron carbonyl (it is also close to a decomposition of the IV–VIII group metal carbonyls) can be represented as follows [52]:

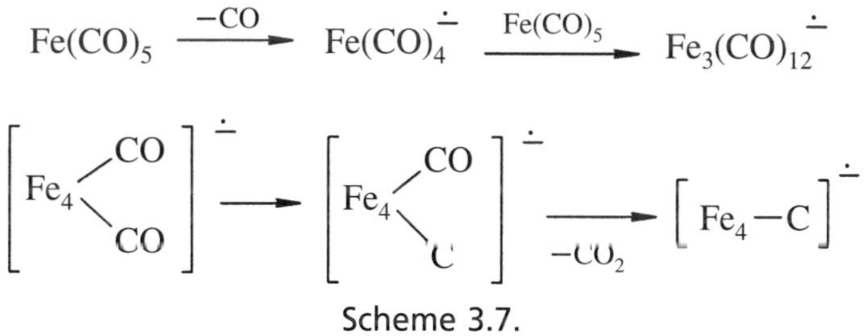

Scheme 3.7.

Under special conditions the metal carbonyls decomposition can be accompanied by formation of not only the metals but also their compounds. For example, one can obtain not only the α-ferrum but also Fe_3O_4 and FeOOH particles and ferric carbides using the following reactions:

$$Fe + CO \longrightarrow Fe_3O_4 + Fe_3C$$
$$Fe_3O_4 + CO \longrightarrow Fe_3C + CO_2$$
$$CO \longrightarrow 0.5CO_2 + 0.5CO$$

Scheme 3.8.

It was shown in [53, 54] that there are two methods to fabricate the homogeneous polymer-immobilized dispersions of colloidal metal particles (Fe, Co, Cr, Mo, W, Mn, Re, Ni, Pd, Pt, Ru, Rh, Os, Ir) using the precursors thermal decomposition. In the former case an "active" polymer solution (containing amino-, amido-, imino-, nitrilo-, hydroxy-, and other functional groups) is used. In an inert solvent a labile metal compound is gradually added to the solution (this operation creates the favorable conditions for the chemisorption interaction) followed by the suspension thermal decomposition at 370–440 K or by radiation.

In the second method a "passive" polymer is used that can react with an initial metal complex after only one ligand is lost (in our systems it is the CO group). Such polymers [e.g., PS (polystyrene), PB (polybutadiene), styrene–butadiene copolymers, etc.] being gradually added to the solution of the inert solvent of the initial complex at the proper temperature leads to the ligand separation, an anion complex bounding with the passive polymer followed by its thermal decomposition.

Some stages of this multistep process (especially the particle growth) are similar to the mechanism of the metal vapor condensation on the polymer [1]. The thermolysis in the presence of polymers was in greater extent studied for carbonyls of cobalt (see, for example, references 56 and 57) and iron [53, 56–58] (Table 3.7) when the process proceeds by a direct $M_a(CO)_b$ vaporization over polymers or by a preliminary adsorption on them.

Particularly the $Fe(CO)_5$ thermolysis in a xylol solution (*cis*-PB or styrene–butadiene copolymer at 408 K, 24 hr, Ar atmosphere) has gone through the successive stages and led to the formation gave rise to the ferrum tricarbonyl(diene) chains $[C_8H_{12}Fe(CO)_3]_n$:

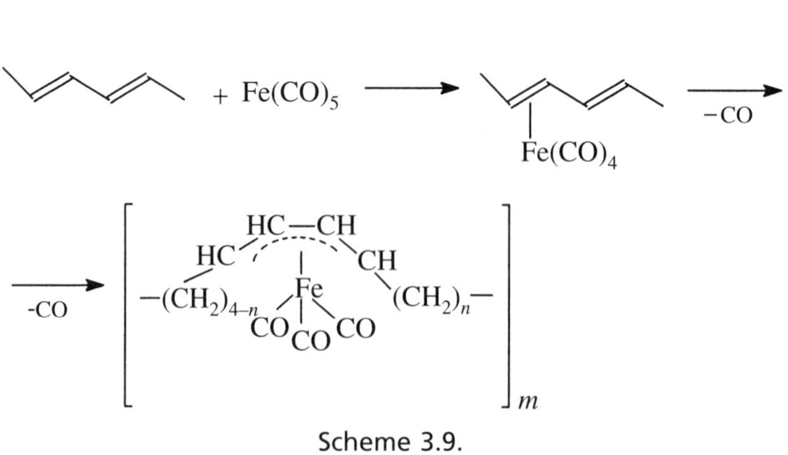

Scheme 3.9.

TABLE 3.7. The Thermal Decomposition of Metal Carbonyls in Polymer Matrices

$M_a(CO)_b$	Polymer Matrix	Thermolysis Conditions	Polymer content in mass (%) and weight and the polymer-immobilized NPs size (in nm)	References
$Co_2(CO)_8$	(Without polymer)	Toluene	>100	55
$Co_2(CO)_8$	PP(atactic)	Toluene	75%; >100	55
$Co_2(CO)_8$	PS	Toluene	75%; 10–30	55
$Co_2(CO)_8$	Polyurethane	Toluene	75%; 5–30	55
$Co_2(CO)_8$	Polychloroprene	Toluene	93%; 30–60	55
$Co_2(CO)_8$	Polyesters	Toluene	75%; 6–20	
$Co_2(CO)_8$	Terpolymer MMA–ethylmethacrylate—vinylpyrrolidone (33:66:1)	Toluene	20–30	25
$Co_2(CO)_8$	Copolymer vinylchloride–vinylacetate spirit (91:6:3)	Benzene chloride	75%; 7–47	55
$Co_2(CO)_8$	Copolymer MMA–vinylpyrrolidone (90:10)	Benzene chloride	75; 6–25	55
$Co_2(CO)_8$	Styrene–acrylonitrile copolymer (88:12)	Toluene	75%; 6–13	59
$Fe(CO)_5$	PB	Decalin, 413–433 K	5–15	53
$Fe(CO)_5$	PB	Decalin, 423 K	~6	58
$Fe(CO)_5$	*Cys*-PB	Xylol, 408 K		57, 58
$Fe(CO)_5$	*Cys*-PB	1% solution dioxane–xylol, 408 K	37% Fe	55
$Fe(CO)_5$	PTFE	DMFA, 413 K	5–15	58
$Fe(CO)_5$	Natural rubber	8% solution dioxane–xylol, 398 K	17% Fe	57
$Fe(CO)_5$	Polystyrene-block-polybutadiene	Dichlorobenzene, 418 K	7–8	53
$Fe(CO)_5$	Styrene–butadiene copolymer (5.8:1)	Decalin, 423 K	~6	60
$Fe(CO)_5$	Styrene-4–vinylpyridine copolymer (1:0.05)	*o*-Dichlorobenzene	2% Fe; ~6	61

TABLE 3.7. Continued

$M_a(CO)_b$	Polymer Matrix	Thermolysis Conditions	Polymer content in mass (%) and weight and the polymer-immobilized NPs size (in nm)	References
$Fe(CO)_5$	Styrene-4–vinylpyridine (1:0.1)	*o*-Dichlorobenzene	1.8 Fe; 16	60
$Fe_3(CO)_{12}$	*Trans*-PB	Benzene, 10% ethanol, 353 K	23% Fe	54
$Fe_3(CO)_{12}$	Styrene–butadiene copolymer (25:75)	Benzene, 10% ethanol, 353 K	8% Fe	62
$Fe_3(CO)_{12}$	PB (81% 1,2-chains)	Benzene-dimethoxyethane and 353 K	16% Fe	53
$Fe(CO)_5$	Isotactic PP (melt)		5% Fe	63–65
$Fe(CO)_5$	Atactic PP (melt)		5–30% Fe	63–65
$Fe(CO)_5$	PE (melt)		1–29% Fe; 1.5–7	63–65
$Cr(CO)_6$	PTFE (fluoroplast-40) (melt)		0.5–4% Cr; 1–5	63–65

At the thermolysis initial stage a very active $Fe(CO)_4^-$ anion is formed by reacting with the isolated double bonds, after which the chain double-bond isomerization proceeds generating π complexes with the ferrotricarbonyl residues. The product composed from the η^4-(butadienyl)ferrotricarbonyl chains with trans–trans and cis–trans units. The ferrotricarbonyl complexes having two double nonconjugated bonds are nonstable, and the intermolecular products in such a situation can be formed. A distinctive feature of the process is that the same type of ferrotricarbonyl complexes connected with allylic fragments were identified [66] at the $Co_2(CO)_8$ or $Fe_3(CO)_{12}$ interactions with polystyrene-polydiene block-copolymers. For polystyrene the $M(CO)_3$ fragments connected through the π-complexes with phenyl ring are formed from $M(CO)_6$ (M = Cr, Mo, W) [67].

Such polymer-immobilized π-allylic complexes undergoing thermal decomposition can form nanoparticles in polymer [68], during which the thermolysis can be carried out even without a solvent (for example, by heating of the formed metallopolymer films using an infrared lamp). The nanoparticle sizes depend on many factors such as the characteristics of the used dispersant polymer, its molecular mass (the optimal value is $M_w \approx 100{,}000$), and the nature of functional groups (L) and solvent. The polymer crystallization proceeds at the solvent removed from the reacting systems, but its supermolecular structure formation depends on the cluster particles concentration and their size distribution. The solvent nature is very important because the $M_a(CO)_b$ complexes in the basic solvents (e.g., in DMF) can disproportionately form the ionic type complexes where a metal formal charge equals +2. The product composition identified by IR spectroscopy depends on the component ratio:

$$nFe(CO)_5 + 6DMF \longrightarrow \begin{cases} Fe^{2+}(DMF)_6[Fe(CO)_4]^{2-} + CO \\ Fe^{2+}(DMF)_6[Fe_2(CO)_1]^{2-} + 7CO \\ Fe^{2+}(DMF)_6[Fe_3(CO)_{11}]^{2-} + 9CO \\ Fe^{2+}(DMF)_6[Fe_4(CO)_{13}]^{2-} + 12CO \end{cases}$$

Scheme 3.10.

In the general case, one has

$$Fe(CO)_5 + Nu \rightarrow [Fe^{2+}(Nu)_x]\,[Fe_n(CO)_m]^{2-}$$

Scheme 3.11.

Here Nu is a nucleophile (Py, *N*-methylpyrrolidone, etc., and $x = 2$–6), $n = 2$ or $m = 8$; $n = 3$, $m = 11$; or $n = 4$, $m = 13$.

The octacarbonyl-dicobalt transformation scheme in these conditions takes the form

$$2Co_2(CO)_8 + 12DMF \rightarrow 2[Co^{2+}(DMF)_6][Co(CO)_4^-]_2 + 8CO$$

Scheme 3.12.

The basic solvents and high temperatures are favorable to metal carbonyl bonding by polymers and allow us to obtain the stable colloidal dispersions (e.g., with the iron particles 5–10 nm in size) by a carbonyl thermolysis in the dilute polymeric solutions. Such nascent nanoparticles are very reactive; in addition, the small particles (<10 nm) are superparamagnetic and the larger ones (10–20 nm) exhibit a magnetic hysteresis [60].

In solvents with a low dielectric permeability (e.g., in toluene), the obtained Co particles size (2–30 nm with a narrow size distribution) can be controlled by the reaction conditions (temperature, reagents concentration, etc.) or by varying the polymer composition (tercopolymer of methylmethacrylate–ethylacrylate–vinylpyrrolidone [56]).

The nucleophilic fragments of macroligands (particularly in styrene-*N*-vinylpyrrolidone copolymer) bring about a polymer-catalyzed decomposition of the metal carbonyl [53] via the scheme

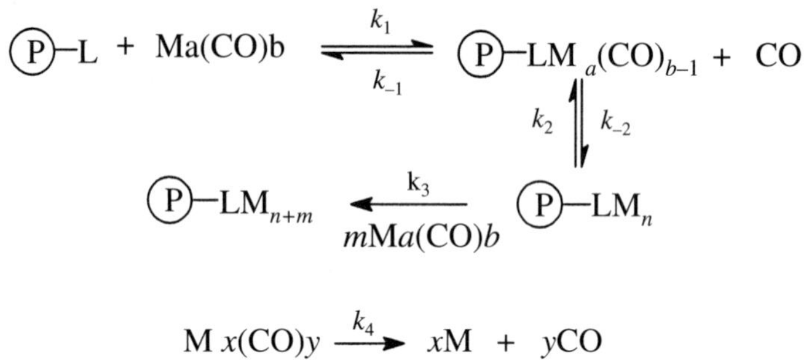

Scheme 3.13.

At the first stage a soluble macromolecular complex is formed with functional groups L. The complex transforms to a polymer-immobilized cluster particle that, after the cleavage of CO groups, undergoes an increase up to a nanoparticle. Actually we see the same development stages as in the above-considered processes of the nanoparticle formation in polymer—that is, the initiation, particle growth, and reaction termination (that can be presented as a simple disproportionation at the particle surface). The main condition for a proper matching of a functionalized polymer is connected with the greater reaction rate on the polymer surface to reduce the decomposition in the solution to a minimum—that is, to provide

$$k_1 + k_2 - k_{-1} - k_{-2} > k_4 \quad (20)$$

$$k_3 > k_4 \quad (21)$$

As a rule, the metal carbonyl decomposition in polar media proceeds by the main reaction routes. First is a main process—that is, the particles growth from the "hot" metal atoms and the formation of nanoparticle having 1–10 nm in size:

$$M_n + M \rightarrow M_{n+1},\ M_n + M_m \rightarrow M_{n+m}$$

Scheme 3.14.

Coincidentally with the forming nanoparticle sizes, a probability of the particles growth termination increases due to their surface noncovalent interactions with macromolecules, although such interactions are weak. First, for the interaction strength on the order of 10^{-4} J/m^2, even the particles of 1–10 nm in size will be captured by macromolecules, and the growth of the particles will be stopped (as a result of an screening effect); that is, the stronger the interaction, the smaller the particle size. For the little and modest-sized particles, the reaction with polymer is described by the following scheme:

$$M_n + \textcircled{P} \longrightarrow \textcircled{P}\!-\!M_n$$

Scheme 3.15.

where $\textcircled{P}$ is the functionalized polymer.

In effect, it is a chain termination, which usually leads to a nonmagnetic material formation. Second, the disproportionation reaction products interact with the polymer chain, giving rise to some new side reactions (the reticulation, the chain destruction or isomerization, and the mononuclear carbonyl complex immobilization). As, for instance, in 1,4-*cis*-PB (containing 92% of 1,4-*cis*-, 4% 1,4-*trans*-, and 4% 1,2-units with $M_w = 246{,}000$) after interacting with $Fe_3(CO)_{12}$ during 2 hr at 350 K, a geometrical isomerization of the chain 1,4-units proceeds and the 1,4-trans units content rises up to 76% [57, 59].

Both of processes run spontaneous, and their competition (connected with the differences of the reacting particles nature and the various reaction conditions) is characterized by a ratio of the thermolysis and the M particles diffusion to the "hot" metal centers in a solid polymer matrix. If the big particles (i.e., big in relation to the distances between the chains, crystalline blocks, or polymer lattice points) are introduced in the system, the polymer structure and its physical and mechanical properties degrade.

In the absence of solvent, the carbonyls thermolysis can be accompanied by some peculiarities. In some systems the Fe, Cr, Co, Ni, and Mn carbonyls can have a catalytic action on the polymer carbonization and graphitization processes—for example, in petroleum, mesogenic pecks (the specific matrices with polycondensed aromatic structures) are able to act as π-ligands stabilizing the metal clusters [70, 71]. Even the 1% content of metal carbonyl catalyzed the mentioned processes, but the iron content rise up to 5% (at 423 K) leads to noticeable growth of the structures similar to that of graphitic layer compounds.

It can be noticed too that polymers having more polar groups promote the finest particles growth, and the same can be said about the metal carbonyls concentration in polymer. For a thermal $Fe(CO)_5$ decomposition in PEHP matrix a bimodal distribution of the ferrum particles size is observed (with the mean sizes of the distribution modes equal to 2–3 and 4–10 nm) and the particles content is ~20% and 80% (in volume) correspondingly as is shown on Figure 3.6 [72]. The majority of the oxide Fe_3O_4 particles has dimensions ~5 nm.

An original method of metallopolymer production by precursors thermal decomposition is to localize the particles being formed due to a fast monomolecular decay of the solutions containing the metal compounds in polymer melts—that is, in the natural voids of the polymer matrix (as PE, PP, PTFE, etc.). Such materials are called "cluspol" [30, 63–65], and for their production it is necessary to provide the most possible melt temperature, which must be considerably above the temperature of the carbonyl decay initiation. For this purpose the carbonyl dilute solutions are used under these conditions, providing the ultimately fast and complete removal of the split out ligand from the reaction system. Such an approach has many advantages because the temperature rise from one side promotes the metal-forming precursor decomposition and from other side decreases the by-products yield. Furthermore, in a melt (as

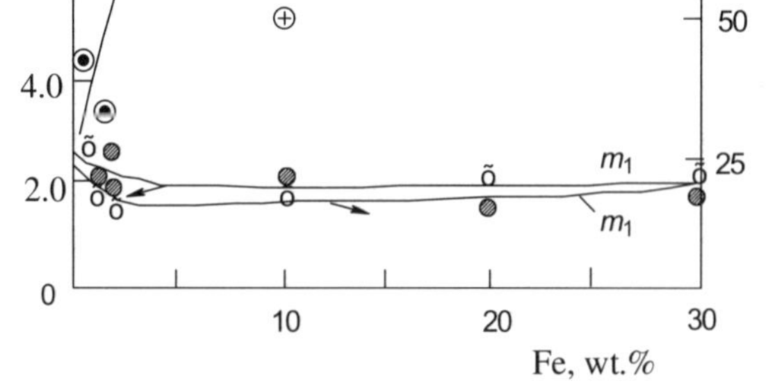

Figure 3.6. The curves of Fe nanoparticles' distribution on the size in a PELD matrix (m_1 and m_2 are the first and second modes of distribution, respectively).

distinct from a solution) a short-range order of the initial polymer structure is preserved, and its voids become readily available for the forming nanoparticles localization.

The obtained polymer-immobilized nanoparticles are characterized by a relatively high dispersion, a uniform distribution through polymer volume, and nonreversible macromolecules sorption on their surface. They are localized primarily in the disordered interspherulite regions of the matrix, between the lamellas or in the spherolite centers. Such a disposition hampers a segmental motion of the amorphous phase for a free volume decrease and a possible interlacing. As a result, the polymer thermostability increases (e.g., the T_m value for atactic PP increases by 50–80°C) and the products represent the monolithic (pseudocrystalline) materials. The spatial periodicity of the ferrum clusters in such structures depends on concentration and averages between 9 and 12 nm for high (20–30% in mass) and 20–22 nm for low (2–3% in mass) concentrations. The particle size distribution is narrow (the half-width is about 1 nm), and it is also possible that the nanoparticles in the matrix intermolecular voids favor the polymer crystalline phase destruction and its transformation to an amorphous state.

It is of interest that neither physical nor chemical properties of such materials allow us to detect a metallic phase for the strong polymer–NP interactions that provide the polycrystalline matter formation. In other words, such products at the same ingredients ratio present the monophase metallopolymers. The nanoparticles' size and distribution depend on the polymer's chemical nature; that is, PEHP, an alternating copolymer of ethylene and carboxide, polyacetonaphthalene, polycarbonate, and so on [72]? With a rise of the nanoparticles' contents in such polymers, naturally their coherent scattering region (CSR) is wider and the crystallinity degrees of PEHP and polycarbonates rise (Table 3.8). At the same time, the polymer matrices cannot preserve the nanoparticles from an oxidation (for example, the ferric nanoparticles in PEHP with the mean size 50 nm after oxidation transforms into Fe_3O_4 particles of the same size).

It is of interest that metal carbonyls such as $Co_2(CO)_8$ form complexes with polyionic ligands (such as ethynylbensene-$Co_2(CO)_6$) and can be decomposed explosively [73] at 513 K (generating CH_4 and H_2) and at 1070 K (forming multilayer nanotubes). It is possible that a metallized polyione pyrolysis can lead to the single-wall nanotube formation too [74], just as the $Fe(CO)_5$ pyrolysis (at 1320 K) in the presence of other carbon sources [75]. The thermal treatment of platinum and cobalt containing poly(phenylene diacetylenes) leads to the Pt (1–3 nm) or Co (5–100 nm) cluster formation into the carboglasses [76].

Another way of the metallocomposite production by the precursor thermolysis is the metal carbonyl decomposition in the halide-containing matrices. Using the above-mentioned classification, such methods can be related to the "passive" ones. The PETF modifications can be performed by Fe^{3+} or Mn^{4+}

TABLE 3.8. X-ray Diffraction Characteristics of Fe-composites Materials

Composite	CSR (nm)	Crystalline Phase Content (%)	Nanoparticle
An alternating copolymer of ethylene and carboxide	20	—	—
+10% Fe	21	—	—
+30% Fe	23	—	—
+50% Fe	23	—	α-Fe, 16.5 nm Fe_3O_4, 5 nm
Polyacenaphthylene	2.5	—	—
+0.1% Fe	2.6	—	—
+1% Fe	2.6	—	—
+10% Fe	2.6	—	Fe_3O_4, 3.5 nm
Polycarbonate	8.0	65	—
+0.5% Fe	9.0	60	—
+1.0% Fe	9.0	60	—
+5.0% Fe	10.0	50	—
+10% Fe	10.0	40	—
PEHP	21.5	30	—
+1.0% Fe	21.5	30	—
+5.0% Fe	26.5	20	—
+10% Fe	25.0	15	Fe_3O_4, 5.5 nm

oxides [77–79] using the sorption of the proper carbonyls and their subsequent decomposition under the action of $KMnO_4$ or H_2O_2. After these reactions the oxides are introduced into the amorphous regions of PETF. The most common situation is that nanoparticles, derived in a polar solvent, are forming the complicated complexes with metal carbonyls that are able to react with fluoropolymers. The main factors affecting a selective interaction of $Fe(CO)_5$ or $CO_2(CO)_8$ with PETF during the thermolysis are generalized in references 80–82. The IR-spectral method allows us to fix the cation–anion $[Fe(DMF)_6]^{2+}[Fe_3(CO)_{11}]^{2-}$–PETF and $[Co(DMF)_6]^{2+}[Co(CO)_4]^{2-}{}_2$–PETF complex formation during adsorption that allows us to perform the kinetic studies of the process as a whole and perform the thermal decomposition in particular. The intermediates thermolysis (413 K, DMF) is a first-order reaction, but these processes' rate constants and activation energies differ essentially (Table 3.9).

Attention is drawn to the fact that the rate constant of polymer-immobilized cobalt carbonyl is very high (its order of magnitude is compar-

TABLE 3.9. The Characterization of Thermolysis of Metal Carbonyls in PETF Matrix

	PETF–$Fe(CO)_5$	PETF–$Fe_3(CO)_{11}^{2-}$	PETF–$Co(CO)_4^-$
$k \times 10^{-5}$, s^{-1}	8.41	12.2	104
E_a (kJ mole^{-1})	142.6	51.62	45.80

able with the value for ferrum carbonyl), which correlates with a higher reactivity of the cobalt carbonyl anion. The very high value of E_a for the neutral $Fe(CO)_5$ decomposition in PETF (as compared with the immobilized compounds) suggests that the reactions have different mechanisms. The Fe and Co carbonyls' thermolysis *in situ* leads to the heterogeneous metal domain formation in a polymeric martix. The polymer-connected carbonyls (both for Co and Fe) generates the systems with uniformly distributed ferromagnetic particles having 5–10 nm in size and mean surface from 80 to 700 nm^2 per particle.

The electron diffraction analysis of such systems [81] have revealed the formation of γ-Fe_2O_3 and FeF_2 particles for initial $Fe(CO)_5$ and revealed Co, Co_2O_3, and CoF_2 for initial $Co_2(CO)_8$. The metal fluoride formation is connected with the reactions involving the polymer chain and is accompanied by the C–F bond splitting:

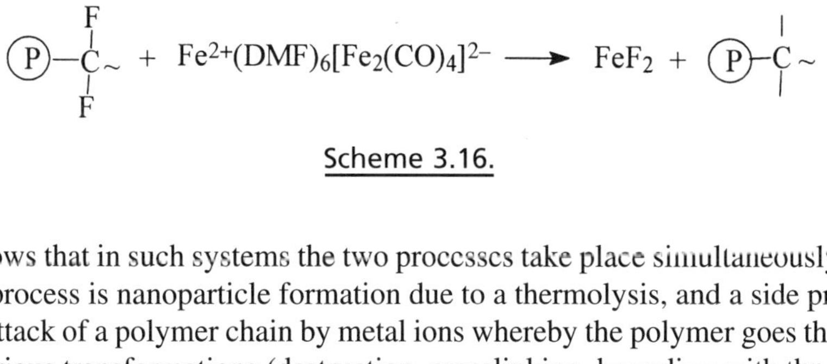

Scheme 3.16.

It follows that in such systems the two processes take place simultaneously. The main process is nanoparticle formation due to a thermolysis, and a side process is an attack of a polymer chain by metal ions whereby the polymer goes through the various transformations (destruction, crosslinking, bounding with the metal complexes, etc.). Moreover, the secondary carbonium ions (formed from PETF) are very active and can be involved in many chemical reactions. As an illustration, atomic zinc interacting with the PETF surface formed ZnF_2 [83, 84]. The characteristics of the bonds between formed clusters and nanoparticles with polymer matrix and their topography are not established conclusively, but it is evident that the polymer–metal particle interactions are defined by the reactivity of metallocenters formed during the decomposition.

It is possible [85] that introducing $Co_2(CO)_8$ into the toluene solution of styrene block-copolymer with 4-VPy also gives rise to the cation–anion complexes localized into a micelle having 4-VPy units (similar to DMF) with composition $[Co(4\text{-}VPy)_6]^{2+}$ $[Co(CO)_4]^{2-}$. A topochemistry of such nanoparticles and their sizes or forms are defined by molar ratio [4-VPy]/[Co], because the bounded (in micelles) cobalt after a fast thermolysis at 383 K forms small particles growing at the expense of dissolved $Co_2(CO)_8$ up to the spheroids (~10 nm in size) or the star-like particles (with the nonregular forms and mean sizes 20–23 nm) built up from the initial anisotropic clusters. Only if $Co_2(CO)_8$ is present in large excess do the nanoparticles begin to originate and grow out of the micelle (cubes with mean size 21 nm). As a rule, such particles are vastly larger than the particles formed, as an example, during $CoCl_2$ reduction by superhydride $Li(C_2H_5)_3BH$ [86].

In the last few years there have been new trends in the metal carbonyl decomposition in the presence of polymer namely the sonochemical reactions [81–91] connected with the known effect of the acoustical cavitation, including the bubbles initiation, growth, and explosive "blow-in" into a high-boiling-point solvent. The process is characterized by a high local temperature (several thousand of degrees) and pressure (150–200 MPa), along with the acoustical emission of the excited particles, called *sonoluminescence*. In our problems the acoustic field helped to generate the high-volatility metal compounds (particularly from metal carbonyls) with their subsequent agglomeration in the presence of organic polymer (e.g., PVP [92]) or inorganic (silica gel) polymer up to the polymer-stabilized nanoparticles. The general scheme of such sonochemical synthesis of nanostructured materials can be present as follows:

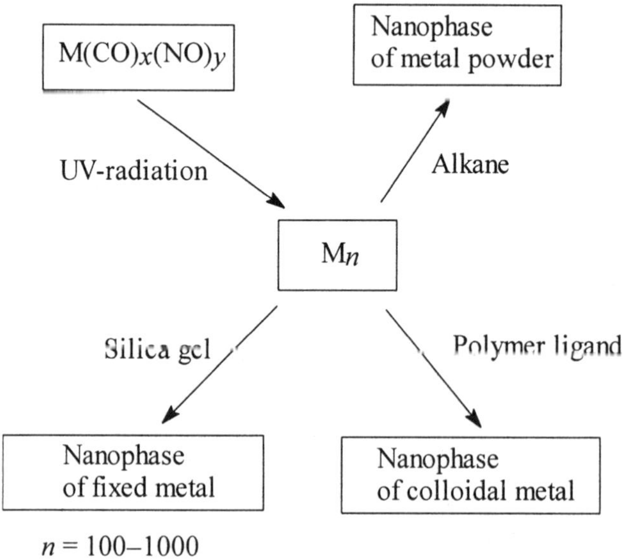

Scheme 3.17.

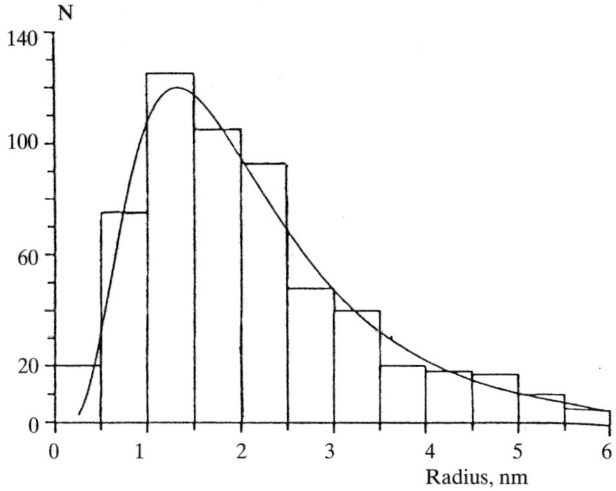

Figure 3.7. The experimental histograms of Fe/PFO nanoparticles' distribution on the size (*N* represents the total number of particles).

Such a reaction of $Fe(CO)_5$ (at 293–363 K, PVP) without ultrasonic radiation proceeds very slowly; and only after few days there, a material is formed with very low Fe content (2%, the isolated particles 2–5 nm in size). It is of interest that the sonochemical decomposition of $Fe(CO)_5$ does not proceed in the presence of PVP if THF is used as the solvent, but the reaction is very effective when anisole is used as the solvent and PFO is used as the polymer matrix [93]. A black product formed contains up to 10% (in mass) of the spheric particles of nonoxidized Fe (mainly γ-Fe, with little content of α-Fe) with 1–12 nm in size (the mean diameter is 3 nm, as shown in Figure 3.7). It is likely that the big particles present the flocks of little ones (~2–2.5 nm). The sonochemical synthesis allows us to produce the functionalized amorphous nanoparticles of ferric oxide with 5–16 nm in diameter [94]. The ultrasonic irradiation in the PFO presence allows us to also produce the stabilized nanoparticles of copper, gold, and so on. In the literature the findings are not about the bimetallic particle formation in the ultrasonic fields by carbonyl metal reduction in the polymer matrices' presence (as, for example, in the case of the carbon-supported Pt–Ru from $PtRu_5C(CO)_{16}$ reduced clusters [95]).

5. POST-THERMOLYSIS OF OTHER-TYPE PRECURSORS IN POLYMER MATRICES

Besides carbonyls, some other metal-forming precursors (usually formiates, acetates, oxalates, or organometallic compounds) are used to produce the polymer-immobilized nanoparticles by thermolysis. For example, a metal–polymer composition can be prepared by thermolysis of a copper formate triethylene–diaminic complex $[Cu(EDA)_3](HCOO)_2$ into PS through a common solvent (DMF) [96]. The complex decays at 443 K and forms metallic copper in a highly dispersed state. It is of interest that the process temperature in PS is at 20° lower than in a bulk copper formate, which may be associated with a polymer catalytic action on the complex thermal decomposition.

The mechanism of the process is that the polymer reactive centers promote the metal nucleation and aggregation, after which the thermolysis occurs and the metal-containing substance is redistributed. The maximum amount of copper being introduced in PS through a common solvent is about 10%. At the same time, the polymer presence increases the temperature of cadmium trihydrate–oxalate decomposition [97], and the decay products increase the initial temperature of PETF intensive destruction. The copper formate thermal decomposition in the highly dispersed PETF presence allows us to produce a metallopolymeric composition (20–34% of copper) where the NP size distribution is maximal at ~4 nm, without any chemical interaction between the components.

At high temperature thermolysis (1270 K) of cobalt acetate with PS, PAA, and PMVK present, the metal clusters that are catalyzing an oxygen electrolytic reduction had formed [98]. As the thermal decomposition of silver trifluoro-acetyl-acetonate at 613 K occurs in polyimide film [99–101], the film becomes metallized (a structure called "film on film") and a nanocomposite is formed with high surface conductivity and light reflection coefficient (above 80%).

There are some examples of the nanoparticle formation in polymers by metal hydroxide decomposition [102]. The thermal decay of silver hydroxide ammonia solution in the isotactic PP melt (temperature range 543–563 K in polymethylsiloxane) occurs according to the following scheme:

$$4[Ag(NH_3)_2]OH \rightarrow 4Ag^0 + 8NH_3 + 2H_2O + O_2$$

Scheme 3.18.

The silver content in PP varies from 3% to 15% (in mass), the particle sizes change in a wide range (2–25 nm), and the small particles (<5 nm in diameter) are roentgen-amorphous but the bigger ones have a crystalline structure.

The Pd particles (with $S_{sp} = 26\,m^2/g$), produced by the decomposition of a palladium hydroxide fine dispersion in PVC, formed the space structures with the nodes where the highly dispersed Pd particles are located [103]. It must be noted that such a structure increases the temperature of PVC complete dehydrochlorination.

An original method of copper-containing composite production on a porous PELP base (obtained by crazing method) is connected with use of a Cu^{2+}–monoethanolamine complex [104]. The complex simultaneously combines the functions of an adsorption active medium, a complex-forming agent, and a copper ion reductant. Under the complex decay (at temperatures from 363 to 398 K), the copper nanoparticles are forming in the porous PE. They have a wide size distribution (from 60 to 320 nm), and the metal mass content is about 8%. It is worthy to note that even a little amount of such filler (1–2 vol. %) improved considerably the polymer mechanical properties. Also, we must note that there is a recently developed universal method for the synthesis of nanocomposites on metal nanoparticles (or their sulphide) embedded in polymer matrices by thermolysis of metal n-alkanthiolates $M_x(SC_nH_{2n+1})$, where $n = 12$, 16, and 18 at moderately low temperatures (390–470 K) in a polystyrene matrix [105]. Metal nanoparticles are homogeneously distributed in a polymer matrix and have mean sizes of 7 nm (Pd) and 10 nm (Ag and Au). The size of particles can be controlled by the concentration of the precursors, as well as by the thermolysis time and temperature.

Some examples are of interest from the very limited findings about nanoparticles formation by organometallic compound decomposition in polymers. The cryochemically synthesized bisarene complexes $(C_6H_5CH_3)_2M$ (M = Ni, Co) were subjected to a fast decomposition in PELD at a temperature higher than that of the complex decay. The metal content in material produced achieved 1.4–4.3% [106]. The cobalt nanoparticles (mean size ~1.6 nm, a narrow size distribution) were obtained by H_2 reduction of organometallic precursor $Co(\eta^3\text{-}C_8H_{13})(\eta^4\text{-}C_8H_{12})$ at PVP presence at 273–333 K [107]. As a precursor, bis(cyclooctatetraen)-iron $Fe(C_8H_8)_2$ can be used too [108]. By contrast, the similar ruthenium complex $Ru(\eta^4\text{-}C_8H_{10})(\eta^6\text{-}C_8H_{12})$ had decomposed slowly even at 0.3 MPa in the presence of PVP. The Mo, Cr, and W hexacarbonyls do not photochemically decompose at high pressures (~35 MPa) of N_2 or H_2 and low temperatures (~30 K) in polyethylene matrix but gave the different substituents only [109] as well as ferrum carbonyls in PE matrix at 190 K [110]. The nanoparticles of Ru (~1.2 nm) [111], Pd, and Pt [112, 113] also can be produced from the organometallic precursors in PVP matrix at 298 K. Nickel dicyclo-octadiene $Ni(COD)_2$ can spontaneously decay (at room temperature in PVP dichloromethane solution), forming the particles with mean size 2–3 nm [114]. The nanocomposites can be obtained in the same manner (by precursor decomposition) based on Nafion-type films of ionic polymer [115–117].

An original method of orgametallic compounds impregnation into the PS, PEHP, and PETF amorphous regions and polyacrylates with their subsequent structural modification and clustering is based on the usage of some liquids in a supercritical state (CO_2, 8–25 MPa, 303–313 K), as was shown for cymanthrene $(\eta^5\text{-}C_5H_5)Mn(CO)_3)$ [118]. In another method the platinum dimethylcyclooctadiene(II) precursor was dissolved in CO_2 and the thin films of poly(4-methyl-1-pentene) or PETF were impregnated by the solution [119, 120]. After reduction (by thermolysis or hydrogenolysis) the polymeric nanocomposites are formed with Pt particles (15–100 nm in size). The kinetic data of impregnation for manganese cyclopentadienyl-tricarbonyl and copper hexafluoroacetylacetonate, dissolved in the supercritical media (CO_2), show that the process is very intense in the polymer amorphous regions, very weak in the partially crystalline regions, and practically absent in crystalline ones [116].

In summary of this section, it must noted that, in spite of numerous studies, nowdays we know very little about carbonyl hydrides and other substituted (mixed) carbonyls thermolysis in polymeric systems, as well as in reactive plastics. For example, in some experiments the decomposing metal carbonyls were placed into an epoxide resin heated up to the nanoparticles deposition on the forming polymer surface [121]. It is possible that the highly reactive metal particles in such systems can initiate the epoxy cycle cleavages followed by a three-dimensional space structure formation. Iron carbonyl being decomposed into polybenzimidazole suspension (in transformer oil at 473 K) forms the ferrum nanoparticles (1–11 nm) capable of polymer thermostabization [122].

It should be stressed that the considerable simple picture of the nanocomposites formation is observed only when for the thermolysis the thermally unstable substances are used (such as the metal-containing precursors); otherwise the polymer matrix thermal destruction (i.e., the reactions connected with the chemical bond breakage and radicals generation) begins to play an important role along with the nanoparticle formation processes. The thermodestruction products present low-molecular-weight volatile compounds of complex type (including monomer) and nonvolatile residue, subsequently becoming a carbonized structure. On the other hand, the highly dispersed metals of alternating valence or their lowest oxides (Fe, FeO, Ni, Cu, etc.) with a defective structure are often introduced into a polymer, especially to inhibit its thermal or thermooxidative destruction processes [123, 124]. For example, the small amount of iron particles in PEHP (0.05–1.0% in mass) increases its thermal stability in comparison with the pure polymer. Of interest is the method of "non-chain" inhibition of thermal oxidative destruction of thermostable polymers [124] based on oxygen binding with highly effective acceptors—that is, metal nanoparticles formed in polymer matrix (i.e., acceptors for oxygen are formed directly in matrix).

It must be noted that the metal-containing precursors and the nanoparticle interactions with the polymer matrix (formed under the pyrolysis), as well as the product topography, are not sufficiently studied, but it is no question that a role of metal centers and their activities are very important for the general picture of the metal particles' interaction with a polymer matrix.

6. THE COMPUTER MODELING OF ASSEMBLING OF METALLOPOLYMER NANOCOMPOSITES DURING CONTROLLED THERMOLYSIS

As shown above, the synthesis of metallopolymer nanocomposites through solid-phase thermal transformation of both macromolecular metallocomplexes and metal-containing monomers—in particular, transition metal carboxylates of unsaturated acids—is a very promise and perspective way. In such processes the formation of nanoparticles, along with their matrix stabilization, occurs simultaneously and is connected with the self-organization of nanoparticles during thermal transformation. This process results in the formation of composites containing (depending on the nature of the metal-containing groups) metal or metal oxide nanoparticles with a quite narrow distribution of size and is homogeneously located in the volume of stabilized matrix.

Although the solid-phase thermal process combines a set of multistage physicochemical transformations [125, 126], on the whole it can be represented by the following simplified scheme of transformations [127].

Thermal decomposition of metal-containing polymer fragment:

$$ML_r \rightarrow M^* + rL \tag{22}$$

Cluster formation and nanoparticle growth:

$$M \xrightarrow{M} M_2 \xrightarrow{M} M_3 \xrightarrow{M} M_{n-1} \xrightarrow{M} [M_n] \tag{23}$$

Polymerization of demetallized polymer ligand (L_r):

$$rL \rightarrow L_r \tag{24}$$

where M is a metal, M* is a metal atom or metal-containing molecule, and L is an unsaturated ligand.

A current wide set of experimental data concerns mainly the yield characteristics of systems such as nanoparticle distribution in the size and matrix space, physical and chemical properties of the matrix formed, and so on. At the same time, the investigations on the mechanism and kinetics of nanocomposite formation and on understanding how these parameters change during thermal transformations are absent because of the difficulties of the experimental studies of the kinetics of such processes.

Computer modeling is the simplest and most productive approach for studying kinetic processes. In recent years the varied variants for solving these problems have been developed [127, 128]. Thus, in the frame of the model of diffusive-restricted aggregation the computer modeling of the kinetics of nanoparticle formation in the solid phase during thermolysis of metal-containing macroligands is performed using the Delphi program in Windows [129]. As the basis of the kinetic model, a polymer medium of different structure—isotropic (globular) and anistropic (layer and fibril) ones)—is used. The polymer chains may contain regular or nonregular metal-containing groups. The products of thermal decomposition of reactive metal-containing groups of polymer chain—monoatomic metal particles or its oxide—can be considered as a source of the aggregated particles. The medium in which the transformations (motion and growth of cluster-forming particles) occur were represented as a three-dimensional lattice consisting of 50 × 50 × 50 cubic-cell-reactive groups (1.25×10^5 centers). The size of the cell ($a = 10^{-9}$–10^{-8} m) is varied and is determined by the size of the specific fragment of a regular polymer chain that contains one reactive decaying metal-containing group.

The computer model is based on the following determinant algorithms:

1. Decomposition of metal-containing groups of polymer with formation of monoatomic metal (or its oxide) particles. The rate of the process W_J is described by the law of first-order reaction $W_J = kC = Ck_0\exp[-E_{aJ}/(RT)]$, where C is the current amount of the reaction centers ($C_0 = 1.25 \times 10^5$ monoatomic centers), E_{aJ} k_0, and T are the varying parameters, namely, the energy of activation, preexponential factor, and temperature of thermolysis. Algorithms take also into account the probability of catalytic decay of reaction centers by diffusing particles.
2. The solid-phase diffusion of monoatomic metal (or its oxide) particles to give multiatomic cluster particles. It is assumed that diffusion processes in the solid phase are activated. The rate of diffusion $W_{D,N} = D_NC^* = C^*D_1 \cdot N^{-1/3}$, where $C^* = C/a$ is the current amount of reaction centers per one cell of the a size, N is the size of N-atomic cluster, $D_1 = D_0\exp[-E_{aD}/(RT)]$ is the diffusion coefficient for a monoatomic particle, E_{aD} and $D_0 = \nu\exp(\Delta S/R)$ are the varying parameters, the energy of activation, and the entropy factor (($\nu \sim 10^{12}\,\text{sec}^{-1}$). Such a model assumes that the coexistence of two or more separate particles in one cell is not possible because they immediately form a single cluster. In the isotropic medium the diffusing particle with a corresponding probability can move in one of the 26 directions.
3. Taking into account the cluster dissociation in the reaction $[M_n] \rightarrow [M_{n-1}] + M$ with formation of multiatomic particles that move to the neighbor cells. The rate of process $W_N = k_NC_N = C_Nk_{0N}\exp[-E_{aN}/(RT)]$ depends on the cluster size N and diminishes with its growth, C_N is the current concentration of cluster particles of the N size, and E_{aN} and k_{0N} are the energy of activation for the N-cluster and preexponential factor ($\sim 10^{13}\,\text{sec}^{-1}$), respectively. The energy of activation can vary and

increase from E_{a2} (the energy of activation for two atomic cluster) to $E_{a,\infty}$ (the energy of activation for nanoparticle). $E_{a,\infty} \approx E_{sub}$ is the energy of sublimation for a bulk metal.

4. Visualization of cluster formation. The algorithm allows us to observe the dynamics of the particle growth and their distribution of size as a function of the initial parameters during the computer experiment. The typical picture of the particle growth scanning is represented in Figure 3.8.

At the end of the experiment the following dependencies are obtained: [ln(C + 1)/ln(C_{max} + 1), t, N] represents the time dependence of cluster distribution on the size, where C_{max} is the maximum amount of clusters in the moment of time t; [J, N] is an ultimate distribution of clusters on the size where J is the amount of N-size cluster.

The time of iteration, $\Delta\tau$, equals $a^2D^{-1}e^{-2}\xi^{-1}$, where ξ is the varying iteration coefficient ($\xi \geq 1$). The time of experiment, t_Σ, equals Rt_{polym}, where t_{polym}

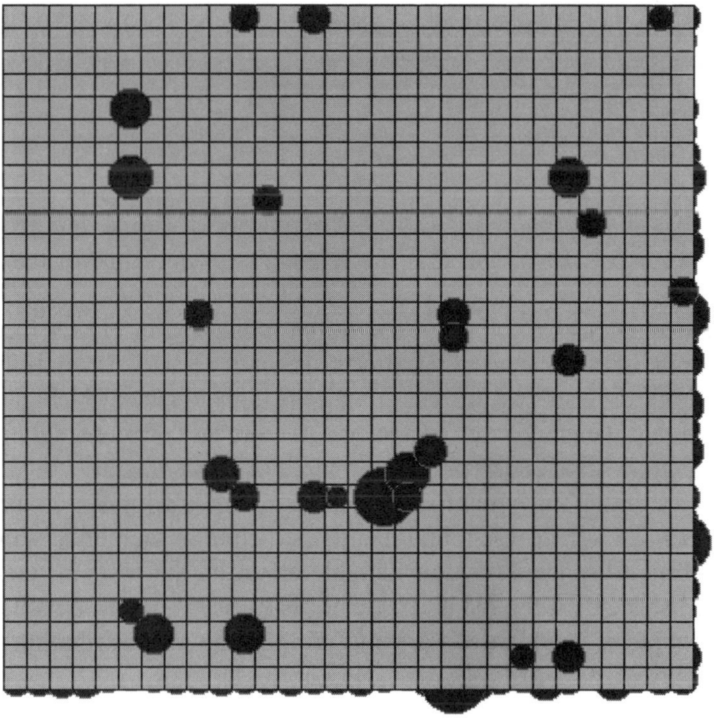

Figure 3.8. A picture fragment of 50 × 50 × 50 cluster growth during the computer experiment.

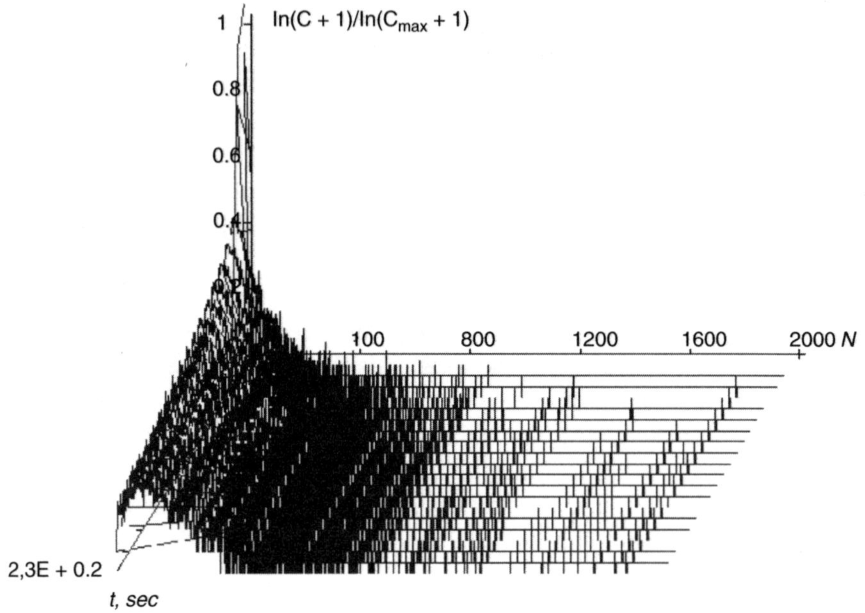

Figure 3.9. The distribution of clusters on the size during the experiment: T = 653 K, $W_{D,N} = 1 \times 10^{-18}$ at·m^2·sec^{-1} (E_{aD} = 250 kJ·mole^{-1}), $W_J = 2.2 \times 10^{25}$ at·m^{-3}·sec^{-1} (E_{aJ} = 167 kJ·mole^{-1}), $W_2 = 1 \times 10^3$ at·m^{-3}·sec^{-1}(E_{a2} = 125 kJ·mole^{-1}), $W_\infty = 4 \times 10^{-14}$ at·m^{-3}·sec^{-1}($E_{a\infty}$ = 330 kJ·mole^{-1}).

$= -k^{-1}\ln(C/C_0)$ is the time of decay of reactive metal-containing groups of polymer ($C/C_0 = 5 \times 10^{-3}$), and R is the relative (varying) time of the cluster formation ($R \geq 1$).

The investigations carried out showed that increasing of cluster formation duration results in the change of spectra of the particle size distribution: With increasing R, the $J(N)$ spectra are shifted into the higher value of N (Figure 3.9). It was found that the conversion is a successive process. For example (Figure 3.10), for the time $t_\Sigma = 228$ sec ($R = 5$) the yield of clusters with $\overline{N} \approx 2$, 37, 65, and 125 passes through maximum, at which the particles with $\overline{N} \approx 65$ have disappeared at $t \approx 200$ sec. It was found that the structure of the polymer medium (layers or fibrils) displayed the considerable effect on the particle size distribution. If the space between the layers or fibrils is inert—that is, outside of the thermolysis zone the nanoparticles are not formed—the size distribution spectra are shifted into a region of lesser sizes (Figure 3.11).

The model considered can be useful for studying the kinetics and mechanism of the formation of metal-containing clusters and nanoparticles in the polymer medium with reactive metal-containing groups. The methodology of

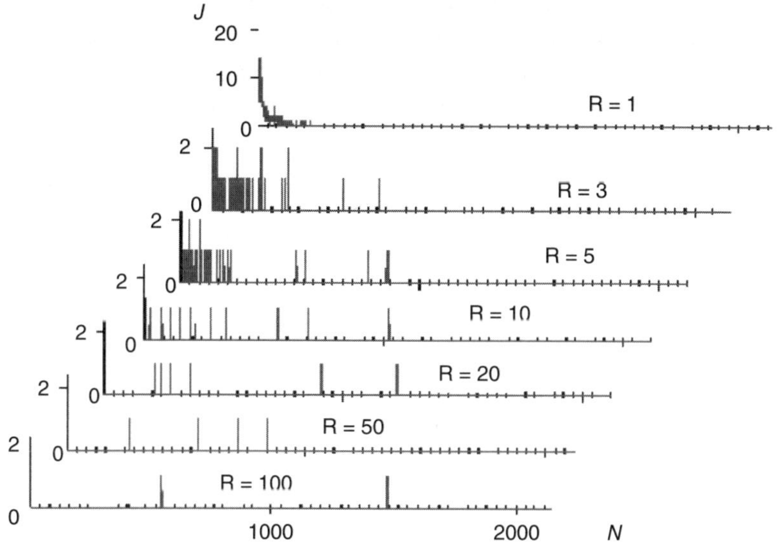

Figure 3.10. The effect of the relative duration of cluster formation on the particle size distribution. Globular structure.

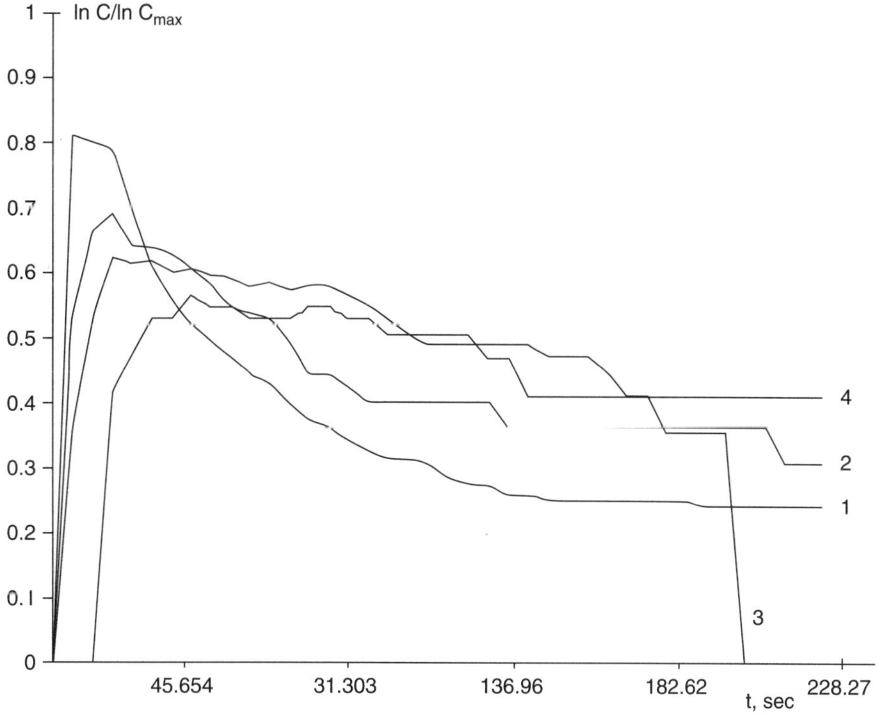

Figure 3.11. The kinetics of the cluster formations. *1*, 1 > N > 4; *2*, 35 > N > 39; *3*, 63 > N > 67; *4*, 123 > N > 127. Globular structure.

this type of computer experiment allows us to carry out the corrections and transformations of the model and thereby to extend the scope of models including the formation of bi- and even polyheterometallic nanoparticles. Most likely, such improved and multiparametric models can be useful for optimization of metallopolymer nanocomposites.

Thus, thermal methods for synthesis of nanoparticles in a polymer matrix possess unlimited opportunities for the construction of polymer nanocomposites and for a choice of the optimal variant. In this chapter we did not analyze sol–gel synthesis of nanocomposites as well as hybrid polymer–inorganic nanocomposites, in preparation of which the controlled thermolysis plays an important role.

Many theoretical problems concerning physicochemistry of generation and growth of nanoparticles and the influence of polymer matrix on their composition and structure as well as the peculiarities of the growth of fractal and fractal filaments, specifics, and limitations of cluster–cluster aggregations are still necessary to solve. All these problems for polymer free nanostructures are mainly studied in detail and analyzed on the quantitative level. There have been attempts to clarify reactivity of the atomic metal regarding the polymer surface from the ionization potential of metal and the energy of low vacancy molecular orbital of polymer. The approaches based on the controlled pyrolysis to obtain polymer-mediated self-organized magnetic nanoparticles have been successfully developed. For example, magnetic materials on the base of binary FePt [130] obtained by thermal reduction of $Pt(AcAc)_2$ and $FeCl_2$ or CoPt [131] as well as PtRu [95] bimetallic nanoparticles are described. Recently, the synthesis of $Fe_{50}Pt_{50}$ using PVP or polyethyleneimine (PEI) [132] is performed. The multilayer self-assembled nanoparticles are localized on the PEI-modified silicon oxide surface. The optimal annealing time and temperature (30 min at 770–870 K) were found when the cubic form of a Pt nanoparticle transforms into hexagonal ones. Such layer-by-layer assembled nanoparticles possess the great potentials.

REFERENCES

1. A. D. Pomogailo, A. S. Rozenberg, and I. E. Uflyand, *Metal Nanoparticles in Polymers*, Khimiya, Moscow (2000), 672 pages.
2. M. J. Hampden-Smith and T. T. Kodas, *Chem. Vap. Deposition* **1**, 8 (1995).
3. A. D. Pomogailo and I. E. Uflyand, *Macromolecular Metal Chelates*, Khimiya, Moscow (1991).
4. E. I. Aleksandrova, G. I. Dzhardimalieva, A. S. Rozenberg, and A. D. Pomogailo, *Russ. Chem. Bull.* **42**, 259 (1993) (English translation).

5. E. I. Aleksandrova, G. I. Dzhardimalieva, A. S. Rozenberg, and A. D. Pomogailo, *Russ. Chem. Bull.* **42**, 264 (1993) (English translation).
6. Yu. M. Shulga, O. S. Roshchupkina, G. I. Dzhardimalieva, I. V. Chernushchevich, A. F. Dodonov, Yu. V. Baldokhin, P. Ya. Kolotyrkin, A. S. Rozenberg, and A. D. Pomogailo, *Russ. Chem. Bull.* **42**, 1661 (1993) (English translation).
7. A. S. Rozenberg, E. I. Alexandrova, G. I. Dzhardimalieva, A. N. Titkov, and A. D. Pomogailo, *Russ. Chem. Bull.* **42**, 1666 (1993) (English translation).
8. A. S. Rozenberg, E. I. Alexandrova, G. I. Dzhardimalieva, N. V. Kir'ykov, P. E. Chizhov, V. I. Petinov, and A. D. Pomogailo, *Russ. Chem. Bull.* **44**, 858 (1995) (English translation).
9. A. S. Rozenberg, G. I. Dzhardimalieva, and A. D. Pomogailo, *Polym. Adv. Technol.* **9**, 527 (1998).
10. S. Madorskii, *Thermal Decomposition of Organic Polymers*, Mir, Moscow (1967).
11. D. Wöhrle and A. D. Pomogailo, *Metal Complexes and Metals in Macromolecules: Synthesis, Structures and Properties*, Wiley-VCH, Weinheim (2003), 667 pages.
12. C. Paal and C. Amberger, *Ber. Dtsch. Chem. Ges.* **37**, 124 (1904).
13. A. D. Pomogailo and D. Wöhrle, in *Macromolecule–Metal Complexes*, edited by F. Ciardelli, E. Tsuchida, and D. Wohrle, Springer, Berlin (1996), pp. 11–129.
14. S. N. Zhurkov, V. A. Zakrevskii, V. E. Korsukov, and V. S. Kuksenko, *Phys. solids.* **13**, 2004 (1971).
15. G. M. Bartenev and E. S. Savin, *Polym. Sci. USSR. Ser. B* **25**, 625 (1982).
16. O. F. Shlenskii, N. V. Afanas'ev, and A. G. Shashkov, *Thermodestruction of Materials*, Energoatomizdat, Moscow (1996).
17. L. V. Ruban and G. E. Zaikov, *Russ. Chem. Rev.* **63**, 373 (1994).
18. R. S. Beer, C. A. Wilkie, and M. L. Mittleman, *J. Appl. Polym. Sci.* **46**, 1095 (1992).
19. I. C. McNeill and R. C. McGuiness, *Polym. Degrad. Stabil.* **9**, 1 (1984).
20. C. A. Wilkie, J. W. Pettergreww, and C. E. Brown, *J. Polym. Sci., Polym. Lett.* **19**, 409 (1981).
21. L. E. Manring, *Macromolecules* **21**, 528 (1988); **22**, 2673 (1989); **24**, 3304 (1991).
22. L. E. Manring, D. Y. Sogah, and G. M. Cohen, *Macromolecules* **22**, 4652 (1989).
23. C. A. Wilkie, J. T. Leone, and M. L. Mittleman, *J. Appl. Polym. Sci.* **42**, 1133 (1991).
24. K. Gurova, C. Uzov, M. Zagortcheva, and G. Gavrilova, *J. Appl. Polym. Sci.* **74**, 3324 (1999).
25. T.-H. Ko and C.-Y. Chen, *J. Appl. Polym. Sci.* **71**, 2219 (1999).
26. A. M. Summan, *J. Polym. Sci., Part A: Polym. Chem.* **37**, 3057 (1999).
27. L. Belfiore, M. P. McCurdie, and E. Ueda, *Macromolecules* **26**, 6908 (1993).

28. O. P. Krivoruchko and V. I. Zaikovskii, *Mendeleev Commun.* 97 (1998).
29. C. McNeill and J. J. Liggat, *Polym. Degrad. Stabil.* **29**, 93 (1990); **37**, 25 (1992).
30. S. P. Gubin and I. D. Kosobudskii, *Russ. Chem. Rev.* **52**, 1350 (1983).
31. N. M. Bravaya, A. D. Pomogailo, and E. F. Vainshtein. *Kinet. Catal.* **25**, 1140 (1984).
32. E. F. Vainshtein and G. E. Zaikov, in *Polymer Yearbook*, Vol. 10, edited by R. A. Pethric, Harwood Academic Publishers, London (1993), p. 231.
33. E. I. Aleksandrova, A. S. Rozenberg, and A. N. Titkiv, *Chem Phys.* **13**(8–9), 83 (1994).
34. D. Dollimore, *Thermochim. Acta* **117**, 331 (1987).
35. I. V. Kumpanenko and N. V. Chukanov, *Russ. Chem. Rev.* **50**, 1627 (1981).
36. A. D. Pomogailo, G. I. Dzhardimalieva, A. S. Rosenberg, and D. N. Muraviev, *J. Nanoparticle Res.* **5**, 497 (2003).
37. A. N. Timofeev, I. Yu. Filatov, V. G. Savostyanov, and K. I. Marushkin, *High Pure Substances* **5**, 45 (1994).
38. A. D. Pomogailo, A. S. Rozenberg, and G. I. Dzhardimalieva, in *Metal-Containing Polymers Materials*, edited by Ch. U. Pittman, Jr., Ch. E. Carraher, Jr., M. Zeldin, J. E. Sheats, and B. M. Culbertson, Plenum Press, New York, (1996), p. 313.
39. M. Ławecka, M. Leonowicz, M. Kopcewicz, A. Ślawska-Waniewska, J. Kozubowski, G. I. Dzhardimalieva, A. S. Rozenberg, and A. D. Pomogailo, *Kompozyty* (*Composites*) **2**, 404 (2002).
40. A. S. Rozenberg, E. I. Alexandrova, N. P. Ivleva, G. I. Dzhardimalieva, A. V. Raevskii, O. I. Kolesova, I. E. Uflyand, and A. D. Pomogailo, *Russ. Chem. Bull.* **47**, 259 (1998) (English translation).
41. A. T. Shuvaev, A. S. Rozenberg, G. I. Dzhardimalieva, N. P. Ivleva, V. G. Vlasenko, T. I. Nedoseikina, T. A. Lyubeznova, I. E. Uflyand, and A. D. Pomogailo, *Russ. Chem. Bull.* **47**, 1460 (1998) (English Translation).
42. I. Skupinska, H. Wilezura, and H. Bonink, *J. Therm. Anal.* **31**, 1017 (1986).
43. A. Gronowski and Z. Wojtezak, *J. Therm. Anal.* **36**, 2357 (1990).
44. P. A. Vasil'ev, A. L. Ivanov, and A. N. Glebov, *Russ. J. Gen. Chem.* **68**, 535 (1998).
45. A. S. Rozenberg, G. I. Dzhardimalieva, and A. D. Pomogailo, *Dokl. Phys. Chem.* **356**, 66 (1997).
46. A. D. Pomogailo, V. G. Vlasenko, A. T. Schuvaev, A. S. Rozenberg, and G. I. Dzhardimalieva, *Colloid J.* **64**, 524 (2002).
47. A. D. Pomogailo, A. S. Rozenberg, G. I. Dzhardimalieva, and M. Leonowicz, *Adv. Mater. Sci.* **1**, 19 (2001).
48. A. D. Pomogailo, G. I. Dzhardimalieva, and A. S. Rosenberg, *Acta Phys. Pol. A*, **102**, 135 (2002).
49. A. D. Pomogailo, A. S. Rosenberg, and G. I. Dzhardimalieva, *Solid State Phenom.* **94**, 313 (2003).
50. V. A. Sergeev, N. I. Bekasiova, M. A. Surikova, E. A. Baryshnikova, Ya. V. Genin, and N. K. Vinogradova, *Dokl. Phys. Chem.* **332**, 601 (1993).

51. V. A. Sergeev, N. I. Bekasov, and M. A. Surikova, *Polym. Sci. Ser. B* **34**, 70 (1992).
52. V. G. Syrkin, *Gas-Phase Metallization Through Carbonyls*, Metallurgy, Moscow (1985).
53. T. W. Smith and D. Wychick, *J. Phys. Chem.* **84**, 1621 (1980).
54. Patent 4252671-4252678, CIIIA.
55. P. H. Hess and P. H. Parker, Jr., *J. Appl. Polym. Sci.* **10**, 1915 (1966).
56. J. R. Thomas, *J. Appl. Phys.* **37**, 2914 (1966).
57. M. Berger and T. A. Manuel, *J. Polym. Sci.* A-1 **4**, 1509 (1966).
58. T. W. Smith and D. J. Luca, *Proceedings, Symposium on Modified Polymers, Las Vegas (Nev.) 1982*. Abstract New York, London (1983), p. 85.
59. A. B. Gilman and V. M. Kolotyrkin, *High Energy Chem.* **12**, 450 (1978).
60. C. H. Griffiths, M. P. O'Horo, and T. W. Smith, *J. Appl. Phys.* **50**, 7108 (1979).
61. H. Biederman, *Vacuum* **34**, 405 (1984).
62. C. E. Kerr, B. E. Eaton, and J. A. Kadue, *Organometallics* **14**, 269 (1995).
63. S. P. Gubin, I. D. Kosobudskii, and G. I. Petrakovskii, *Dokl. AN USSR* **260**, 655 (1981).
64. S. P. Gubin and I. D. Kosobudskii, *Dokl. AN USSR* **272**, 1155 (1983).
65. I. D. Kosobudskii, S. P. Gubin, V. P. Piskorskii, G. A. Petrakovskii, L. V. Kashkina, V. N. Kolomiichuk, and N. M. Svirskaya, in *Electronics of Organic Materials*, edited by A. A. Ovchinnikov, Nauka, Moscow (1985), p. 62.
66. L. M. Bronshtein, P. M. Valetskii, S. V. Vinogradova, A. I. Kuzaev, and V. V. Korshak, *Polym. Sci. USSR* **29**, 1694 (1987).
67. C. U. Pittman, Jr, P. L. Grube, O. E. Ayers, S. P. McManus, M. D. Rausch, and G. A. Moser, *J. Polym. Sci.*, A 1, **10**, 379 (1972).
68. L. M. Bronstein, S. P. Solodovnikov, E. Sh. Mirzoeva, E. Yu. Baukova, and P. M. Valetsky, *Proc. ACS Div. Polym. Mater. Sci. Eng.* **71**, 397 (1994).
69. M. Berger and D. J. Buckley. *J. Polym. Sci.*, A-1, **1**, 2945 (1963).
70. R. A. Arents, Yu. V. Maksimov, I. P. Suzdalev, and Yu. B. Amerik, *Hyperfine Interactions*, **56**, 167 (1990).
71. Yu. B. Amerik, Yu. M. Korolev, and V. N. Rogovoi, *Neftekhimiya*, **36**, 304 (1996).
72. Yu. M. Korolov, A. L. Bykova, and Yu. B. Amerik, *Polym. Sci. Ser. B*, **39**, 1856 (1997).
73. P. I. Dosa, C. Erben, V. S. Iyer, K. P. C. Vollhard, and I. M. Vasser, *J. Am. Chem. Soc.* **121**, 10430 (1999).
74. C.-H. Kiang and W. A. Goddard, *Phys. Rev. Lett.* **76**, 2515 (1996).
75. R. Sen, A. Govindaraj, and C. N. R. Rao, *Chem. Mater.* **9**, 2078 (1997).
76. T. X. Neenan, M. R. Callstrom, and O. J. Schueller, *Macromol. Symp.* **80**, 315 (1994).
77. F. Galembeck, C. C. Chironi, and C. A. Ribeiro, et al. *J. Appl. Polym. Sci.* **25**, 1427 (1980).

78. R. Baumhardt-Neto, S. E. Galembeck, I. Joekes, and F. Galembeck, *J. Polym. Sci., Polym. Chem. Ed.* **19**, 819 (1981).
79. F. Galembeck, *J. Polym. Sci.: Polym. Lett. Ed.* **15**, 107 (1977); *J. Polym. Sci.: Polym. Chem. Ed.* **16**, 3015 (1978).
80. R. Tannenbaum, C. L. Flenniken, and E. P. Goldberg, in *Metal-Containing Polymeric Systems*, edited by C. Carraher, C. Pittman, and J. Sheats, Plenum, New York (1985), p. 303.
81. R. Tannenbaum, C. L. Flenniken, and E. P. Goldberg, *J. Polym. Sci.: Part B: Polym. Phys.* **28**, 2421 (1990).
82. S. Reich and E. P. Goldberg, *J. Polym. Sci., Phys. Ed.* **21**, 869 (1983).
83. A. I. Pertsin and Yu. M. Poshutin, *Polym. Sci. Ser. B* **38**, 919 (1996).
84. A. I. Pertsin and I. O. Volkov, *Polym. Sci. Ser. B* **38**, 1249 (1996).
85. L. M. Bronshtein, P. M. Valetskii, and M. Antonietti, *Polym. Sci. Ser. A* **39**, 1847 (1997).
86. O. A. Platonova, L. M. Bronstein, S. P. Solodovnikov, I. M. Yanovskaya, E. S. Obolonkova, P. M. Valetsky, E. Wenz, and M. Antonietti, *Colloid Polym. Sci.* **275**, 426 (1997).
87. K. S. Suslick (editor). *Ultrasound: Its Chemical, Physical, and Biological Effects*, VCH Press, New York (1988).
88. K. S. Suslick, *Science* **247**, 1439 (1990).
89. K. S. Suslick, S. B. Choe, A. A. Cichowlas, and M. W. Grinstaff, *Nature* **353**, 414 (1991).
90. G. Kataby, T. Prozorov, Yu. Koltyp, C. N. Sukenik, A. Ulman, and A. Gedanken, *Langmuir* **13**, 615 (1997); **15**, 1702 (1999).
91. K. V. P. M. Shafi, A. Gedanken, R. Prozorov, and J. Balogh, *Chem. Mater*. **10**, 3445 (1998).
92. K. S. Suslick, T. Hyeon, M. Fang, and A. A. Cichowlas, *Mater. Sci. Eng.* **A204**, 186 (1996); *Chem. Mater.* **8**, 2172 (1996).
93. D. Caro, T. O. Ely, A. Mari, B. Chaudret, E. Snoeck, M. Respaud, J.-M. Broto, and A. Fert, *Chem. Mater.* **8**, 1987 (1996).
94. K. V. P. M. Shafi, A. Ulman, X. Yan, N.-L. Yang, C. Estournes, H. White, and M. Rafailovich, *Langmuir* **17**, 5093 (2001).
95. M. S. Nashner, A. I. Frenkel, D. Somerville, C. W. Hills, J. R. Shapley, and R. G. Nuzzo, *J. Am. Chem. Soc.* **120**, 8093 (1998).
96. T. Yu. Ryabova, A. S. Chirkov, L. S. Radkevich, and N. V. Evtushok, *Ukr. Chem. J.* **59**, 1329 (1993).
97. A. I. Savitskii, Sh. Ya. Korovskii, and V. I. Prosvirin, *Colloid J. USSR*, **41**, 88 (1979); **42**, 998 (1980).
98. M. C. Alves and G. Tourillon, *J. Phys. Chem.* **100**, 7566 (1996).
99. J. E. Millburn and M. J. Rosseinsky, *Chem. Mater.* **9**, 511 (1997).
100. R. E. Southward, D. W. Thompson, and A. K. St. Clair, *Chem. Mater.* **9**, 501 (1997).

101. R. E. Southward, D. S. Thompson, D. W. Thompson, and A. K. St. Clair, *Chem. Mater.* **9**, 1691 (1997).
102. S. P. Pogorelov, G. M. Plavnik, A. P. Tikhonov, and T. P. Puryaeva, *Colloid J.* **61**, 100 (1999).
103. Yu. M. Khimchenko and L. S. Radkevich, *Plast. Massy* **1**, 53 (1975).
104. N. I. Nikanorova, E. S. Trofimchuk, E. V. Semenova, A. L. Bolynskii, and N. F. Bakeev, *Polym. Sci. Ser. A* **42**, 1298 (2000).
105. G. Carotenuto, B. Martorana, P. Perlo, and L. Nicolais, *J. Mater. Chem.* **13**, 2927 (2003).
106. A. Yu. Vasilkov, P. V. Pribytko, E. A. Fedorovskaya, A. A. Slinkin, A. S. Kogan, and V. A. Sergeev, *Dokl. Phys. Chem.* **331**, 179 (1993).
107. J. Osuna, D. Caro, C. Amiens, B. Chaudret, E. Snoeck, M. Respaund, J.-M. Broto, and A. Fest, *J. Phys. Chem.* **100**, 14571 (1996).
108. A. Carbonaro, A. Greco, and G. Dall'asta, *J. Organomet. Chem.* **20**, 177 (1969).
109. S. E. J. Goff, T. F. Nolan, M. W. George, and M. Poliakoff, *Organometallics* **17**, 2730 (1998).
110. A. I. Cooper and M. Poliakoff, *Chem. Phys.* Lett. **212**, 611 (1993).
111. J. S. Bradley, J. M. Millar, E. W. Hill, S. Behal, B. Chaudret, and A. Duteil, *Faraday Discuss. Chem. Soc.* **92**, 255 (1991).
112. A. Duteil, R. Queau, B. M. Chaudret, C. Roucau, and J. S. Bradley, *Chem. Mater.* **5**, 341 (1993).
113. J. S. Bradley, E. W. Hill, S. Behal, C. Klein, B. M. Chaudret, and A. Duteil, *Chem. Mater.* **4**, 1234 (1992).
114. D. Caro and J. S. Bradley, *Langmuir* **13**, 3067 (1997).
115. F. Waller, *J. Catal. Rev. Sci. Eng.* **28**, 1 (1986).
116. A. M. Hodges, M. Linton, A. W.-H. Mau, K. J. Cavell, J. A. Hey, and A. Seen, *J. Appl. Organomet. Chem.* **4**, 465 (1990).
117. M. Pehnt, D. L. Schulz, C. J. Curtis, K. M. Jones, and D. S. Ginley, *Appl. Phys. Lett.* **67**, 2176 (1995).
118. E. E. Said-Galeev, L. N. Nikitin, Yu. P. Kudryavtsev, A. L. Rusanov, O. L. Lependina, V. K. Popov, M. Polyakoff, and S. M. Houdl, *Khim Phizika*, **14**, 190 (1995).
119. J. J. Watkins and T. J. McCarthy, *Chem. Mater.* **7**, 1991 (1995).
120. J. J. Watkins and T. J. McCarthy, *Polym. Mater. Sci. Eng.* **73**, 158 (1995).
121. E. M. Natanson and Z. P. Ulberg, *Colloid Metals and Metallopolymers*, Naukova Dumka, Kiev (1971).
122. D. M. Mognonov, V. G. Samsonova, V. V. Khakinov, and I. N. Pinchuk, *Polym. Sci. Ser. B* **39**, 1250 (1997).
123. G. P. Gladyshev, O. A. Vasnetsova, and N. I. Mashukov, *Mendeleev. Chem. J.* **34**, 575 (1990).
124. G. P. Gladyshev, N. I. Mashukov, and S. A. Eltsin, *Polym. Sci. USSR, B* **28**, 62 (1986).

125. N. D. Zhuravlev, V. I. Raldugin, and A. P. Tikhonov, *Colloid J.* **61**, 322 (1999).
126. S. Kan, S. Yu, and X. Peng, *J. Colloid. Interface Sci.* **178**, 673 (1966).
127. A. S. Rozenberg, A. A. Rozenberg, G. I. Dzhardimalieva, and A. D. Pomogailo, in *Program and Abstracts, 10th IUPAC Internernational Symposium on Macromolecule–Metal Complexes, Moscow, May 18–23, 2003*, edited by E. A. Karakhanov and A. L. Maximov, MAX Press, Moscow (2003), p. 101.
128. D. E. Ulberg, V. V. Il'in, N. V. Churaev, and Yu. V. Nizhnik, *Colloid J.* **54**, 151 (1992).
129. A. S. Roseberg, A. A. Rosenberg, A. V. Lankin, G. I. Dzhardimalieva, and A. D. Pomogailo, *Dokl. Phys. Chem.* **393**, 361 (2003).
130. S. Sun, C. B. Murray, D. Weller, L. Folks, and A. Moser, *Science*, **287**, 1989 (2000).
131. J. H. Park and J. Cheon, *J. Am. Chem. Soc.* **123**, 5473 (2001).
132. S. Sun, S. Anders, H. F. Haman, J.-U. Thiele, J. E. E. Baglin, T. Thomson, E. E. Fullerton, C. B. Murray, and B. D. Terris, *J. Am. Chem. Soc.* **124**, 2884 (2002).

4

NANOSTRUCTURED POLYMERIC NANOREACTORS FOR METAL NANOPARTICLE FORMATION

L. M. Bronstein

Department of Chemistry, Indiana University, Bloomington, Indiana

1. INTRODUCTION

Formation of metal nanoparticles in a polymer matrix became a popular tool for design of new metal–polymer nanocomposites, the properties of which can be greatly altered compared to those of pure polymers [1–4]. Incorporation of nanoparticles into a polymeric system may impart magnetic, semiconductor, catalytic, or sensing properties, depending on the kind of nanoparticles, formed inside a polymer, and nanoparticle characteristics. To obtain metal–polymer nanocomposites with well-defined and well-reproducible properties, one should carry out a subtle control over nanoparticle growth, particle size distribution, and particle-interface interactions. These are key issues to the desired properties and targeted applications of polymeric materials. Narrow particle size distribution is especially crucial, since optical, magnetic, sensing, and even catalytic properties strongly depend on the precise control over particle size. One of the prominent ways to control nanoparticle size and morphology is to employ functional polymeric nanostructures with well-defined interfaces [5–7]. These interfaces can be generated via hydrophobic–hydrophilic microphase separation; or, due to nanopores or nanocavities, they can be formed within a

Metal-Polymer Nanocomposites, Edited by Luigi Nicolais and Gianfranco Carotenuto
ISBN 0-471-47131-3

polymer during synthesis. The presence of functional groups able to interact with metal compounds allows incorporation of metal species into the functional nanophase, while subsequent reduction or thermal (or other) treatment results in nanoparticle formation within this restricted area. The characteristics of the functional nanophase, metal compound loading, type of reducing agent, and other parameters are responsible for metal nanoparticle and metal–polymer nanocomposite characteristics. This chapter describes our major accomplishments in the field of metal–polymer nanocomposites, illustrating synthetic approaches, structural characterization, and properties. Comprehensive review articles describing "Nanoparticles in Nanostructured Polymers" and "Polymer Colloids and Their Metallation" are presented in the *Encyclopedia of Nanoscience and Nanotechnology* (American Scientific Publishers, 2004) and the *Dekker Encyclopedia of Nanoscience and Nanotechnology* (Marcel Dekker, 2004). In this chapter, all nanostructured polymeric systems are divided into two large groups—solid or heterogeneous (insoluble in any media) and soluble or microheterogeneous (colloidally soluble)—and will be discussed in separate sections.

2. SOLID POLYMER NANOSTRUCTURES

Among solid polymer nanostructures, two types of systems will be described: (a) complexes of polyelectrolyte gels with surfactants and (b) crosslinked polymers containing nanopores or nanocavities.

2.1. Polyelectrolyte Gel–Surfactant Complexes

Polyelectrolyte gel–surfactant complexes are formed when a polyelectrolyte gel is placed in the solution of oppositely charged surfactants. Surfactant molecules easily penetrate a gel due to hydrophobic interactions and form complexes with the polyelectrolyte charged groups. At the same time, surfactant molecules self assemble within the gel, and their ordering even exceeds the ordering of surfactants in solutions, as was reported in a number of publications [8–12]. Since a polyelectrolyte gel–surfactant complex contains counterions in well-ordered surfactant areas within a weakly crosslinked gel, these counterions can be replaced with ions of interest (transition metal ions). The subsequent treatment (reduction, thermolysis, photolysis, etc.) should lead to formation of nanoparticles in a well-ordered nanostructured medium. Indeed, metal ions are absorbed by polyelectrolyte gel–surfactant complexes when these ions are oppositely charged toward polyelectrolyte charged groups [13–16]. However, incorporation of the ions does not destroy the surfactant ordering: Metal compounds form clusters located between charged groups of a polyelectrolyte chain and surfactant head groups. During further reduction, these clusters play the role of nuclei for metal nanoparticle formation.

The use of small-angle X-ray scattering (SAXS) and anomalous SAXS (ASAXS) [15] allowed us to quantitatively characterize the influence of the type of the reducing agent and of the structure of the metal compound on size distributions of Pt nanoparticles formed in polyelectrolyte gel–surfactant complexes. We studied three collapsed gel–surfactant complexes: a cationic gel of polydiallyldimethylammonium chloride (PDADMACl) with two anionic surfactants, sodium dodecyl sulfate (SDS) and sodium dodecylbenzene sulfonate (SDBS); and an anionic gel of poly(methacrylic acid) with a cationic surfactant, cetylpyridinium chloride (PMA/CPC). For all the cationic gel–anionic surfactant complexes, the sluggish reducing agent ($N_2H_4 \cdot H_2O$) leads to much larger nanoparticles than does the fast reducing agent ($NaBH_4$), yet the metal compound structure plays a crucial role in nanoparticle formation. The particles obtained from the planar $PtCl_4^{2-}$ ions are smaller (the mean radius of a small particle fraction is about 1 nm) and have a narrower size distribution; moreover, particles larger than 15–20 nm are absent (Figure 4.1). In contrast, the octahedral $PtCl_6^{2-}$ ion-based materials yield larger particles. In so doing, the gel–surfactant complex derived from H_2PtCl_6 (low pH) yields particles with the radii up to 40 nm. Even larger particles (with the radii up to 80 nm) are formed in the anionic gel–cationic surfactant complex PMA/CPC where the particle growth is not controlled by the internal gel structure [15]. Presumably, the formation of larger particles in the PDADMACl–SDS system from $PtCl_6^{2-}$ ions than from $PtCl_4^{2-}$ ions is due to formation of larger precursor (salt) clusters from ions with octahedral geometry.

Analysis of the Bragg peaks in the scattering data permitted us to also quantitatively characterize the ordering in the gel–surfactant complexes. In Table 4.1 one can see the structural parameters characterizing the internal order in the collapsed polyelectrolyte gel–surfactant complexes calculated from the Bragg peaks in the SAXS and ASAXS profiles. The mean long-range order dimension L estimating the size of quasi-crystalline zones in the sample, as well as the radius of interaction r_m yielding maximum separation between spatially correlated structural motifs, should increase with increasing order in the system. In contrast, the relative mean square deviation of the distance between the neighboring periodic motifs $\Delta/\bar{d}$ (degree of disorder) should decrease if the system becomes more ordered, while the Bragg peaks nearly disappear if $\Delta/\bar{d} > 0.25$.

A surprising result was obtained for both cationic gel–anionic surfactant systems: The metal-containing samples appear not less ordered than the initial polyelectrolyte gel–surfactant complexes despite the diminished amplitudes of the peaks. For the PDADMACl–SDBS system, the average distance between

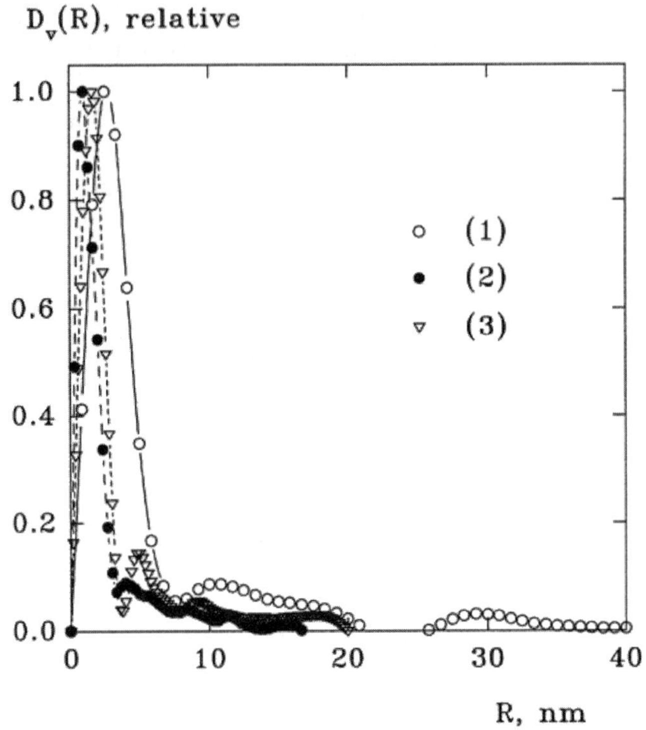

Figure 4.1. Volume distribution functions evaluated by GNOM for the PDADMACl–SDS gel containing Pt nanoparticles derived from H_2PtCl_6 (1), $(NH_4)_2PtCl_4$ (2), and Na_2PtCl_6 (3). (Reprinted with permission from reference 15. Copyright 2000, American Chemical Society.)

the ordered structure motifs $\bar{d}$ increases from 2.7 to about 3.4 nm after the addition of the metal compounds (compared to the initial complex), but the parameters describing the degree of order remain practically unchanged. The internal order is even enhanced after the metal reduction with both sluggish and fast reducing agents. The gel–surfactant ordering increases locally during the nanoparticle formation while the volume fraction of the ordered structures in the samples decreases. This indicates that the highly ordered surfactant areas within hydrogels concentrate around the growing nanoparticles. In contrast, when the $PtCl_6^{2-}$ ions are incorporated into the PMA–CPC complex, the ordering changes due to new micellar structures. The latter are formed as a result of the interaction of the $PtCl_6^{2-}$ ions with the cetyl pyridinium cations. The reduction of a metal compound is accompanied by restoration of the initial PMA–CPC ordering; however, an increase of the local ordering similar to the PDADMACl/SDS(SDBS) systems does not occur.

TABLE 4.1. Structural Characteristics of the Hydrogel–surfactant Complexes Obtained from the SAXS and ASAXS Data

Sample	s_{max} (nm^{-1})	$\overline{d}$ (nm)	L (nm)	$\Delta/\overline{d}$
	PDADMACl–SDBS			
Collapsed gel	2.34	2.68	27	0.100
Gel + $H_2Pt(OH)_2Cl_4$	1.88	3.34	38	0.094
Gel + $H_2Pt(OH)_2Cl_4$ + $NaBH_4$	1.87	3.36	74	0.066
Gel + $H_2Pt(OH)_2Cl_4$ + N_2H_4	1.94	3.24	60	0.074
	PDADMACl–SDS			
Collapsed gel	1.53	4.10	31	0.116
Gel + H_2PtCl_6	1.52	4.13	29	0.118
Gel + $(NH_4)_2PtCl_4$	1.51	4.17	32	0.114
Gel + Na_2PtCl_6	1.52	4.13	38	0.105
Gel + $H_2Pt(OH)_2Cl_4$	1.58	3.98	39	0.101
Gel + Na_2PtCl_6 + $NaBH_4$	1.56	4.02	98	0.064

Source: Reprinted with permission from reference 15. Copyright 2000, American Chemical Society.

As can be seen from the discussion above, the polyelectrolyte gel–surfactant complexes present interesting hybrid metal–polymer nanocomposites, allowing a vast variety of incorporated metals and metal–polymer–surfactant structures. The limitations of these systems are their heterogeneous character (insoluble in any media) and excessive sensitivity to external parameters (pH, temperature, etc.).

2.2. Hypercrosslinked Polystyrene

Hypercrosslinked polystyrene (HPS) [17, 18] is a unique polymer network consisting of nano-sized rigid cavities (pores) with a size in the 2- to 3-nm range (Figure 4.2). HPS is normally synthesized by incorporating methylene groups between adjacent phenyl rings in a dissolved polystyrene homopolymer or a gelled poly(styrene-*r*-divinylbenzene) copolymer in the presence of ethylene dichloride (a good solvent), yielding a very high degree of crosslinking density (may exceed 100%). HPS is able to swell in a wide variety of different solvents, even thermodynamically poor ones (e.g., water) that facilitate incorporation of various organometallic compounds into the nanostructured HPS matrix. The first use of HPS as a medium for Co nanoparticle formation was reported by us in 1999 [19]. Cobalt nanoparticles have been selected due to the established correlation between size and shape of Co particles and their mag-

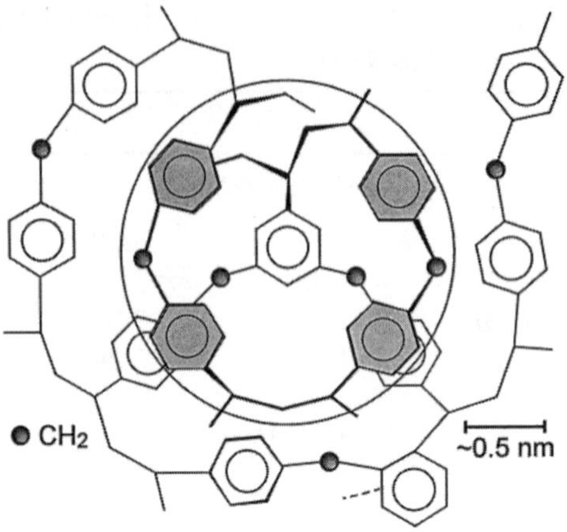

Figure 4.2. Schematic illustration of the internal network of hypercrosslinked polystyrene (HPS). The speckled phenyl rings reside in a different plane relative to the unspeckled ones in the crosslinked material, and each circle identifies a postulated cavity in which Co nanoparticles could grow. (Reprinted with permission from reference 19. Copyright 1999, American Chemical Society.)

netic properties [20]—that is, characteristics of their ferromagnetic resonance (FMR) spectra. Co particles with diameters below 1 nm are nonmagnetic, while the particles with diameters in the range 1–10 nm are superparamagnetic. Larger Co particles are ferromagnetic. In addition, the width of the FMR signal (ΔH) for spherical particles correlates with the size of the Co nanoparticles [20, 21], while the position corresponding to the zero signal (H_0) bears information regarding the shape of the Co nanoparticles. This allows one to determine the general size and shape characteristics of Co nanoparticles solely from FMR spectra.

The Co compounds, $Co_2(CO)_8$ or $[Co(DMF)_6]^{2+}[Co(CO)_4]^-{}_2$ complex, were incorporated in HPS by wet impregnation in 2-propanol or dimethylformamide (DMF), respectively (Table 4.2). The subsequent thermolysis at 200°C results in the formation of Co nanoparticles, the loading and characteristics of which were investigated by X-ray fluorescence (XRF) spectroscopy, FMR spectroscopy, and transmission electron microscopy (TEM). The FMR data presented in Figure 4.3 and Table 4.2 confirmed the formation of spherical nanoparticles [19]. A 2–8 wt.% Co loading the magnitude of the FMR line width indicates that the mean Co nanoparticle diameter is about 2 nm, agreeing well with the mean particle diameter obtained from TEM. When Co content increases above 8 wt. %, the mean particle diameter also increases due to an increase in the population of large Co nanoparticles up to 15 nm in diameter. Here, controlled nanoparticle growth over a wide range of Co contents is

TABLE 4.2. FMR Data of HPS–Co Samples Thermolyzed at 200°C and Various Durations

Co Content (wt. %)	Time of Thermolysis (hr)	H_0 (Oe)	ΔH (Oe)
	$Co_2(CO)_8$		
0.9	2	3115	170
0.9	6	3100	250
3.0	2	3040	200
3.0	6	3050	200
3.0	10	3035	200
	$[Co(DMF)_6]^{2+}[Co(CO)_4]_2^-$ in DMF		
1.2	2	3100	200
1.2	6	3050	200
1.2	10	3060	220

Source: Reprinted with permission from reference 19. Copyright 1999, American Chemical Society.

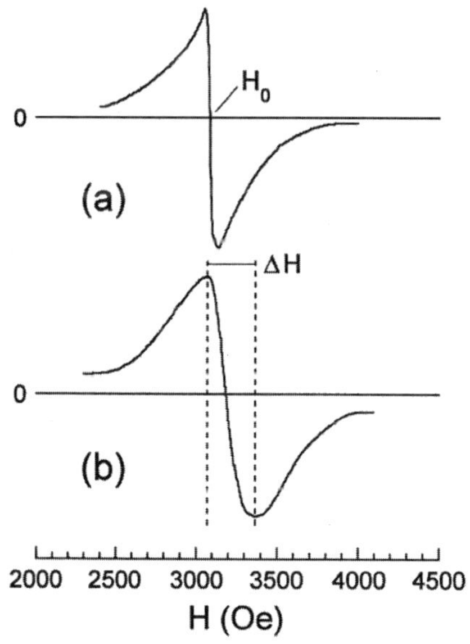

Figure 4.3. FMR spectra of the HPS–Co specimens containing **(a)** 5.07 and **(b)** 10.08 wt. % Co and obtained by thermolysis at 200°C for 2 hr. Labeled in panel **a** is the definition of H_0, the magnitude of which is related to the shape of the Co nanoparticles. The vertical dashed lines in panel **b** identify the line width corresponding to ΔH, which provides a measure of particle size. (Reprinted with permission from reference 19. Copyright 1999, American Chemical Society.)

attributed to nanoscale HPS cavities, physically restricting the size of growing particles. It is worth mentioning that the HPS used in this work contains no functional groups, the interaction of which with particle interface might restrict the particle growth and stabilize the particles. Moreover, the nanoscale cavities in HPS are highly interconnected so they should *not* obstruct the migration of Co atoms or even small Co clusters. A steric limitation arising from the predominance of phenyl rings represents the most probable reason for controlled Co nanoparticle growth in HPS; phenyl rings may interact with metal surfaces and nonspecifically stabilize metal nanoparticles. When the Co content in the HPS–Co composite exceeds 10wt.% (likely a saturation limit), a small fraction of large nanoparticles appears. Their size might correlate with the statistics of the pore size distribution of the HPS matrix.

As can be seen from the above discussion, HPS is a robust, commercially available (by Purolite Int.) nanostructured polymer, yielding control over nanoparticle formation. We believed that this control could be executed for a vast variety of nanoparticles. Following this path, we synthesized Pd and Pt nanoparticles in HPS, using reduction of Pd and Pt compounds instead of thermolysis for the final step of metal–polymer nanocomposite synthesis. Impregnation of HPS with tetrahydrofuran (THF) solution containing platinic acid (H_2PtCl_6) results in the formation of Pt(II) complexes within the nanocavities of HPS [22]. Hydrogen reduction of these metal–polymer nanocomposites yields stable Pt nanoparticles with a mean diameter of 1.3nm (Figure 4.4). One might recall that for Co-containing HPS, the mean particle size was about 2nm. This discrepancy in particle size can be explained by different mechanisms of Co and Pt nanoparticle formation. Since Co nanoparticle formation

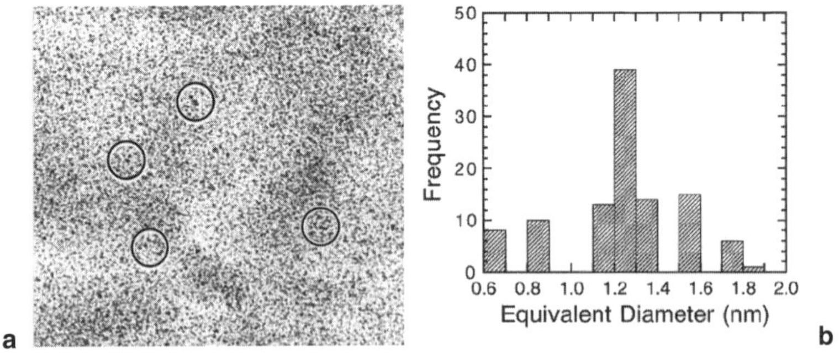

Figure 4.4. Transmission electron micrograph **(a)** and particle size histograms **(b)** of HPS–Pt–THF after H_2 reduction. Groups of single Pt nanoparticles in *a* are highlighted by circles. (Reprinted with permission from reference 22. Copyright 2001, American Chemical Society.)

occurs at 200°C, Co clusters could easily migrate between the HPS pores, yielding Co nanoparticles, whose sizes match the pore size. By contrast, H_2 reduction of Pt nanoparticles occurs at a room temperature, and migration of the Pt clusters or atoms or the Pt precursor molecules from pore to pore is unlikely. This results in the formation of Pt nanoparticles, sizes of which correspond to the size of the Pt precursor (salt or complex) filling the pores [22]. Thus, HPS controls the Co nanoparticle size by physical limitation of their growth within the pore ($D = 2$ nm), while for Pt nanoparticles, pores limit the amount of precursor filling the single pore so when the precursor shrinks during reduction due to the increase of the material density (from Pt complex to Pt metal), the Pt nanoparticle occupies a smaller volume. In both cases, HPS provides strong control over nanoparticle growth. The simplicity of this approach and commercial availability of HPS make this polymer system valuable for broader use. However, the drawback of this system is in the crosslinked nature of HPS so no films or coatings can be made, which limits prospects for HPS use in optical or magnetic materials. When HPS is used for heterogeneous catalysis, this limitation becomes an advantage since this catalyst does not require an additional support. Another drawback of the metallated HPS is the easy removal of metal nanoparticles from the polymer in any media where PS swells. To prevent metal loss, HPS-based catalysts should be used only in poor (for PS) solvents (water is the best); thus swelling is minimized.

2.3. Nanocavities in Polyoctadecylsiloxane

To enhance the stability of nanoparticles inside polymer nanocavities, one should use functional polymers providing the functionalization of the nanocavities. As an example, here we discuss polyoctadecylsiloxane (PODS) containing nanocavities formed by a siloxane bilayer [23, 24]. This polymer is obtained by hydrolytic polycondensation of octadecyltrichlorosilane in water. The resultant condensate consists of highly uniform microcrystallites in which the inorganic siloxy backbones form periodic layers, each containing a monomolecular layer of intercalated water, separated by crystalline assemblies of alkyl chains [23]. The siloxy bilayers form nanocavities with surface silanol groups. When this polymer is placed in an aqueous solution of a metal salt (such as K_2PtCl_6, K_2PtCl_4, K_2PdCl_4, etc.), it absorbs the salt into its hydrophilic layer [25, 26]. Subsequent reduction yields regularly spaced metal nanoparticles, as depicted in Figure 4.5. For metallated PODS samples, metal compound loading greatly depends on the compound type. Incorporation of all divalent ions is driven by entropy: Hydrogen bond formation between the chlorine atoms of metal anions and the silanol groups leads to replacement of several water molecules with each $PtCl_4^{2-}$ or $PdCl_4^{2-}$ ion. In so doing, planar ions ($PdCl_4^{2-}$, $PtCl_4^{2-}$) easily penetrate thin siloxy bilayers, while incorporation of bulkier octahedral $PtCl_6^{2-}$

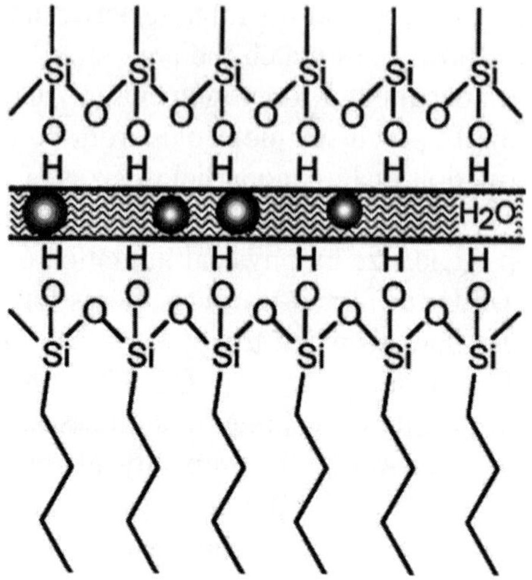

Figure 4.5. Schematic representation of nanoparticle formation in the PODS siloxy bilayer. (Reprinted with permission from reference 25. Copyright 2000, American Chemical Society.)

ions is less successful and the degree of metallation is lower. Reduction of metal compounds with PODS silanol groups (for $PtCl_4^{2-}$) or added reducing agents (for other ions) yields metal nanoparticles located in the siloxy bilayers (Figure 4.6).

As assessed by ASAXS and TEM, in these metal–polymer nanocomposites the particle size distribution is narrow (Figure 4.7) and the metal nanoparticle size does not depend on the reducing agent type, on the metal compound loading, and on the metal compound type. This indicates a "cage" effect, limiting the particle growth by the cavity size. In Figure 4.6 one can see that particles are located solely in the siloxy bilayers. These nanoparticles measure about 1–2 nm in diameter and possess a narrow size distribution likely due to limited volume availability within the siloxy bilayers. Thus, the cavity size of the nanostructured polymer controls the nanoparticle growth. However, the degree of metallation in this material is low due to low fraction of silanol groups. To enhance functionality of PODS's, we impregnated the ordered polymer with surfactants (cetylpyridinium chloride, CPC) or with other long-chain functionalized molecules, incorporation of which inside the PODS hydrocarbon layers was driven by the hydrophobic–hydrophobic interaction in a hydrophilic media (water–ethanol and water–acetone). After this functionalization, we were able to load the hydrophobic layers of the modified PODS's

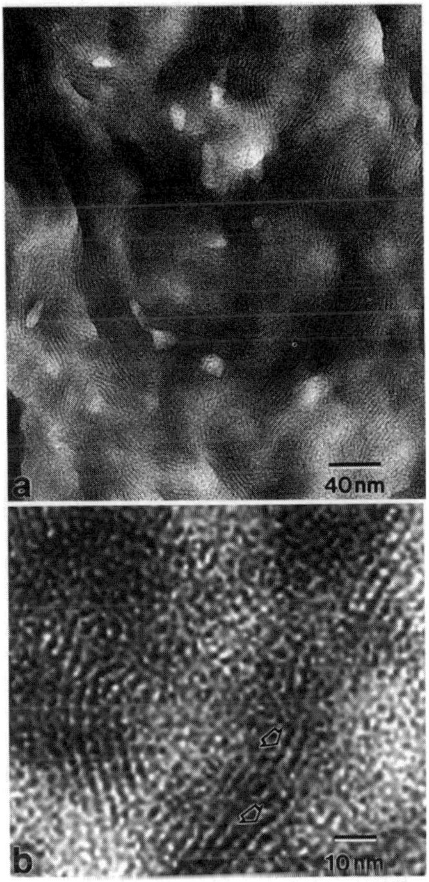

Figure 4.6. **(a)** A TEM image of PODS-Pt1, derived from K_2PtCl_4 and reduced by the silanol groups of PODS. According to elemental analysis, this system contains 1.20 wt. % Pt. **(b)** An enlargement of **(a)** to permit closer examination of the Pt nanoparticles (arrowheads) within the PODS nanostructure. (Reprinted with permission from reference 25. Copyright 2000, American Chemical Society.)

with a higher fraction of metal ions. In so doing, metal nanoparticle formation was not restricted to the siloxy bilayer and was not controlled by cavity size. Thus, PODS allows regular spatial positioning of the nanoparticles and control over the nanopraticle growth (Figure 4.7), but the material is crosslinked and metal loading is low. The functionalization of PODS with functional long-chain hydrocarbons yields a higher degree of metallation, but nanoparticle growth becomes dependent on the reaction conditions (for example, the type of a reducing agent).

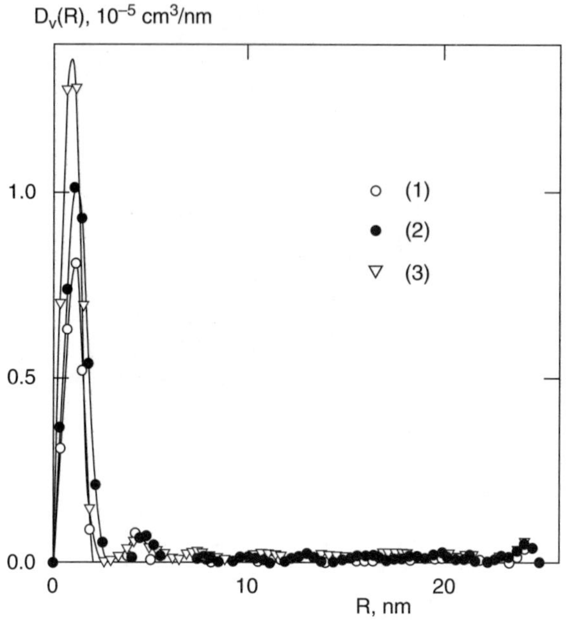

Figure 4.7. Volume distribution functions of the Pt nanoparticles in PODS. Curves 1–3 correspond to the samples based on K_2PtCl_4 after self-reduction and $H_2PtCl_6 \times 6H_2O$ after H_2 and $NaBH_4$ reduction, respectively. (Reprinted with permission from reference 26. Copyright 2000, American Chemical Society.)

3. SOLUBLE POLYMER NANOSTRUCTURES

The existence of nanostructures in polymeric systems assumes the presence of interfaces, so if these systems are solubilized in any solvent without loss of nanostructure (interfaces), they are not genuinely homogeneous, but rather nanoheterogeneous. In this case, interfaces are formed within a nanometer size volume, while these systems are colloidally soluble. Examples of such systems are various kinds of block copolymer micelles, microgels, and polymer colloids.

3.1. Nanoparticles in Block Copolymer Micelles

3.1.1. Cores of Block Copolymer Micelles. In selective solvents (a good solvent *only* for one block), amphiphilic block copolymers form micelles, the characteristics of which (size and shape) depend on the chemical structure and molecular weight of each block and on the solvent type [27–29]. If the core-forming block contains functional groups, which are able to react with metal

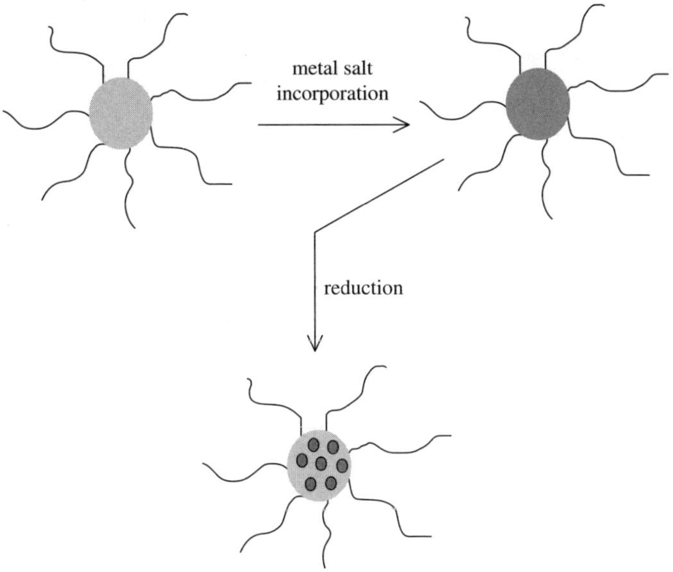

Figure 4.8. Schematic image of metallation of the block copolymer micelle cores.

compounds forming complexes or salts, the micelle core can be charged with a metal compound (Figure 4.8). A block containing no functional groups but providing solubility and micelle stability in a respective solvent should form the micelle corona. Then any further transformations of metal compounds (reduction, thermolysis or other), leading to metal nanoparticle formation, are contained in the core as in a nanoreactor (Figure 4.8). Although micelles are colloidally soluble in the selective solvent, the micelle core can be regarded as quasi-solid, forming the interface between core and corona. In 1995 we described formation of gold nanoparticles in the micelle cores of polystyrene-*block*-poly-4-vinylpyridine (PS-*b*-P4VP) [30]. Practically simultaneously, synthesis of metal and semiconductor nanoparticles in the cores of amphiphilic block copolymer micelles was reported by several research groups [30–35]. As was shown for Au, Pd, and Pt nanoparticles formed in the PS-*b*-P4VP micelles, the nanoparticle morphology strongly depends on the type of reducing agent. A sluggish reducing agent yields one nanoparticle per micelle ("cherry-like" morphology) or a few large particles (Figure 4.9) [35]. If the micelle is crosslinked [36]—that is there is no exchange of macromolecules between micelles—the morphology with a single particle per micelle is feasible. A fast reducing agent yields a so-called "raspberry-like" morphology, containing many small particles per micelle [30]. The latter morphology is thought to be preferable for catalytic applications (Figure 4.9) [35]. In a similar fashion, we

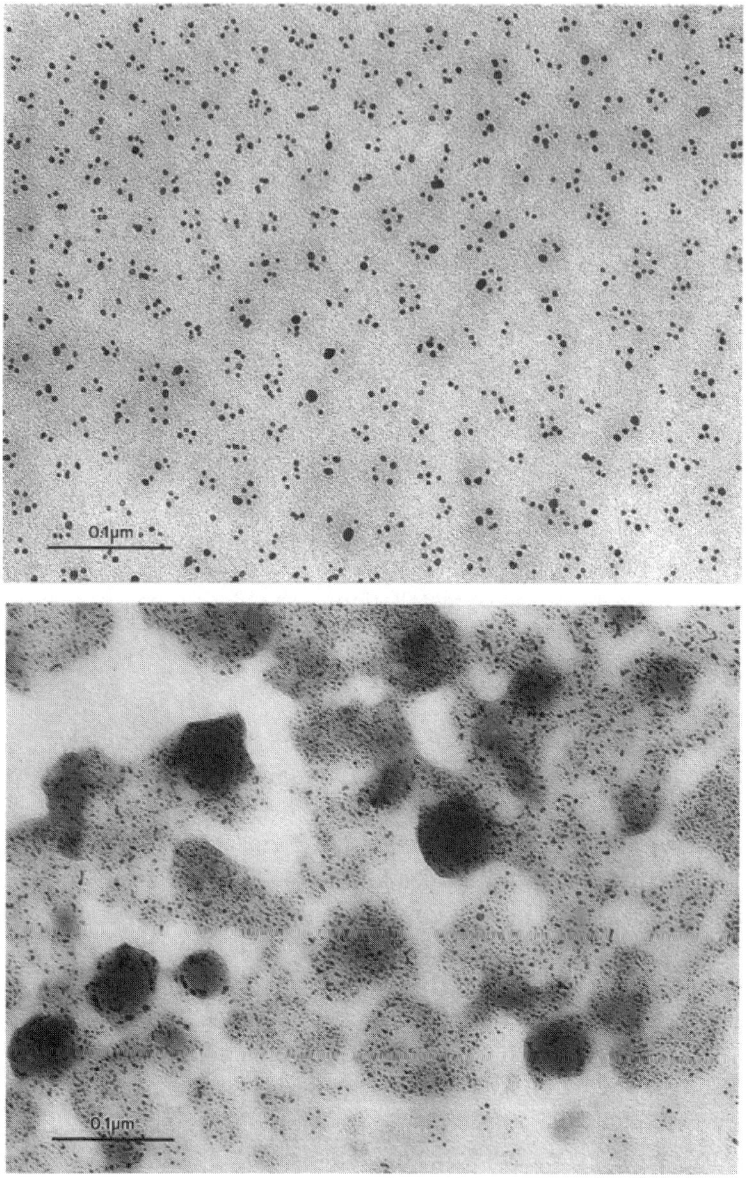

Figure 4.9. Electron micrographs of the Pd colloids synthesized in the PS-*b*-P4VP block copolymer via reduction with hydrazine **(top)** and $NaBH_4$ **(bottom)**. (Reprinted with permission from reference 35. Copyright 1997, American Chemical Society.)

have synthesized bimetallic nanoparticles (PdAu, PdPt, and PdZn) in the PS-*b*-P4VP micelles, yet bimetallic particle morphology depends on the metal pair [37]. FTIR spectra of CO adsorbed onto bimetallic nanoparticles demonstrate the presence of solely Pd atoms on the PdAu nanoparticle surface and the existence of a single type of active centers, while the PdPt and PdZn bimetallics display both Pd and Pt (or Zn) atoms on the nanoparticle surface and the presence of two types of active centers on Pd. The Pd monometallic and PdAu, PdPt, and PdZn bimetallic colloids formed in the PS-*b*-P4VP block copolymer micelles were studied in the hydrogenation of dehydrolinalool (DHL) to linalool (a fragrant substance and a precursor of vitamins A, E, and K). The catalytic activities of bimetallic catalysts were found to be higher than that of the Pd nanoparticles due to the modifying influence of the second metal. For all the catalysts, the optimal conditions were found, providing high selectivity of hydrogenation (up to 99.8%), which is most likely determined by the modifying influence of pyridine groups in the P4VP cores.

Similar to noble metal nanoparticles, Co nanoparticles can be prepared by incorporation of $CoCl_2$ (which is almost insoluble in toluene) in the PS-*b*-P2VP micelles followed by reduction. This results in very small spherical particles (below 1 nm in diameter), the thermal treatment of which at 200°C for 2 hr yields spherical nanoparticles with diameters in the range of 3–5 nm [38]. In the solid state these metal–polymer nanocomposites display extraordinarily high magnetization value at comparatively low Co content; that is, we obtained a tenfold increase of the specific magnification density.

Another suitable Co source is $Co_2(CO)_8$. Despite the fact that this compound is well-soluble in toluene, its preferable incorporation into the micelle core occurs due to formation of cationic–anionic complexes involving 4-VP units [38]. The shape and size of the Co nanoparticles formed after thermolysis of the $Co_2(CO)_8$-filled micelles can he controlled by the ratio N/Co and can be varied from the spherical particles (at low Co loadings) to a star-like and cubic morphology at higher $Co_2(CO)_8$ loading [38]. In this case, both superparamagnetic and ferromagnetic nanocomposites can be obtained. The coercive force for the latter varies from 250 to 450 Oe, depending on the Co content and block ratio.

Recently, block copolymers micelles filled with MoS_x nanoparticles (well miscible with mineral oil) were synthesized in heptane using interaction of $Mo(CO)_6$ with polystyrene-*block*-polybutadiene (PS-*b*-PB) and polystyrene-*block*-polyisobutylene (PS-*b*-PIB) followed by H_2S treatment [39]. To position MoS_x nanoparticles in the PS core, the reaction between $Mo(CO)_6$ and block copolymer should be carried out in argon atmosphere. This yields arene Mo tricarbonyl complexes while olefin Mo carbonyl complexes do not form. By contrast, to place MoS_x nanoparticles in the PB corona, complexation with $Mo(CO)_6$ should be carried out in the CO atmosphere. This suppresses formation of arene $Mo(CO)_3$ complexes and ensures oleftin $Mo(CO)_x$ complexes in the PB corona. MoS_x composition can be influenced by varying the sulfiding temperature: Increase of sulfiding temperature to 98°C results in the species whose elemental analysis matches to MoS_3 or MoS_2. At all compositions, MoS_x nanoparticles are amorphous even when the nanoparticle diameter reaches 4.5 nm. For the PS-*b*-PB block copolymer, the location of MoS_x species in the micelle corona makes them more accessible to working surfaces and allows better antifrictional properties than those for MoS_x species situated in the micelle core. By contrast, for PS-*b*-PIB, whose overall micelle density is low, location of MoS_x nanoparticles in the micelle core also leads to a low friction coefficient and a high critical load. On top of that, addition of block copolymer micelles filled with MoS_x nanoparticles improves antiwear properties. These characteristics make block copolymer micelles filled with MoS_x nanoparticles, prospective additives to lubricating oils.

Among all the block copolymers, forming micelles in organic solvents, PS-*b*-P4VP is the most versatile one, since interaction of pyridine units with nearly any metal compound is feasible. In addition, metallated block copolymer micelles are very stable in toluene solutions (for months and years); this often makes PS-*b*-P4VP a polymer of choice. However, from the environmental point of view, one would prefer using less hazardous solvents and water is a preferred solvent. However, when water is concerned as a solvent, there are very few suitable block copolymers—that is, those forming micelles in water and containing functional groups in the micelle core. A few examples of such block copolymers include PEO-*b*-P2VP and PEO-*b*-PB [40, 41]; the micellization of the former block copolymer depends on pH [42]. At pH below 5, PFO-*b*-P2VP becomes molecularly soluble in water. However, if a metal compound is already incorporated into the P2VP core, lowering the pH of the PEO-*b*-P2VP micellar solution does not lead to the micelle decomposition: Coordination with the metal compound keeps micelles intact. Despite the fact that all metal salts studied are soluble in water, the metal compound was found to be completely bound to the micelle cores as shown by ultracentrifugation [40]. The driving force for interaction with a metal compound and the rate of a metal uptake depend on the type of the metal salt. Cations (Pd^{2+}) coordinate directly with the pyridine units of the core-forming block, and this process is completed within 15 min. In the case of salts containing the noble metal in the anion (Na_2PtCl_6, Na_2PtCl_4, Na_2PdCl_4), the interaction of micelles with metal salts takes place due to ligand exchange of Cl^- for pyridine units, and this process can require days. Protonation of P2VP units either by preliminary acidification with HCl followed by addition of a neutral metal compound or by using metal-containing acids (such as H_2PtCl_6 or $HAuCl_4$) yields immediate metal binding and micelle formation due to electrostatic interaction and hydrogen bonding. Although the salt is immediately bound, structural equilibration of such micelles demands longer times. Reduction of the metal ions embedded in the PEO-*b*-P2VP micelles results in the formation of well-defined noble metal nanoparticles, the size and stability of which depend on the metal loading and on the micelle structure. The higher the micelle core density, the larger the particles (if all othet condition are identical), suggesting that the nanoparticle size depends on a local metal concentration while diffusion limitations of metal ions are not crucial.

Palladium nanoparticles formed in PEO-*b*-P2VP micelles were studied in selective hydrogenation of DHL to linalool by varying solvent composition ("isopropanol : water" ratio) and the pH of the reaction medium [43]. As was established with TEM and AFM, isopropanol fraction and KOH loading control the micellar characteristics, which in turn govern the catalytic properties. The larger and denser the micelles, the slower the reaction due to internal diffusion limitations within the micelles. At the same time, denser micelle cores provide better modification of the Pd nanoparticle surface with pyridine units and higher selectivity. The highest selectivity (99.4%) was obtained at pH of 9.4 and 95 vol. % of isopropanol. The highest observed TOF (turnover frequency) value was found to be 24.4 sec^{-1} at pH of 13.0 and 70 vol. % of isopropanol.

Unlike PEO-*b*-P2VP, the PEO-*b*-PB block copolymer micelles in water are very dense, so they successfully fulfill two roles: They serve as nanoreactors for Pd, Pt, and Rh nanoparticle formation and as metal-particle-containing templates for mesoporous silica casting [41]. Figure 4.10 presents PEO-*b*-PB micelles (M^{PB} = 13,400; M^{PEO} = 19,200) filled with Pd nanoparticles and mesoporous silica cast over this block copolymer template. Variation of block lengths and composition of the PEO-*b*-PB block copolymers enables the adjustment of the pore size and pore structure of mesoporous silica along with the simultaneous size control of metal nanoparticles. By varying the metal compound type or the way of metal particle formation and treatment, we were able to influence the metal particle size and particle size distribution. It is believed that any amphiphilic block copolymers, forming dense, functionalized micelles in water, can be used for templating of mesoporous metal oxides with metal nanoparticles.

3.1.2. Coronas of Block Copolymer Micelles. If the nanoparticles are synthesized in the corona of amphiphilic block copolymer micelles, they might be more accessible to the substrate in catalytic reactions and become more efficient catalysts. However, synthesis of nanoparticles in the micelle corona is not straightforward. If the micelle corona contains functional groups able to interact with metal compounds, these micelles will precipitate during complexation. If the micelle corona is not functionalized, then nanoparticle stabilization is only governed by hydrophobic interactions with the hydrophobic micelle core, and the stability of the soluble metal–polymer nanocomposites will be inferior [44]. To improve stabilization in the micelle coronas, we suggested using hybrid

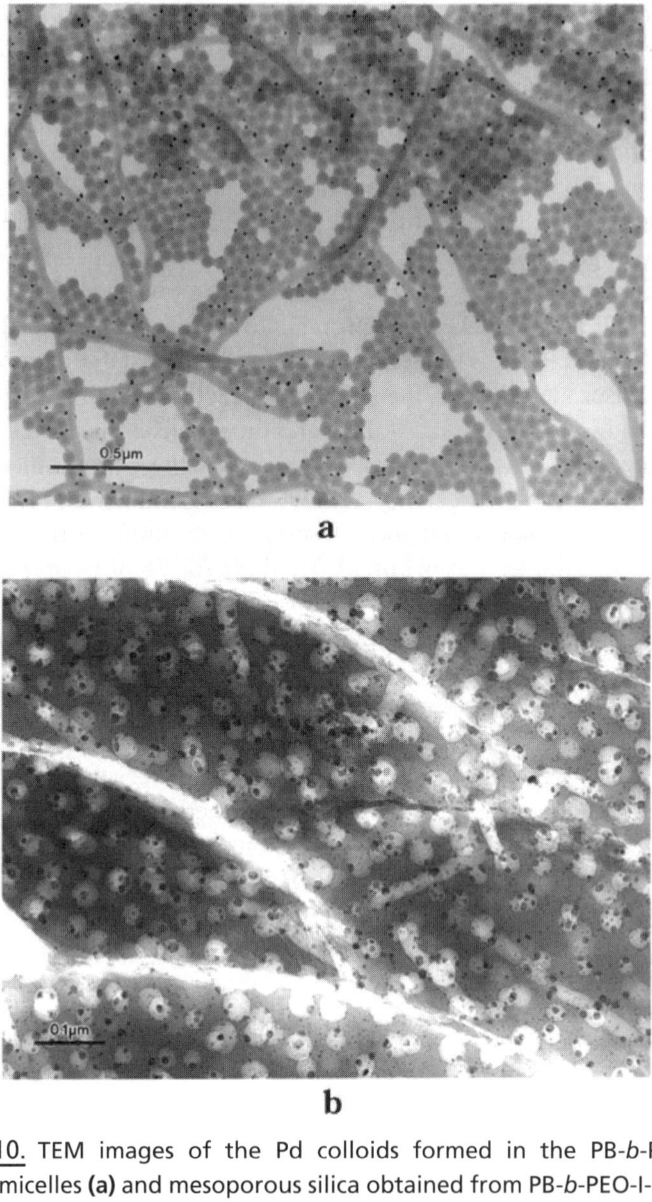

Figure 4.10. TEM images of the Pd colloids formed in the PB-*b*-PEO-I block copolymer micelles **(a)** and mesoporous silica obtained from PB-*b*-PEO-I-Pd(0) as the template after calcination **(b)**. (Reprinted with permission from reference 41. Copyright 1999, American Chemical Society.)

micelles consisting of PS-*b*-PEO and cationic or anionic surfactants [45–47]. We assumed that when surfactant is loaded in an aqueous solution containing block copolymer micelles, the favorable path for surfactant molecules will be penetration into the hydrophobic micelle cores. Then surfactant head groups will be located on the core–corona interface or in its vicinity. Ion exchange of surfactant counterions for the ions of choice (Pt, Pd, or Rh) was expected to saturate the corona with the given ions. Combination of dynamic light scattering (DLS), sedimentation in an ultracentrifuge (UC), and ^{1}H NMR spectroscopy allowed us to prove comicellization of block copolymer macromolecules and cationic or anionic surfactants. Ion exchange of the surfactant counterions in the PS-*b*-PEO/CPC system for $PtCl_6^{2-}$ or $PdCl_4^{2-}$ ions followed by reduction of metal-containing hybrid micellar systems (PS-*b*-PEO/CPC/MX_n) with $NaBH_4$ or H_2 yields nanoparticles, the morphology and stability of which depend on the metal compound loading and the type of the reducing agent [45]. $NaBH_4$ reduction leads to decomposition of micellar clusters and formation of micelles with embedded nanoparticles (Figure 4.11). These systems display exceptional stability (for years), if metal salt loading does not exceed 1.24×10^{-2} M. Hydrogen reduction results in metal nanoparticle formation both in micelles and micellar clusters so colloidal solutions are stable at metal salt concentration below 3.36×10^{-3} M (Figure 4.12). Unlike many other systems [30, 40], here the nanoparticle size does not depend on the reducing agent type, but depends on the metal type [45–47]. We surmise that strong interaction of surfactant head

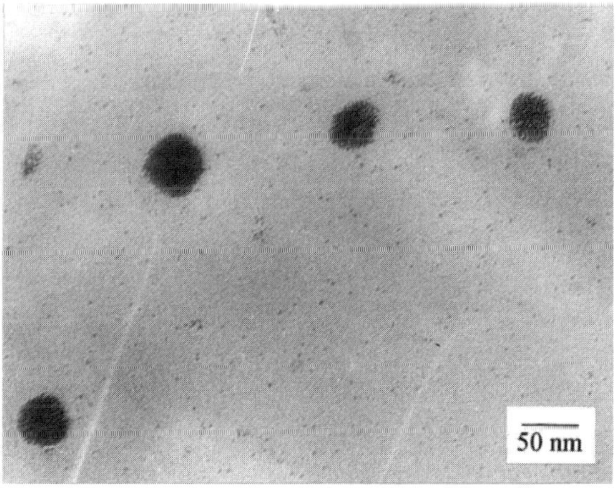

Figure 4.11. TEM micrograph of the Pt nanoparticles stabilized in the PS-*b*-PEO/CPC hybrid micelles in water and obtained via reduction of $H_2PtCl_6{\cdot}6H_2O$ (1.24×10^{-2} mol/L) with $NaBH_4$. (Reprinted with permission from reference 45. Copyright 2000, American Chemical Society.)

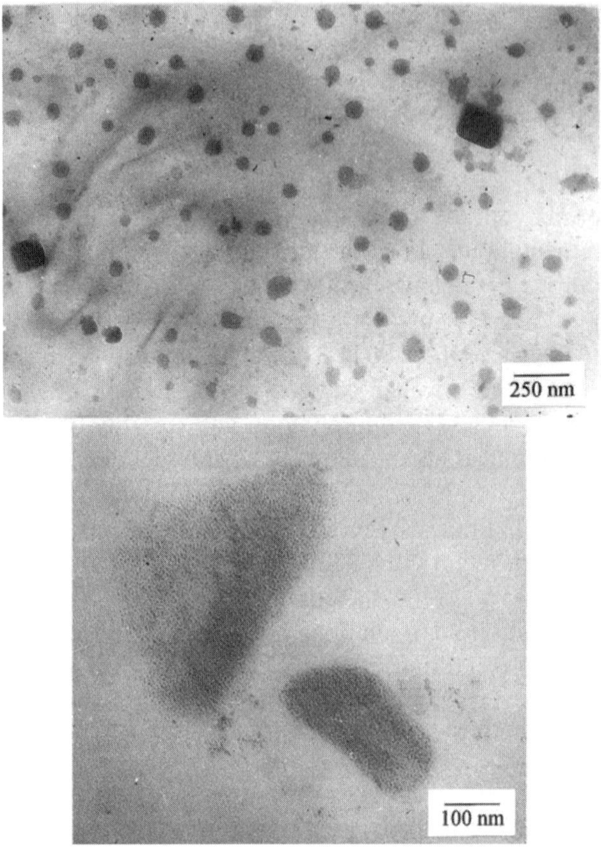

Figure 4.12. TEM micrographs of Pt nanoparticles stabilized in the PS-*b*-PEO/CPC hybrid micelles in water and obtained via reduction of $H_2PtCl_6 \cdot 6H_2O$ (3.36×10^{-3} mol/L) with molecular hydrogen. (Reprinted with permission from reference 45. Copyright 2000, American Chemical Society.)

groups with growing nanoparticles along with hydrophobic interactions with the PS core control the nanoparticle size.

3.1.3. Micelle Formation via Complexation. Micellization of molecularly soluble block copolymers due to interaction with metal compounds was observed in both organic media and water, if one of the two blocks (for a diblock copolymer) is inert while the other is able to form complexes with metal compounds. Micellization of Pd-, Pt-, and Rh-containing polymers derived from PS-*b*-PB with a short PB block was first reported in 1998 [48]. The crosslinks formed due to complexes between metal atoms and PB blocks of different macromolecules were shown to cause micellization. By contrast, Fe carbonyl complexes formed in PS-*b*-PB intramolecularly (no crosslinks) do not yield micelles, and macromolecules remain in the solution. In a similar way, polystyrene-*block*-poly-*m*-vinyltriphenylphosphine (PS-*b*-PPH) was reacted with $(CH_3CN)_2PdCl_2$ or $(PPh_3)_2PdCl_2$ [49]. This resulted in the complexation of triphenylphosphine groups of the PPH block with palladium compounds, inducing micellization instead of gel formation. The morphologies of the Pd-containing PS-*b*-PPH micelles were found to depend on the molecular weight of the diblock copolymer, the block length, and the metal compound type. Interaction of $(CH_3CN)_2PdCl_2$ with PS-*b*-PPH of medium molecular weight (M_w = 29,000) results in the formation of different structures: spherical aggregates, multilamellar vesicles, disklike micelles, and unilamellar vesicles (Figure 4.13) [49]. By contrast, the interaction of $(PPh_3)_2PdCl_2$ with the same block copolymer induces the formation of large multivesicular structures of high uniformity.

The micelles filled with Pd nanoparticles were prepared by reduction of Pd-containing PS-*b*-PPH with $N_2H_4 \times H_2O$, $LiB(C_2H_5)_3H$, or $LiAlH_4$ and analyzed by TEM, X-ray photoelectron spectroscopy (XPS), and X-ray diffraction (XRD) techniques. The size and architecture of Pd nanoparticles formed in the PS-*b*-PPH micellar aggregates was shown to depend on the type of the reducing agent and molecular structure of the Pd complex produced in block copolymer.

In a similar way, micellization via complexation in an aqueous medium was carried out using PEO-*b*-PEI (polyethyeleneimine) [50, 51], since PEI, being a polydentate ligand, easily interacts with any metal compounds yielding metal complexes [52]. Complexation results in aggregation of the PEI blocks and

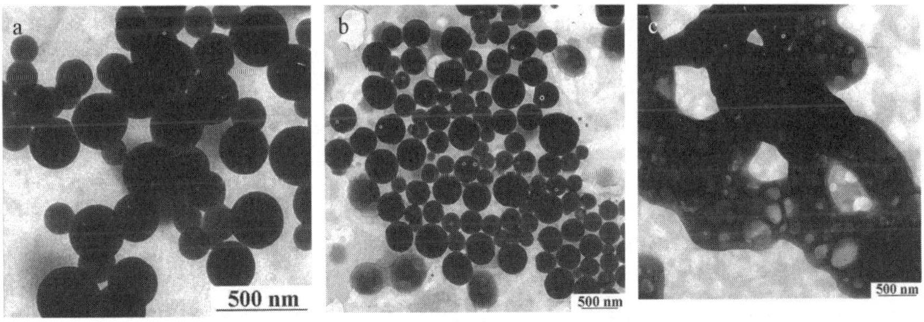

Figure 4.13. **(a)** TEM micrograph of the Pd-containing polymer derived from PS-*b*-PPH1 and $(CH_3CN)_2PdCl_2$. Molar ratio P/Pd = 2/1. **(b)** TEM micrograph of the Pd-containing polymer derived from PS-*b*-PPH1 and $(CH_3CN)_2PdCl_2$. Molar ratio P/Pd = 4/1. **(c)** TEM micrograph of the Pd-containing polymer derived from PS-*b*-PPH1 and $(PPh_3)_2PdCl_2$. Molar ratio P/Pd = 2/1. (Reprinted with permission from reference 49. Copyright 2000, American Chemical Society.)

micelle core formation, while PEO blocks form coronas and ensure colloidal solubility. After nanoparticle formation via reduction, PEO-*b*-PEI micelles remain intact, while changing their hydrodynamic parameters. This suggests strong interaction between the PEI units and a metal nanoparticle surface. Similar to amphiphilic block copolymer micelles in selective solvents, the complex-induced micelles also furnish control over nanoparticle nucleation and growth, although micellar and metal particle characteristics are more dependent on the reaction conditions. Since micelle formation is induced solely by complexation with metal compounds, the micelle core densities are low, so the cores here are not quasi-solid any more. Metal nanoparticle morphologies in these systems can be affected in the same way as in amphiphilic block copolymer micelles by varying the reducing agent type, the metal compound loading, and the pH.

3.2. Nanoparticles in Polyelectrolyte Microgels

Polyelectrolyte (PE) microgels are gel spherical particles having diameters in the nanometer range, bearing cationic or anionic groups, and forming colloidal solutions [53, 54]. The microgel stability in the solutions is ensured by crosslinking. Charged groups of PE microgels provide solubilization in water and can be used for ion exchange. When PE microgels are loaded with metals, they can serve as nanoreactors for metal nanoparticle formation. Gold nanoparticles were formed in microgels based on sulfonated PS [55]. Morphologies of nanoparticles are governed by the degree of microgel crosslinking. The higher the crosslinking density, the higher the probability of spherical particles embedded in the microgels. The type of reducing agent is the other key factor. Fast reduction with $NaBH_4$ results in small gold nanoparticles (4.5 nm) located in microgels. When $NaBH_4$ is added in alkaline solution (0.1 N NaOH), the nucleation is slower, resulting in 7-nm nanoparticles forming long "threads." Yet, only 20% of microgels contain nanoparticles; thus, slow nucleation allows Au clusters and ions to migrate out of microgel areas where aggregation easily occurs.

Microgels filled with Pd and Pt nanoparticles have been prepared in a similar way. These microgels were then used as co-templates (along with amphiphilic PS-*b*-PEO block copolymers) for templating of mesoporous silica [56], aluminosilica, and alumina [57]. Here microgels play a dual role: They are nanoreactors for nanoparticle formation and pore-forming templates when mesoporous material is formed. It is noteworthy that aluminas with metal nanoparticles templated both over cationic and anionic microgels consist of an interpenetrating pore system and alumina nanowires (2–3 nm in diameter and about 40 nm in length) along with Pd or Pt nanoparticles (Figure 4.14). This combination yields higher mesoporosity than that of aluminosilicas. Catalytic

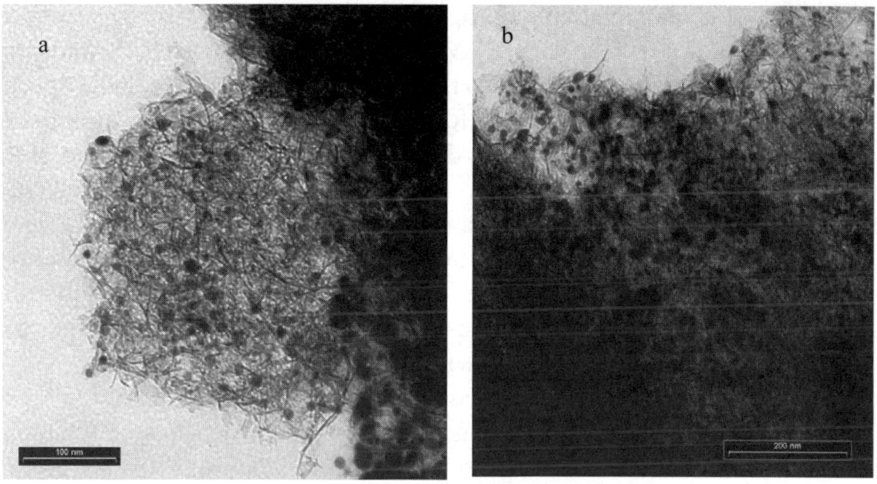

Figure 4.14. TEM images of alumina samples containing Pd nanoparticles and templated over cationic **(a)** and anionic **(b)** microgels. (Reprinted with permission from reference 57. Copyright 2003, American Chemical Society.)

properties of Pd- and Pt-containing aluminosilicas and aluminas were studied in selective hydrogenation of long-chain acetylene alcohols (C_5, C_{10}, C_{20}). Aluminosilica samples showed no appreciable activity or selectivity for either substrate, which is probably caused by high diffusion limitations within pores. Indeed, the shorter the substrate, the higher the activity, so the highest activity was achieved for alcohol C_5. Mesoporous alumina cast over cationic microgels with Pd nanoparticles displayed a high selectivity toward a longest acetylene alcohol C_{20}, along with satisfactory activity. Since normally it is difficult to achieve, we contributed this effect to a perfect correlation between pore, nanoparticle, and substrate sizes.

3.3. Nanoparticles In Multifunctional Polysilsesquioxane Colloids

A new family of polysilsesquioxane colloids was synthesized by hydrolytic condensation of different functional amphiphilic silanes without any additional surfactants [58 60]. To choose a suitable monosubstituted silane, we looked for the combination of a functional group, which would be able to interact with metal compounds, and a hydrophobic part, which would self-assemble in an aqueous medium. We expected that a sol–gel reaction of such a precursor in water would yield multifunctional materials with internal ordering if the hydrophobic tail is sufficiently long. In this way, polysilsesquioxane colloids based on *N*-(6-aminohexyl)aminopropyltrimethoxysilane (AHAPS) were synthesized and studied using SAXS, TEM, liquid- and solid-state NMR, and other methods [58, 59]. Poly-AHAPS (PAHAPS) can be obtained as soluble species (colloids) in water—a poor solvent for AHAPS precursor—or as crosslinked material in THF (a good solvent for AHAPS). Internal lamellar ordering of the AHAPS tails was obtained when the amino groups of the AHAPS molecules were not protonated (no pH adjustment) and aqueous medium was used (Figure 4.15). The size of the PAHAPS colloids depends on the reaction conditions including the dish material where evaporation occurs. Evaporation of a PAHAPS solution on a Teflon dish instead of glass results in much larger colloids because of lack of interaction with the dish surface.

When metal compounds are added to the solution containing PAHAPS colloids, the size of the latter may change, and this change depends on the molar ratio [PAHAPS unit] : metal. For example, when Pt loading is low (40 : 1), no change (additional aggregation) occurs, while at tenfold increase of the loading, colloid size increases. In so doing, PAHAPS colloid size greatly depends on the Pt compound geometry: A planar $PtCl_4^{2-}$ ion induces higher aggregation during metallation than does an octahedral $PtCl_6^{2-}$ ion. Metal particle formation depends on the metal compound geometry as well [59]. For the planar $PtCl_4^{2-}$ ion (from K_2PtCl_4), metal particle size is determined by the reducing

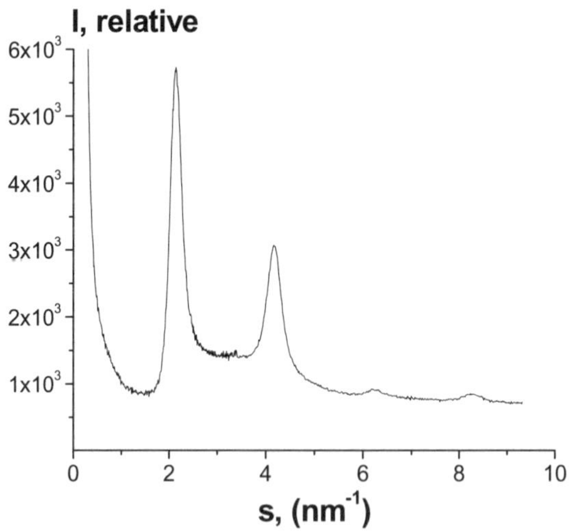

Figure 4.15. SAXS profiles of PAHAPS-H4 in which intensity is shown as a function of scattering vector (s) given by $4\pi \sin(\theta/\lambda)$, where 2θ is the scattering angle and λ (= 0.15 nm) is the wavelength of the incident radiation. (Reprinted with permission from reference 59. Copyright 2003, American Chemical Society.)

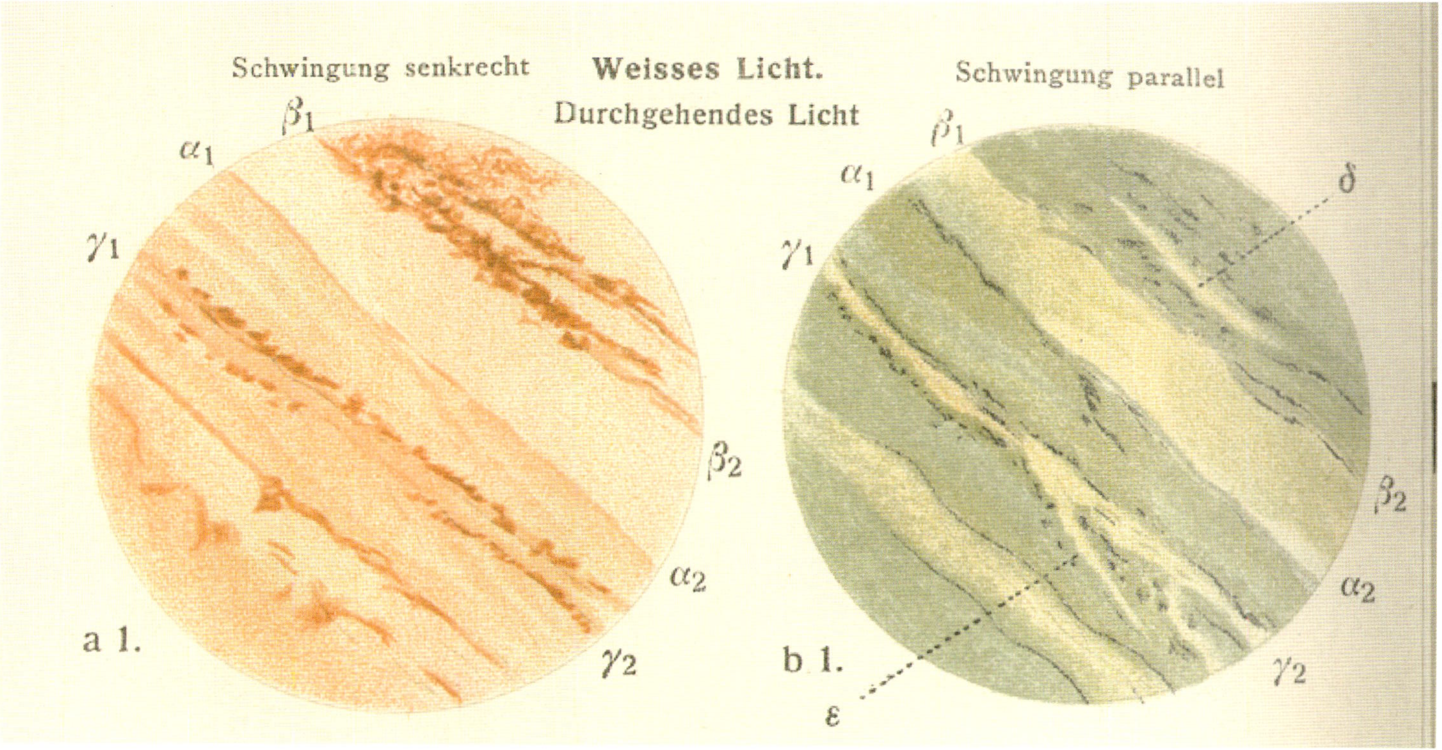

Figure 9.1. Microscope image of a dichroic nanocomposite consisting of gold particles embedded in wood (probably fir), published in 1905 [36]. The polarization plane of the light is parallel to the orientation axis of the wood in the left image and perpendicular in the right image.

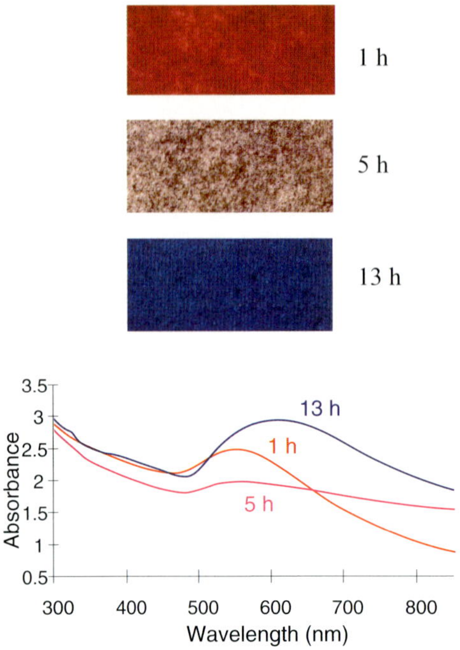

Figure 9.7. Colors and UV–Vis spectra of dispersions of gold particles (average diameter 2.2 nm, covered with a layer of 1-dodecanethiol) kept in a poly(ethylene) solution in *p*-xylene at 130°C for 1 hr, 5 hr, and 13 hr, respectively.

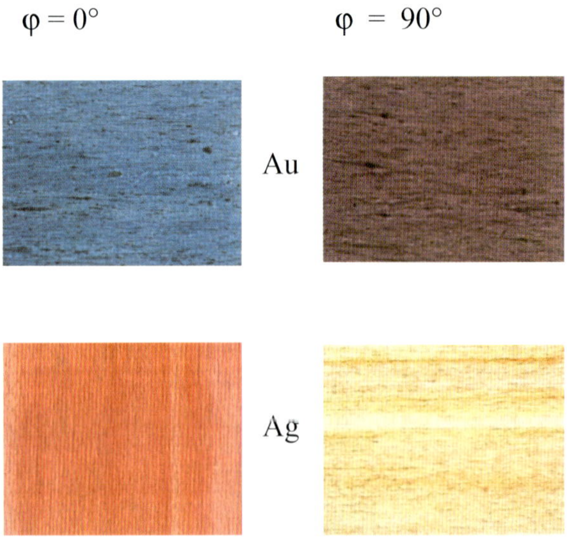

Figure 9.9. Colors of a drawn poly(ethylene)–gold and a poly(ethylene)–silver nanocomposite in polarized light with the polarization plane of the incident light and the drawing direction of the nanocomposite parallel ($\varphi = 0°$) and perpendicular ($\varphi = 0°$), respectively.

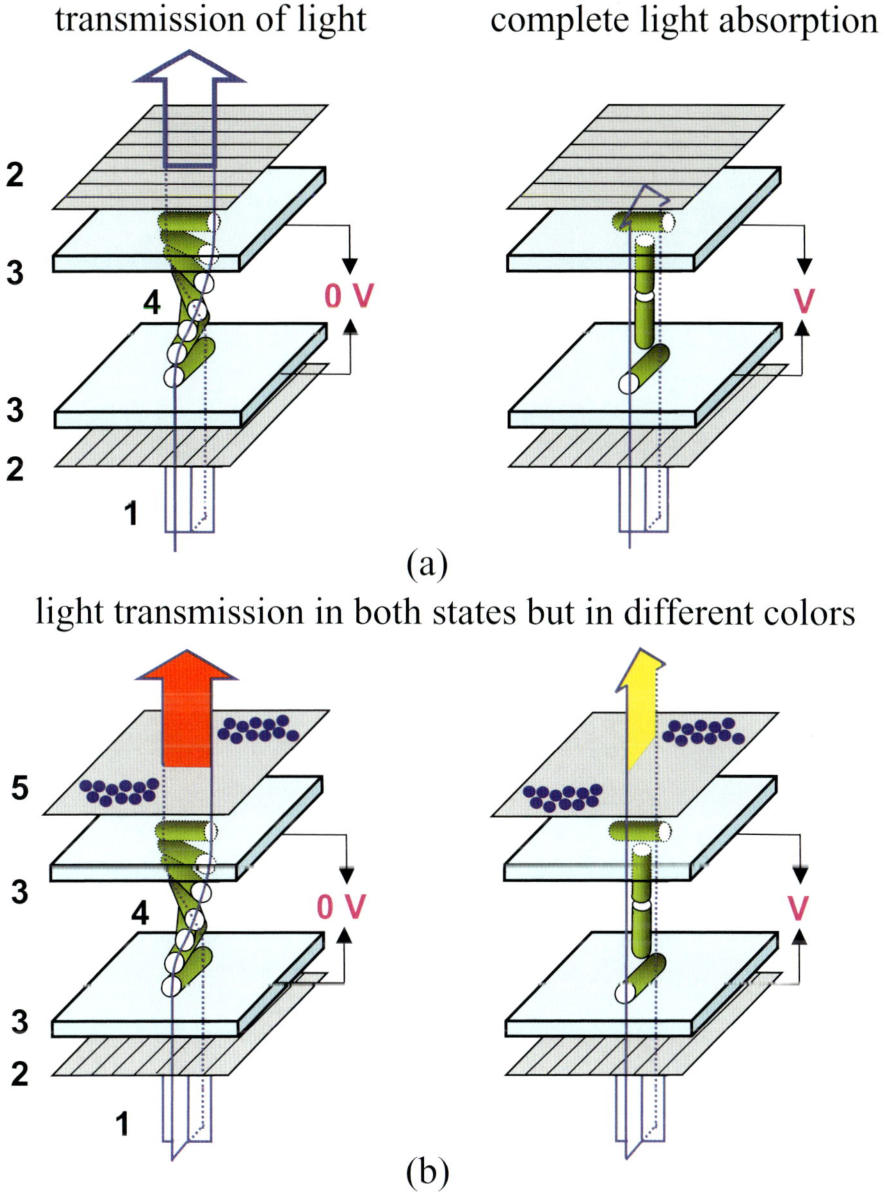

Figure 9.12. Schematic representation of **(a)** a standard twisted-nematic liquid crystal display and **(b)** a related display equipped with a drawn nanocomposite. 1, incoming light (unpolarized); 2, polarizer; 3, glass plate coated with an electrode layer and an orientation layer for the liquid-crystalline molecules; 4, liquid-crystalline molecules forming a 90° helical twist in absence of a voltage or a linear array parallel to an electric field in presence of an electric field; 5, nanocomposite with oriented arrays of metal particles.

Figure 9.13. A twisted-nematic liquid crystal display (LCD) equipped with a poly(ethylene)–silver nanocomposite that had been annealed at 180°C for 15 hr and subsequently drawn as described in the text. The drawing axis of the nanocomposite is oriented parallel to the polarizer in the left image and perpendicular in the right image.

agent type, while for K_2PtCl_6, particle diameter is fairly independent of the reducing agent (Figure 4.16). However, when H_2PtCl_6 is used, PAHAPS becomes partially protonated and this yields larger particles, although particle size distribution remains very narrow. Since amino groups can interact with the majority of transition-metal compounds, PAHAPS colloids can be used as templates for synthesis of magnetic, semiconductor, metal, or metal oxide nanoparticles with a small variation in the synthetic procedure. A schematic image of PAHAPS metallation is presented in Figure 4.17.

Another interesting feature of these metallated polymer colloids is that they can form dendrites with different size and shape from aqueous solutions containing PAHAPS colloids loaded with metal salts or metal nanoparticles (Figure 4.18) [58]. This phenomenon has been reported for the first time for colloidal polymer particles, although formation of dendrites is common for a number of salts or metals [61, 62]. Because the size and shape of the dendrites can be easily controlled by concentration of polymer solutions and metal loading, metallated PAHAPS may be suggested as catalytically active membranes. This is appropriate in those cases when select surface coverage is important. Another possible application for these dendrites loaded with metal nanoparticles is a conductive layer between two surfaces, as described elsewhere [63] for pure Pd dendrite crystals. Unlike pure metals, advantages of metal-loaded PAHAPS are the lower cost and the greater structural stability.

Polymer colloids with ordered interior were synthesized by sol–gel reaction of octadecyldimethyl(3-trimethoxysilylpropyl)ammonium chloride (ODMACl) and a mixture of ODMACl and the trisodium salt of the triacetic acid *N*-(trimethoxysilylpropyl)ethylenediamine (TANED) [60]. The structure and morphology of these colloids were studied with SAXS, TEM, NMR, sedimentation in ultracentrifuge, and other methods. When polymer colloids are obtained from a single precursor (ODMACl), their local structure, molecular weight characteristics, and morphology greatly depend on the pH value and on the sequence of pH adjustment, while lamellar ordering remains nearly unaffected with Bragg spacing d of about 3.6 nm. This Bragg spacing is significantly smaller than two layers of extended ODMACl tails, so one can assume either tilting of ODMACl chains (they are not perpendicular to the lamellar surface) or intercalation of the tails of the two layers (or both). As we observed no vesicles and the PODMACl colloids are much larger than twice the d spacing, they should have a multilamellar structure.

Hydrolytic condensation of an equimolar mixture of cationic and anionic silanes (ODMACl and TANED) yields zwitterionic copolymer colloids with two-dimensional hexagonal packing. Interaction of the ODMACl quaternary ammonium groups with the three carboxy groups of TANED leads to replacement of sodium and chloride ions and formation of gegenions, resulting in a molar ratio ODMACl : TANED = 3 : 1 (each TANED molecule contains three

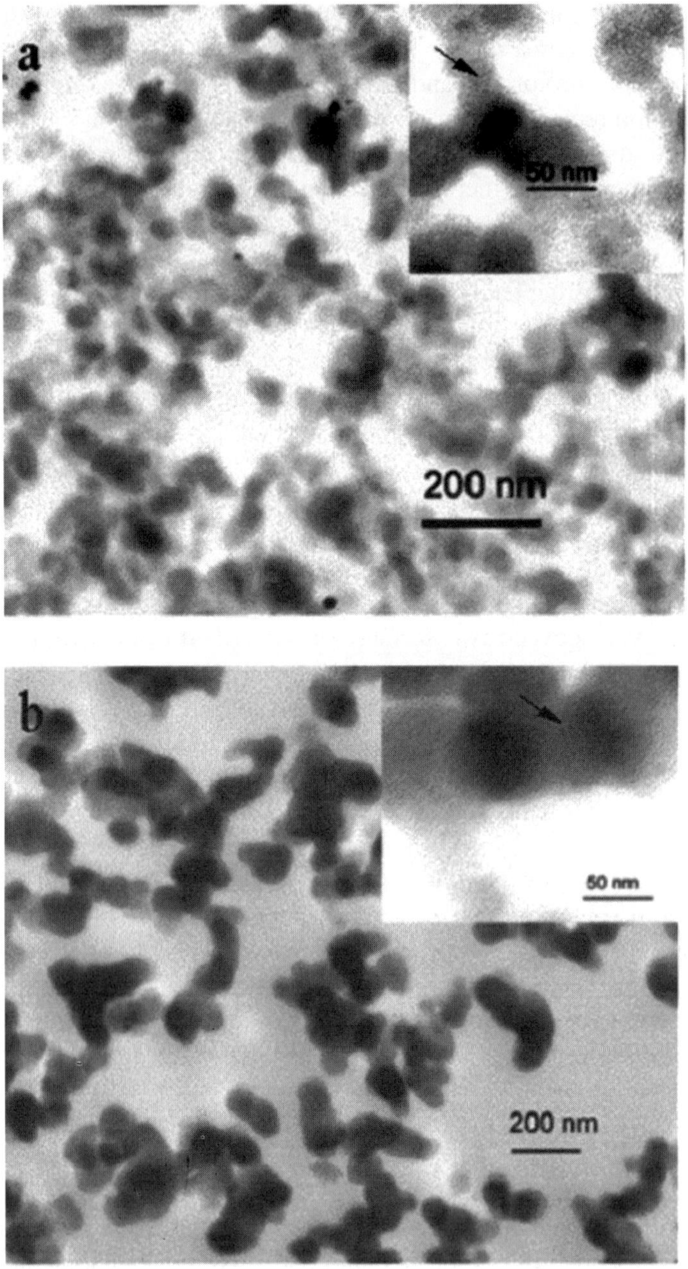

Figure 4.16. TEM images of PAHAPS-F1 colloids prepared by interaction with K_2PtCl_6 at molar ratio [PAHAPS] : Pt = 4 : 1 followed with $N_2H_4 \times H_2O$ **(a)** or $NaBH_4$ **(b)** reduction. Pt nanoparticles appear like dark dots and are shown by arrows. (Reprinted with permission from reference 59. Copyright 2003, American Chemical Society.)

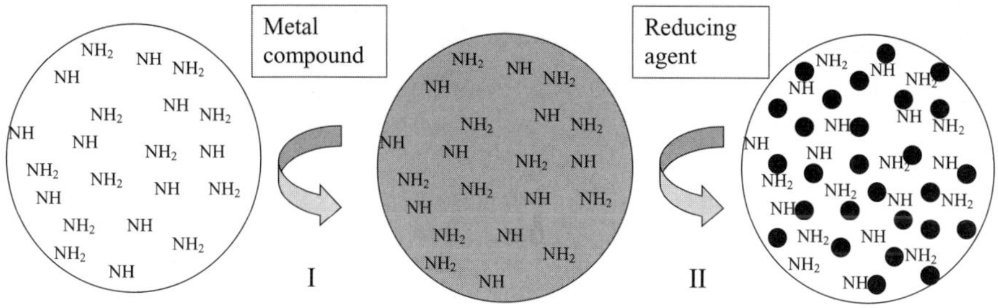

Figure 4.17. Schematic image of PAHAPS colloid metallation. Step I is incorporation of metal compounds inside the PAHAPS colloids due to interaction with amino groups. Step II is metal particle formation due to reduction of metal species. (Reprinted with permission from reference 59. Copyright 2003, American Chemical Society.)

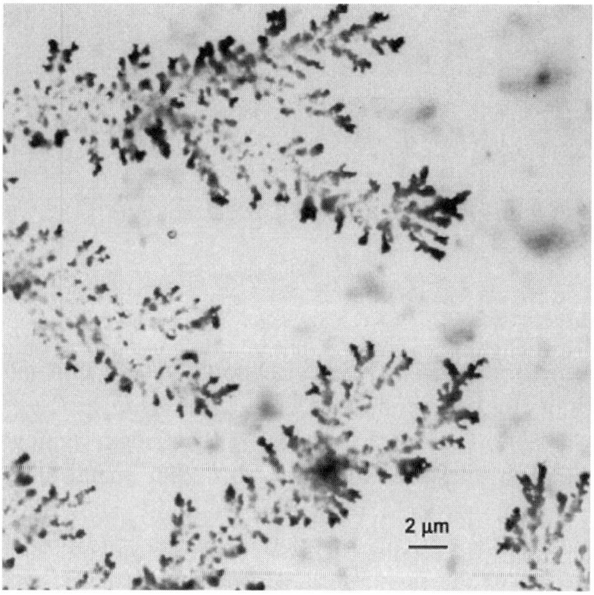

Figure 4.18. TEM image of dendrites formed by PAHAPS colloids containing Pd nanoparticles.

carboxy groups) nearly independently of the initial ODMACl : TANED ratio in the reaction medium. Due to their ordered interior, PolyODMACl (PODMACl) and PODMACl–TANED colloids can be used as templates for controlled positioning of nanoparticles within these colloids. Indeed, since lamellar ordering is preserved during metal reduction, it controls Pt nanoparticle formation within PODMACl colloids providing Pt nanoparticle alignment within the lamellar

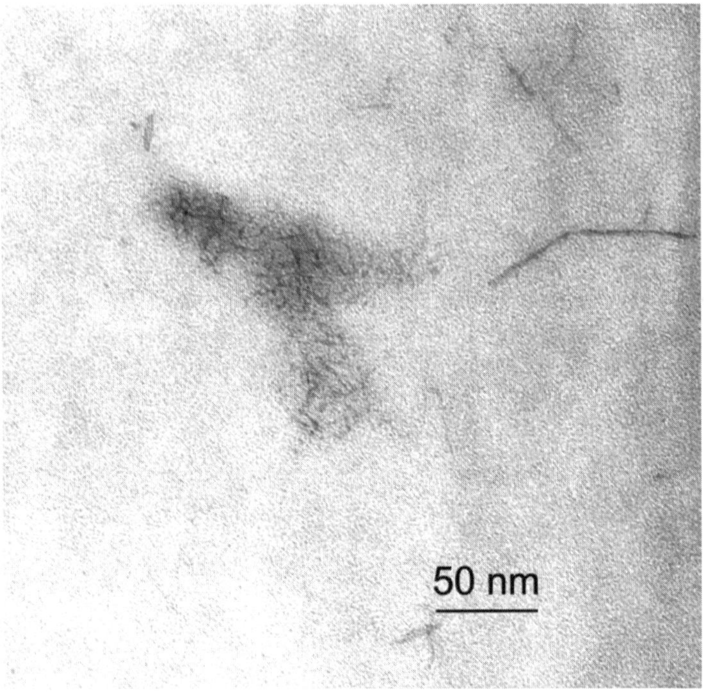

Figure 4.19. TEM image of PODMACl–TANED colloids filled with iron oxide nanoparticles.

structure. Loading of PODMACl–TANED colloids with iron salts followed by pH increase results in the formation of iron oxide nanoparticles located within PODMACl–TANED cylinders (Figure 4.19). This suggests new opportunities for aligning nanoparticles derived from different metal, metal oxides, or semiconductors within ordered polymeric colloids. These materials can be promising for tailoring optical, magnetic, and electric properties of metal nanocomposites by regular positioning of the nanoparticles.

4. CONCLUSIONS AND OUTLOOK

This chapter illustrates our approaches in the design of metal nanocomposites, both in bulk and in solution. The key advantages of soluble systems, especially if they retain solubility after nanoparticle formation, are their potential use for microheterogeneous catalysis and magnetic liquids or formation of thin deposited or free-standing films (the latter were obtained with block copolymer micelles and some functional polysilsesquioxane colloids). These opportunities make promising a number of important applications for nanolithography and catalysis in optical and magnetic materials. On the other hand, for catalytic applications, insoluble metal nanocomposites are often favored, because they do not require an additional support in heterogeneous catalysis. It is noteworthy that the nanoparticle properties can be greatly influenced by designing the polymer nanoenvironment via modification of the nanoparticle surface with polymer functional groups. Hence, the choice of the polymer type and the internal ordering of the nanostructured polymer play a crucial role in the material properties and should be taken into consideration for a specific material application.

Varying the polymer type and characteristics allows development of sophisticated metal–polymer nanocomposites with tunable properties and promise for future applications. Yet, one can expect vigorous development of this field for years to come so that new polymeric systems yielding better control over nanoparticle formation and properties will be designed.

REFERENCES

1. W. Heffels, J. Friedrich, C. Darribere, J. Teisen, K. Interewicz, C. Bastiaansen, W. Caseri, and P. Smith, *Rec. Res. Dev. Macromol. Res.* **2**, 143 (1977).
2. M. W. Ellsworth and D. L. Gin, *Polym. News* **24**, 331 (1999).
3. R. Gangopadhyay and A. De, *Chem. Mater.* **12**, 608 (2000).
4. G. Carotenuto and L. Nicolais, *Rec. Res. Dev. Mater. Sci.* **3**, 303 (2002).
5. C. Sanchez, G. J. de Soler-Illia, F. Rihot, T. Lalot, C. R. Mayer, and V. Cabuil, *Chem. Mater.* **13**, 3061 (2001).
6. V. Sankaran, C. C. Cummins, R. R. Schrock, R. E. Cohen, and R. J. Silbey, *J. Am. Chem. Soc.* **112**, 6858 (1990).
7. M. W. Ellsworth and D. L. Gin, *Polym. News* **24**, 331 (1999).
8. F. Yeh, E. L. Sokolov, A. R. Khokhlov, and B. Chu, *J. Am. Chem. Soc.* **118**, 6615 (1996).
9. A. R. Khokhlov, S. G. Starodubtsev, and V. V. Vasilevskaya, *Adv. Polym. Sci.* **109**, 123 (1993).
10. A. R. Khokhlov, E. Y. Kramarenko, E. E. Makhaeva, and S. G. Starodubtsev, *Macromol. Theory Simul* **1**, 105 (1992).
11. Y. V. Khandurina, A. T. Dembo, V. B. Rogacheva, A. B. Zezin, and V. A. Kabanov, *Vysokomol. Soedin., Ser. A Ser. B* **36**, 235 (1994).
12. H. Okuzaki and Y. Osada, *Macromolecules* **28**, 380 (1995).
13. L. M. Bronstein, O. A. Platonova, A. N. Yakunin, I. M. Yanovskaya, P. M. Valetsky, A. T. Dembo, E. E. Makhaeva, A. V. Mironov, and A. R. Khokhlov, *Langmuir* **14**, 252 (1998).

14. L. M. Bronstein, O. A. Platonova, A. N. Yakunin, I. M. Yanovskaya, P. M. Valetsky, A. T. Dembo, E. S. Obolonkova, E. E. Makhaeva, A. V. Mironov, and A. R. Khokhlov, *Colloids Surf A* **147**, 221 (1999).
15. D. I. Svergun, E. V. Shtykova, M. B. Kozin, V. V. Volkov, A. T. Dembo, E. V. J. Shtykova, L. M. Bronstein, O. A. Platonova, A. N. Yakunin, P. M. Valetsky, and A. R. Khokhlov, *J. Phys. Chem. B.* **104**, 5242 (2000).
16. L. M. Bronstein, D. I. Svergun, and A. R. Khokhlov, in *Polymer Gels and Networks*, edited by Y. Osada and A. R. Khokhlov, Marcel Dekker, New York, (2002), p. 103.
17. V. A. Davankov and M. P. Tsyurupa, *React. Polym.* **13**, 27 (1990).
18. M. P. Tsyurupa and V. A. Davankov, *J. Polym. Sci. Polym. Chem. Ed.* **18**, 1399 (1980).
19. S. N. Sidorov, L. M. Bronstein, V. A. Davankov, M. P. Tsyurupa, S. P. Solodovnikov, P. M. Valetsky, E. A. Wilder, and R. J. Spontak, *Chem. Mater.* **11**, 3210 (1999).
20. S. V. Vonsovskii, *Magnetism,* Nauka, Moscow, 1971.
21. C. P. Bean, J. D. Livingston, and P. S. Rodbell, *Acta Metall* **5**, 682 (1957).
22. S. N. Sidorov, I. V. Volkov, V. A. Davankov, M. P. Tsyurupa, P. M. Valetsky, L. M. Bronstein, R. Karlinsey, J. W. Zwanziger, V. G. Matveeva, E. M. Sulman, N. V. Lakina, E. A. Wilder, and R. J. Spontak, *J. Am. Chem. Soc.* **123**, 10502 (2001).
23. A. N. Parikh, M. A. Schivley, E. Koo, K. Seshadri, D. Aurentz, K. Mueller, and D. L. Allara, *J. Am. Chem. Soc.* **119**, 3135 (1997).
24. W. R. Thompson and J. E. Pemberton, *Langmuir* **11**, 1720 (1995).
25. L. M. Bronstein, D. M. Chernyshov, P. M. Valetsky, E. A. Wilder, and R. J. Spontak, *Langmuir* **16**, 8221 (2000).
26. D. I. Svergun, M. B. Kozin, P. V. Konarev, E. V. Shtykova, V. V. Volkov, D. M. Chernyshov, P. M. Valetsky, and L. M. Bronstein, *Chem. Mater.* **12**, 3552 (2000).
27. I. U. Hamley, *The Physics of Block Copolymers*, Oxford University Press, Oxford, UK (1998).
28. S. Förster and M. Antonietti, *Adv. Mater.* **10**, 195 (1998).
29. N. Hadjichristidis, S. Pispas, and G. A. Floudas, *Block Copolymers Synthetic Strategies, Physical Propertie, and Applications*, John Wiley & Sons, Hoboken, NJ (2003).
30. M. Antonietti, E. Wenz, L. Bronstein, and M. Seregina, *Adv. Mater.* **7**, 1000 (1995).
31. M. Antonietti and S. Henz, *Nachr. Chem. Technol. Lab.* **40**, 308 (1992).
32. H. Saito, S. Okamura, and K. Ishizu, *Polymer* **33**, 1099 (1992).
33. M. Moffitt, L. McMahon, V. Pessel, and A. Eisenberg, *Chem. Mater.* **7**, 1185 (1995).
34. J. P. Spatz, A. Roescher, and M. Möller, *Adv. Mater.* **8**, 337 (1996).
35. M. V. Seregina, L. M. Bronstein, O. A. Platonova, D. M. Chernyshov, P. M. Valetsky, J. Hartmann, E. Wenz, and M. Antonietti, *Chem. Mater.* **9**, 923 (1997).
36. L. Bronstein, M. Antonietti, and P. Valetsky, in *Nanoparticles and Nanostructured Films*, edited by J. H. Fendler, Wiley-VCH Verlag: Weinheim, (1998), p. 145.

37. L. M. Bronstein, D. M. Chernyshov, I. O. Volkov, M. G. Ezernitskaya, P. M. Valetsky, V. G. Matveeva, and E. M. Sulman, *J. Catal.* **196**, 302 (2000).
38. O. A. Platonova, L. M. Bronstein, S. P. Solodovnikov, I. M. Yanovskaya, E. S. Obolonkova, P. M. Valetsky, E. Wenz, and M. Antonietti, *Colloid Polym. Sci.* **275**, 426 (1997).
39. T. P. Loginova, Y. A. Kabachii, S. N. Sidorov, D. N. Zhirov, P. M. Valetsky, M. G. Ezernitskaya, L. V. Dybrovina, T. P. Bragina, O. L. Lependina, B. Stein, and L. M. Bronstein, *Chem. Mater.* 2003, *submitted.*
40. L. M. Bronstein, S. N. Sidorov, P. M. Valetsky, J. Hartmann, H. Coelfen, and M. Antonietti, *Langmuir* **15**, 6256 (1999).
41. L. Bronstein, E. Kraemer, B. Berton, C. Burger, S. Foerster, and M. Antonietti, *Chem. Mater.* **11**, 1402 (1999).
42. T. J. Martin, K. Prochazka, P. Munk, and S. E. Webber, *Macromolecules* **29**, 6071 (1996).
43. N. V. Semagina, A. V. Bykov, E. M. Sulman, V. G. Matveeva, S. N. Sidorov, L. V. Dubrovina, P. M. Valetsky, O. I. Kiselyova, A. R. Khokhlov, B. Stein, and L. M. Bronstein, *J. Mol. Cat. A: Chem.* (2003).
44. A. B. R. Mayer, J. E. Mark, and R. E. Morris, *Polym. J.* **30**, 197 (1998).
45. L. M. Bronstein, D. M. Chernyshov, G. I. Timofeeva, L. V. Dubrovina, P. M. Valetsky, E. S. Obolonkova, and A. R. Khokhlov, *Langmuir* **16**, 3626 (2000).
46. L. M. Bronstein, D. M. Chernyshov, G. I. Timofeeva, L. V. Dubrovina, P. M. Valetsky, and A. R. Khokhlov, *J. Colloid Interface Sci.* **230**, 140 (2000).
47. L. M. Bronstein, D. M. Chernyshov, E. Vorontsov, G. I. Timofeeva, L. V. Dubrovina, P. M. Valetsky, S. Kazakov, and A. R. Khokhlov, *J. Phys. Chem. B* **105**, 9077 (2001).
48. L. M. Bronstein, M. V. Seregina, O. A. Platonova, Y. A. Kabachii, D. M. Chernyshov, M. G. Ezernitskaya, L. V. Dubrovina, T. P. Bragina, and P. M. Valetsky, *Macromol. Chem. Phys.* **199**, 1357 (1998).
49. D. M. Chernyshov, L. M. Bronstein, H. Boerner, B. Berton, and M. Antonietti, *Chem. Mater.* **12**, 114 (2000).
50. L. M. Bronstein, S. N. Sidorov, A. Y. Gourkova, P. M. Valetsky, J. Hartmann, M. Breulmann, H. Colfen, and M. Antonietti, *Inorg. Chim. Acta* **280**, 348 (1998).
51. S. N. Sidorov, L. M. Bronstein, P. M. Valetsky, J. Hartmann, H. Colfen, H. Schnablegger, and M. Antonietti, *J. Colloid Interface Sci.* **212**, 197 (1999).
52. M. Michaelis and A. Henglein, *J. Phys. Chem.* **96**, 4719 (1992).
53. M. Antonietti, *Angew. Chem.* **100**, 1813 (1988).
54. M. Antonietti, W. Bremser, and M. Schmidt, *Macromolecules* **23**, 3796 (1990).
55. M. Antonietti, F. Gröhn, J. Hartmann, and L. Bronstein, *Angew. Chem. Int. Ed.* **36**, 2080 (1997).
56. N. T. Whilton, B. Berton, L. Bronstein, H.-P. Hentze, and M. Antonietti, *Adv. Mater.* **11**, 1014 (1999).

57. L. M. Bronstein, D. M. Chernyshov, R. Karlinsey, J. W. Zwanziger, V. G. Matveeva, E. M. Sulman, G. N. Demidenko, H.-P. Hentze, and M. Antonietti, *Chem. Mater.* **15**, 2623 (2003).
58. L. M. Bronstein, C. Linton, R. Karlinsey, B. Stein, D. I. Svergun, J. W. Zwanziger, and R. J. Spontak, *Nano Lett.* **2**, 873 (2002).
59. L. M. Bronstein, C. Linton, R. Karlinsey, E. Ashcraft, B. Stein, D. I. Svergun, M. Kozin, I. A. Khotina, R. J. Spontak, U. Werner-Zwanziger, and J. W. Zwanziger, *Langmuir* **19**, 7071 (2003).
60. L. M. Bronstein, C. Linton, R. Karlinsey, B. Stein, G. I. Timofeeva, D. I. Svergtun, P. I. Konarev, M. Kozin, J. Tomaszewski, U. Werner-Zwanziger, and J. W. Zwanziger, *Langmuir* **20**, 1100 (2004).
61. H. Honjo and S. Ohta, *Phys. Rev. A* **45**, R8332 (1992).
62. H. C. Zeng and L. C. Lim, *J. Mater. Res.* **13**, 1426 (1998).
63. W. E. Bernier and E. G. Bundga, US 6331119.

5

METAL–POLYMER NANOCOMPOSITE SYNTHESIS: NOVEL *EX SITU* AND *IN SITU* APPROACHES

G. Carotenuto and L. Nicolais

Institute of Composite and Biomedical Materials, National Research Council, Naples, Italy

B. Martorana and P. Perlo

Fiat Research Center, Orbassano, Italy

1. INTRODUCTION

Currently, there is a very strong interest in using nano-sized metals (i.e., metal clusters containing from 100 to 100,000 atoms) as advanced additives in plastics functionalization, and considerable research activities are being done in this novel field of composite science [1]. Because of surface effects and the dramatic changes in properties occurring when critical lengths governing some physical phenomenon (magnetic, structural, etc.) become comparable with size, nano-sized metals have unique properties (e.g., plasmon absorption, near-IR photoluminescence, superparamagnetism, Coulomb staircase, etc.). The embedding of nanoscopic metal structures into polymeric matrices represents the most simple way to take advantage of some of these novel physical characteristics. Polymer-embedded nanotructures are potentially useful for a

Metal-Polymer Nanocomposites, Edited by Luigi Nicolais and Gianfranco Carotenuto
ISBN 0-471-47131-3

number of technological applications, especially as advanced functional materials (e.g., high-energy radiation shielding materials, microwave absorbers, optical limiters, polarizers, sensors, hydrogen storage systems, etc.) [1]. In addition to the intrinsic nanoscopic material properties and the possibility to make transparent metal–polymer combinations, these materials are interesting also because the presence of a very large filler–matrix interface area can significantly affect the polymer characteristics (e.g., glass transition temperature, crystallinity, free volume content, etc.), allowing the appearance of further technologically exploitable mechanical and physical properties (e.g., fire resistance, low gas diffusivity, etc.).

The control of nanoparticle morphology becomes a very important aspect, since morphology profoundly influences the material performance. As a long-term goal the development of synthesis schemes able to control particle size, shapes, and composition independently from one another is very important, in order to allow tuning of nanocomposite properties.

Metal–polymer nanocomposites can be obtained by two different approaches, namely, *in situ* and *ex situ* techniques. In the *in situ* methods, metal particles are generated inside a polymer matrix by decomposition (e.g., thermolysis, photolysis, radiolysis, etc.) or chemical reduction of a metallic precursor dissolved into the polymer. In the *ex situ* approach, nanoparticles are first produced by soft-chemistry routes and then dispersed into polymeric matrices. Usually, the preparative scheme allows us to obtain metal nanoparticles whose surface has been passivated by a monolayer of *n*-alkanethiol molecules (i.e., $C_nH_{2n+1}\text{-}SH$). Surface passivation has a fundamental role since it avoids aggregation and surface oxidation/contamination phenomena. In addition, passivated metal particles are hydrophobic and therefore can be easily mixed with polymers. The *ex-situ* techniques for the synthesis of metal/polymer nanocomposites are frequently preferred to the *in situ* methods because of the high optical quality that can be achieved in the final product.

2. *EX-SITU* SYNTHESIS OF Au-BASED NANOCOMPOSITES

Polymer-embedded gold nanoparticles have been extensively studied [1]. Because of unique physical characteristics, gold–polymer nanocomposites are potentially useful for a number of advanced functional applications, especially in the optical and photonic fields. In particular, these materials can be used as light-stable color filters [2], polarizers [3, 4], ultra-low refractive index materials [5], nonlinear optical devices [6], optical sensors [7], and so on. However, still limited are the chemical routes that allow us to obtain monodispersed thiol-derivatized gold nanoparticles with controlled size to be embedded into polymeric matrices. At present, the only method available for the synthesis of thioaurite compounds is that developed by Brust et al. [8]. This technique yields thiol-derivatized gold clusters of moderate polydispersity. However, owing to solubility issues, the method can be applied only to a little number of alkane– or arene–thiol compounds. Such a fundamental restriction prevents the full process exploitation. In addition, the gold particle size can be only lightly modified by adjusting the amount of gold salt. Cluster properties are size-dependent, and the possibility to fine-tune particle dimension is of a fundamental importance in developing functional additives for polymers.

A new method for the controlled synthesis of alkane–thiol-derivatized gold clusters to be used as filler for polymeric nanocomposites has been recently developed by us [9]. This scheme for thiol-derivatized gold clusters preparation involved the mixing under stirring of a poly(*N*-vinyl pyrrolidone) (PVP, Aldrich, M_w = 10,000 amu) solution in ethylene glycol (stabilized at temperatures ranging from 60°C to 110°C) with a little quantity of a concentrated tetrachloroauric acid ($HAuCl_4$, Aldrich) solution in ethylene glycol. For example, in a typical reactive mixture composition, the first solution was prepared by dissolving 4 g of PVP into 20 mL of ethylene glycol, and the second one was prepared by dissolving 5 mg of $HAuCl_4$ in 1 mL of ethylene glycol. After gold nanoparticle formation, the solution had a ruby-red color. In order to end the reaction and to separate nanoparticles, the reactive mixture was cast into a large amount of acetone (250 mL) and the system was sonicated for a few minutes. The Au–PVP nanocomposite system was achieved after flocculation. The product was washed several times with pure acetone and then dried at room temperature under vacuum. The Au–PVP nanocomposite was used as precursor for the thiol-derivatized gold preparation. In particular, the Au–PVP system was dissolved into a dilute ethanol solution of dodecanethiol (DDT, $C_{12}H_{25}$-SH, Aldrich, 98%). DDT acted as a surface passivating agent for gold clusters since it was able to produce a continuous and uniform organic monolayer around them. Thiol molecules bonded the gold surface much stronger than PVP side groups, and therefore the polymeric stabilizer was completely removed from cluster surface, according to a ligand-exchange process:

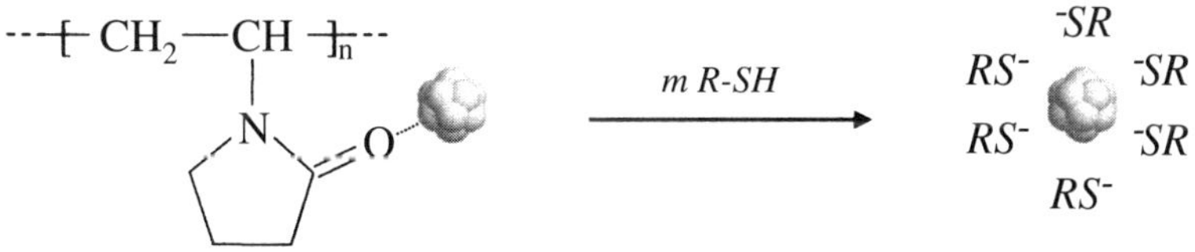

(1)

After two hours the exchange reaction was considered as complete and the hydrophobic thiol-coated nanoparticles were separated from the PVP–ethanol

Figure 5.1. Dry thiol-derivatized gold cluster sample.

solution by centrifugation. A waxy, purple solid product with golden shading (see Figure 5.1) was obtained which was accurately washed with ethanol, in order to remove all impurities. This material was stable for months in air at room temperature. Gold–DDT clusters dissolve in hydrocarbons, chlorine solvents, and ethers because of their hydrophobic nature. Clusters were purified by a dissolution/re-precipitation treatment using chloroform and ethanol, respectively.

In order to produce nanocomposite films, thiol-derivatized gold nanoparticles were dispersed in a chloroform solution of polystyrene (PS, Aldrich, M_w = 230,000 amu). High-transparent purple-colored nanocomposite films resulted by solution casting process (see Figure 5.2).

3. Au–PS NANOCOMPOSITE CHARACTERIZATION

The microstructure of both gold clusters embedded into PVP matrices and gold clusters derivatized by dodecanethiol was imaged by transmission electron microscopy (TEM). Transmission electron micrographs were obtained by a Philips EM208S microscope, using an accelerating voltage of 100 kV. Au-clusters–PVP nanocomposites were dissolved in ethanol, and the solution deposited on graphitized Formvar films supported on copper mesh grids for TEM. Heptane was used to disperse alkanethiol-derivatized gold clusters. As visible in Figure 5.3a, all particles had a pseudo-spherical shape which in some cases corresponded exactly to a polyhedron, and aggregates were not present in the samples. The particle size was moderately polydispersed (standard devi-

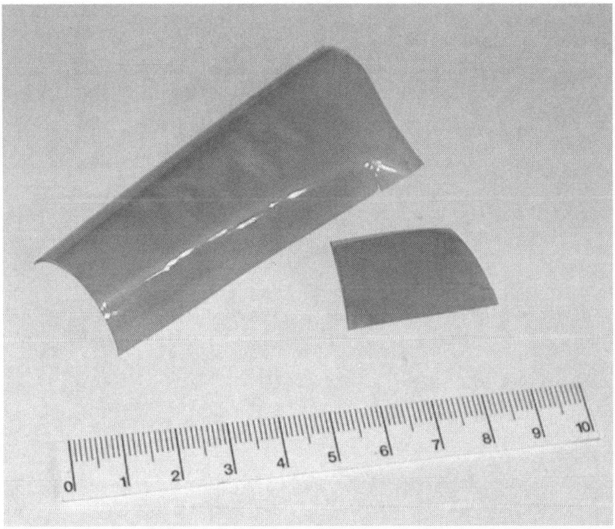

Figure 5.2. Au–PS nanocomposite film.

ation of ~10%) with a monomodal distribution approximately described by a Gaussian function. The particle size distribution was dependent on the reaction conditions (i.e., temperature, time, reactive mixture composition, etc.). However, the most simple way to control the average particle size was by varying the PVP/$HAuCl_4$ weight ratio. In particular, Au clusters with average diameters of 10, 4, and 2 nm were obtained using PVP/$HAuCl_4$ ratios of 400, 800, and 1600 respectively (all reactions were done at 60°C using 21 mL of ethylene glycol and 5 mg of $HAuCl_4$; the reaction time was of 5 min). Therefore, the particle size decreased significantly with increasing of PVP/$HAuCl_4$ weight ratio, allowing us to obtain by the same technique both gold nanoparticles and very small clusters. PVP had a nucleating effect, and therefore higher nucleation rates were obtained by increasing the PVP amount, whereas the cluster growth process was reduced. Sample polydispersity decreased with increasing of PVP amount. Particle size was also dependent on reaction time and temperature; however, such a dependence turned out to be much less sensible and was difficult to control experimentally. Heptane suspensions of gold particles passivated by dodecanethiol spontaneously produced self-organized submicron-sized domains of close-packed nanoparticle arrays by solvent evaporation (see Figure 5.3b). Coalescence aggregates produced by the sintering of the metallic cores were not contained in the final product, thus proving that the ligand-exchange process was really effective and complete.

Figure 5.4 shows the XRD pattern of gold nanoparticles embedded into a PVP matrix. X-ray diffraction measurements were performed with a Rigaku

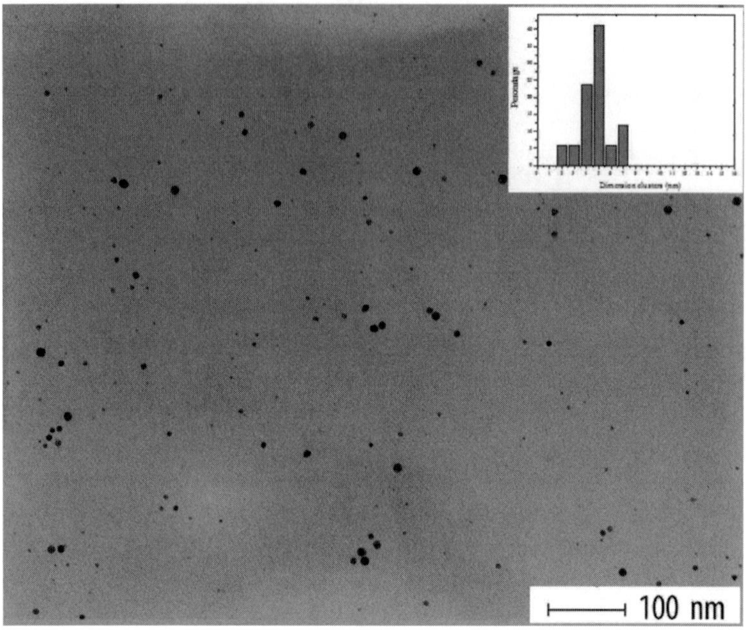

(a)

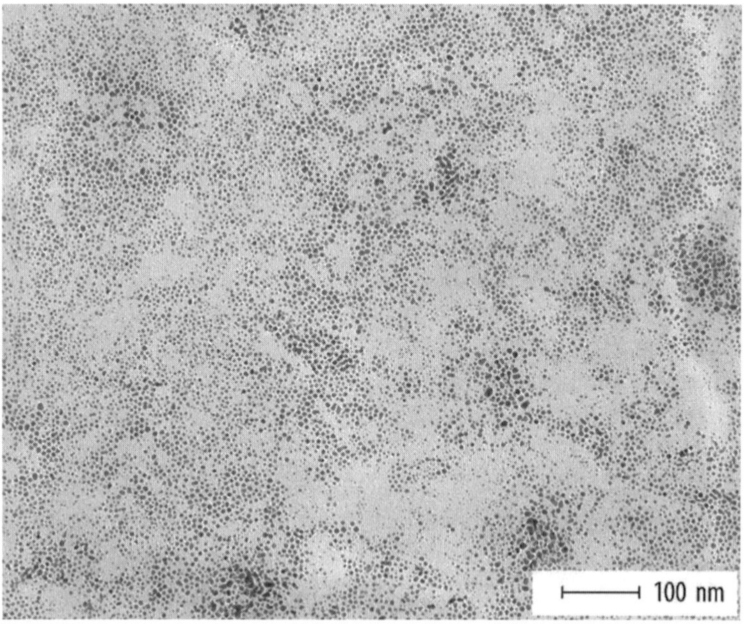

(b)

Figure 5.3. TEM micrographs of (**a**) PVP-embedded gold clusters and (**b**) dodecanethiol-derivatized gold clusters.

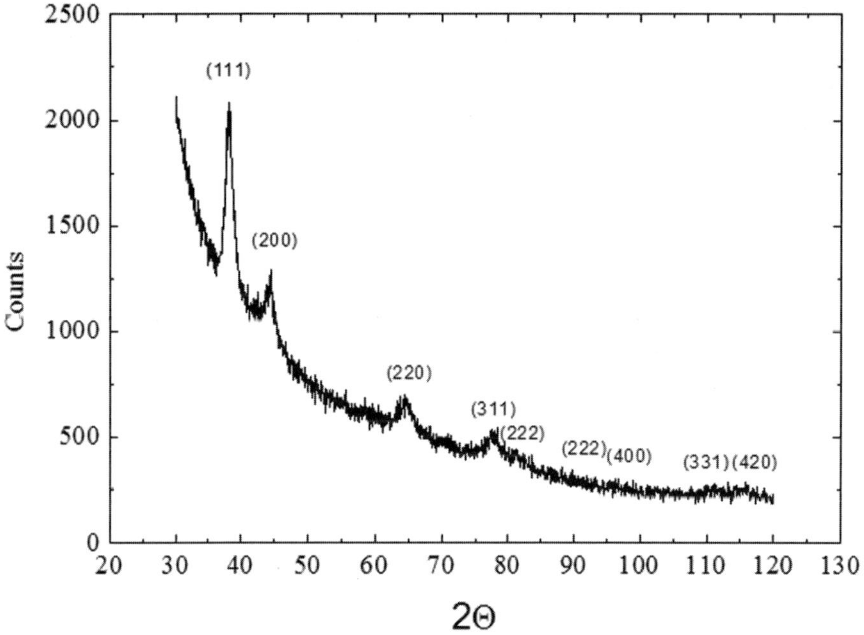

Figure 5.4. XRD pattern of gold nanoparticles embedded into a PVP matrix.

DMAX-IIIC goniometer using Cu-K_α radiation (λ = 0.154056 nm) and a pyrolytic graphite monochromator in the diffracted beam. The goniometer was operated in the standard Bragg–Brentano θ/2θ parafocusing geometry. The spectrum included two broad peaks at 2θ = 11.5° and 21.1° (not visible in Figure 5.4) attributed to the diffraction of the noncrystalline polymer matrix. The spectrum revealed several significantly broad scattering peaks of low intensity; the most evident signals were at 38.2° and 44.6° that can be assigned to the (111) and (200) planes, respectively. The obtained pattern indicated that gold particles were in the face-centered cubic (fcc) structure. The value of the lattice constant calculated from the XRD pattern was a = 4.0770(7) Å, which was consistent with the value for pure bulk gold. The significant broadness of the scattering peaks was caused by the very small size of crystallites. The average size of Au crystals was calculated by Scherrer's equation, and it was determined to be 11.0 ± 0.03 nm for a sample with PVP/$HAuCl_4$ weight ratio of 400. Such value was in good accordance with the corresponding TEM measurement, thus proving a monocrystalline nature of gold clusters.

The formation-growth of gold clusters was monitored by UV–Vis spectroscopy, by looking at the very intensive surface plasmon absorption characterizing nano-sized gold. Optical spectra were recorded at different reaction times by a UV–Vis spectrophotometer (HP-8453 UV–Vis Spectrophotometer), equipped with a Peltier apparatus to control the reaction temperature with an accuracy of ±0.1°C. The system employed a magnetic stirrer to allow for the homogenization of the reactive mixture and far-UV quartz cuvettes. The temporal size development of clusters was monitored by automatic recording of spectral data on a personal computer connected to the spectrophotometer. The method was very sensitive and allowed for a fast and accurate identification of nanometric solid-phase formation and sizing of resulting spherical particles. The temporal evolution of the UV–Vis spectrum of the reactive mixture is shown in Figure 5.5a. As visible, the spectum is given by the overlap of three different signals: (i) an absorption band at 350 nm produced by the Au^{3+} ions, (ii) a broad absorption band produced by interband transition of gold (from 400 nm to 600 nm), and (iii) the surface plasmon absorption band of gold clusters located at 560 nm. Since the Au^{3+} ions concentration decreases with time being reduced to the zero-valence state, the peak intensity lowered during reaction, assuming negative values. Such a behavior is in accordance with bleaching of yellow color characteristic of gold solutions at the beginning of the reaction. The spectrum convolution by Lorentz's functions (see Figure 5.5b) allows us to accurately measure the features of the plasmon peak (i.e., maximum absorption value and wavelength, full width at half-maximum, integral area, etc.). The analysis of the plasmon peak may give morphological–topological information on the produced cluster system, and in particular the average size of nanoparticles. According to the literature [10], the maximum absorption value, A_{max}, is related to the number of absorption centers, N, and to the average particle radius, R, by the following semiempirical expression:

$$A_{max} = k \cdot N \cdot P^{2.8} \qquad (2)$$

where k is a proportionality constant and 2.8 is a semiempirical factor (the theoretical value is 3). Therefore, if the number of gold particles, N, is a constant during the metal phase-separation process, the cubic root of the maximum absorption value is proportional to cluster radius and such information can be used to study the mechanism of cluster growth. The number of metal clusters can be considered as a constant when the nucleation rate is high enough to give separation between nucleation and growth stages [11]. In other words, all cluster nuclei are generated at same time and successively grow together. At temperatures higher than 70°C for a PVP/$HAuCl_4$ weight ratio of 400, the nucleation of gold is fast enough; consequently, nucleation and growth are separated stages of cluster formation, and the number of particles is practically a constant during the radial development of clusters. In this case, a correct evaluation of cluster size can be obtained by using equation (2). However, since the proportionality constant, k, is unknown, TEM calibration, measurements should

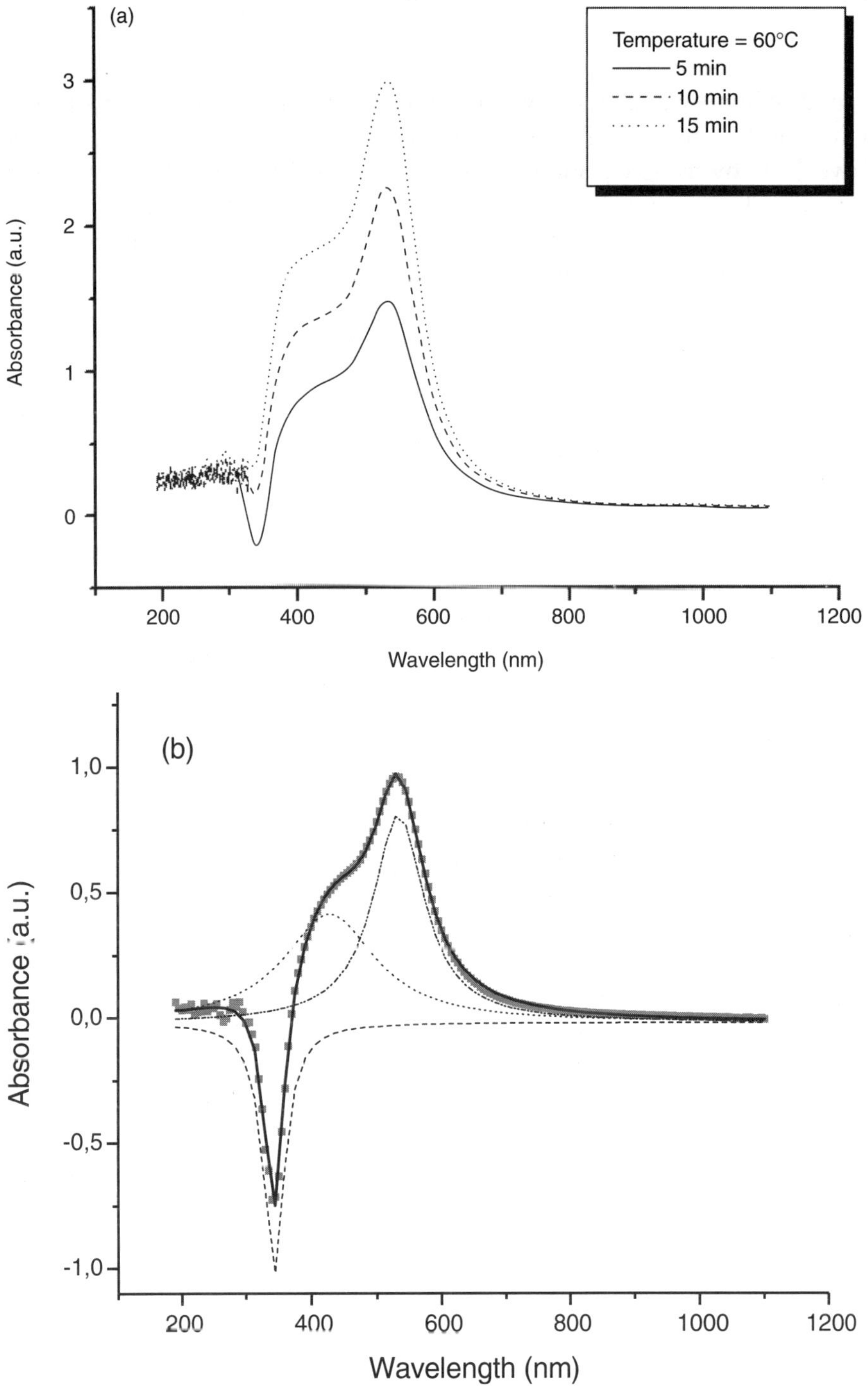

Figure 5.5. Absorption spectra (UV–Vis–NIR) of the reactive $HAuCl_4$–ethyleneglycol–PVP mixture: (**a**) Temporal evolution at 60°C and (**b**) spectrum convolution by Lorentz's functions.

be used. Without calibration, only the profile of the temporal evolution of cluster radius can be obtained. The results of the analysis of spectral data obtained above 70°C by using equation (2) are shown in Figure 5.6a. As visible above 70°C, the cluster radius increased according to a linear function of time. The linear behavior of cluster radius with time is related to the special mechanism of cluster growth that for this system seems to follow a surface-deposition controlled kinetic [12]. Such analysis is in accordance with literature studies for the gold cluster formation by a formaldehyde-based reduction [12]. Figure 5.6b shows the temporal evolution of the maximum plasmon absorbance at temperatures below 70°C. The curve behaved according to a sigmoidal function. The deviation of the cluster radius from the linear behavior and the sigmoidal shape of absorbance versus time plot was probably due to an increase of the cluster number (i.e., the number of absorption centers, N) with time as the main phenomena. Consequently, below 70°C, nucleation and growth are not separated stages of the cluster formation process, but they happen simultaneously during the reaction (for a PVP/$HAuCl_4$ weight ratio of 400).

Spectral data showed a partial resolution of interband and plasmon absorption with increasing of reaction temperature. This phenomenon was caused by the thermal expansion of the metal phase which red-shifted the plasmon absorption band.

Samples of pure gold clusters passivated by dodecanethiol were characterized by an endothermic transition well visible in the DSC thermograms. Differential scanning calorimetry (DSC, TA-Instrument Mod.Q100) tests were carried out at 10°C/min under inert atmosphere of dry nitrogen. A sharp endothermic peak appeared at ~30°C, and it was produced by the melting of small crystallites originated from the co-crystallization of interdigitated alkanethiol chains [13]. The melting temperature of interdigitated alkanethiol crystals was much higher than the melting point of pure dodecanethiol (–7°C). When gold clusters derivatized by dodecanethiol molecules were accurately dispersed into a polystyrene matrix, the DSC thermogram of nanocomposite specimens did not show such endothermic transition even for high percentages of gold filler (i.e., up to 30%). The absence of an endothermic transition in the nanocomposite samples indicated the absence of a co-crystallization process among alkanethiol chains of neighbor gold clusters which, according to TEM investigations, proved the contact-free dispersion of passivated gold clusters in polymer.

The thermogravimetric analysis (TGA, TA-Instrument Mod.Q500) of gold clusters derivatized by dodecanethiol showed a weight loss at temperatures above 280°C. The weight loss was produced by the degradation of the organic layer coating the gold cluster surface, which probably occurred according to a dehydrogenation mechanism [14]. The percentage of thiol molecules absorbed on the surface of a monodispersed cluster sample with an average size of

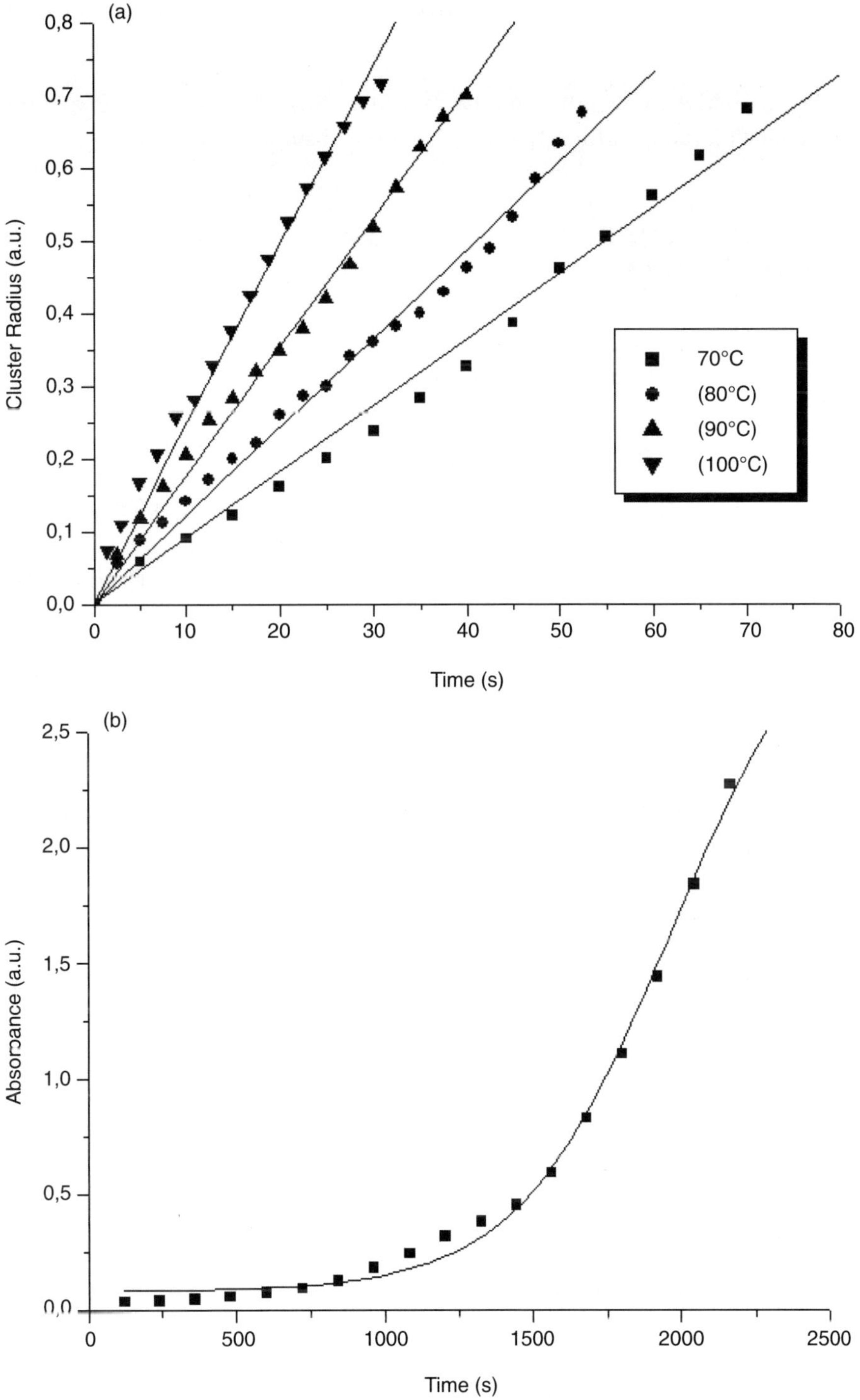

Figure 5.6. Behavior of the cluster radius with time at temperatures above 70°C (**a**), and temporal evolution of absorbance maximum at 25°C (**b**) (in both cases the PVP/$HAuCl_4$ weight ratio was 800).

10nm was ~80% by weight. Finally, thiol-derivatized gold cluster compounds can be considered as a high-thermally stable substance at dry state.

4. PREPARATION OF NANOCOMPOSITES BASED ON OTHER METALS

Alcoholic reduction and polyol process are well-known and useful chemical routes for the synthesis of nano-sized metal particles [15]. These techniques have been widely used for the preparation of very small clusters of noble metals (e.g., Ag, Au, Pd, Pt, Rh, Ru, Os, Ir, Re, etc.), light transition metals (e.g., Co, Ni, Cu, etc.), and easily reducible metals (e.g., Pb, Sn, Cd, W, Hg). A variety of layered core/shell and alloyed clusters can also be obtained by these approaches [15]. The use of hydrophilic polymers as steric stabilizers (e.g., gelatine, poly(vinylalcohol), poly(acrylic acid), poly(methyl vinyl ether), poly(vinyl pyrrolidone), etc.) are particularly effective, owing to the intensive short-range steric repulsions that these polymers are able to produce. In addition, the use of a polymeric stabilizer is convenient since it allows us to isolate and store clusters in a stable polymer-embedded form, simply by co-precipitation with nonsolvent liquids (e.g., acetone, THF). The resulting composite solid is a hydrophilic material that can be dissolved into a variety of polar liquids. Alcohols are very good solvents for PVP because of the polar nature and low molecular weight (M_w = 10,000 amu). On the other hand, alcohols can dissolve all types of thiol molecules, provided that alcohols with larger hydrocarbon chains are used with increasing of alkane-thiol molecular weight.

Because of the very strong affinity of thiol groups (SH) for the gold surface, the PVP-embedded gold clusters can be used as an excellent precursor material for the thiol-derivatized cluster preparation. Polymer is removed from the gold cluster surface according to a ligand-exchange process. Thiol molecules bond the gold surface by a covalent interaction (i.e., thiol chemisorption) generated from a reduction–oxidation reaction (i.e., $Au_n + mRSH = Au_n(SR)_m + m/2H_2$), and therefore the physically adsorbed nucleophylic groups of PVP are completely removed from the metal surface. Because of the high solubility of PVP and thiols in polar solvents, a homogeneous-phase reaction is usually involved. Both very small clusters and nanoparticles can be obtained simply by varying the PVP/$HAuCl_4$ weight ratio, and therefore a very general preparative technique that can offer complete control over the metal filler structure results.

Thiol-derivatized gold clusters are materials stable for months in air at room temperature which can be handled like simple chemical compounds. They have a hydrophobic nature and consequently are very soluble in aliphatic and aromatic hydrocarbons (e.g., heptane, cyclohexane, benzene), chlorine solvents (e.g., chloroform, carbon tetrachloride, dichloroethane), and ethers (e.g., diethyl ether and tetrahydrofuran), only moderately soluble in styrene, and practically insoluble in esters (e.g., methyl methacrylate), acetone, alcohol, and water. Therefore, the product can be easily dispersed into organic solvents able to dissolve hydrophobic polymers like polystyrene. The resulting cluster–polystyrene solutions can be utilized to prepare composite films by solution-casting technology. Because of the small size of gold particles and the amorphous nature of polystyrene, the obtained films do not produce light-scattering phenomena. The elevated transparency of these nanocomposites makes them ideal materials for optical applications.

This novel synthesis scheme has been also investigated with alloyed gold–silver and pure-silver clusters. Ag–Au-alloyed clusters obtained by simultaneous reduction and co-precipitation of two metals give results completely similar to those of the pure-gold clusters. For example, dodecanethiol completely removed PVP from the alloyed cluster surface, producing the thiol derivative. However, the ligand-exchange reaction was not complete in the case of pure-silver clusters embedded in PVP, and consequently the thiolate product isolation–purification was not possible. The success of this chemical technique with other PVP-embedded metal clusters has not be tested yet.

5. *IN SITU* METAL–POLYMER NANOCOMPOSITE SYNTHESIS

A large-scale production of polymer-embedded nano-sized metals should be necessarily based on the thermal decomposition of metal precursors directly added to polymers during their hot–processing stage. A number of organic precursors have been studied for this application [16]; however, any of the tried materials has shown a completely satisfying behavior. We have recently discovered that homoleptic mercaptides (i.e., $Me_x(SR)_y$) can be used as metal precursors in the industrial production of metal–polymer nanocomposites [17]. Mercaptides of transition metals are covalent organic salts characterized by simple synthesis, high compatibility with polymers for the hydrophobic nature, and adequate thermolysis characteristics. These compounds have been occasionally used as polymer additives—for example, for the thermal stabilization of PVC [18]. In particular, mercaptides are quite thermally stable at room temperature, and therefore they can be handled and stored without special care. However, these chemical compounds quantitatively decompose at temperatures that are a little higher (110–180°C), producing zero-valence metal or metal sulfides in addition to organic by-products. The solventless thermolysis of pure mercaptides gives rise mainly to metal sulfides as product [19]. However, mercaptide decomposition in polystyrene matrices frequently produces metal clusters, and the formation of a metal sulfide phase is observed only in presence of high electropositive metals [17]. Such a behavior can be explained on the basis of a decomposition mechanism of mercaptide in polymeric environments different from that involved in solventless thermolyses.

In this part of the chapter we discuss (a) the controlled thermolysis of thiolate solutions in polystyrene matrix at temperatures above the polymer glass transition temperature and (b) the reaction mechanism in the case of silver–polystyrene nanocomposite systems. However, the same reaction mechanism is probably involved in the thermolysis of other mercaptide–polystyrene systems. This technique has proven to be an excellent new preparative scheme for the generation of both metal and sulfide clusters in polymers. In particular, high-molecular-weight *n*-alkanethiolates have shown to be the most effective compound class since the low volatility of thermolysis by-products avoids film foaming during the annealing process.

6. METAL PRECURSOR SYNTHESIS

Usually, high-molecular-weight *n*-alkanethiolates are not commercially available products, but their synthesis is very simple, and only common chemical reagents are required. In particular, thiolates of different metals (e.g., Co, Pd, Pb, Cd, Cu, Au, Ag, etc.) were obtained according to the following general preparative scheme. Metal salts quite soluble in alcohols (e.g., acetates) were selected for this reaction. A few milliliters of alkanethiol (e.g., C_{12}, C_{16}, or C_{18} thiols, Aldrich 98%) were dissolved in ethanol and added dropwise (under stirring, in ~15 min) to a stoichiometric quantity of a metal salt solution in ethanol. The reaction was performed at room temperature, and reagents were used without purification. During thiol solution addition, the system becomes a slurry because of thiolate precipitation. Reaction yields close to 100% resulted, owing to the very low thiolate solubility in the alcoholic medium. In particular, for the reductive nature of thiols, changes in the metal oxidation number were also possible before precipitation. When there were no changes in the metal oxidation number (e.g., with Cu^{2+}, Pb^{2+}, and Pd^{2+} ions), the precipitation promptly occurred; whereas precipitation required some time to take place in the case of an oxidation number variation (e.g., Au(I)–S–R formation from Au(III) salts). Depending on the metal nature and therefore on the ability for the metal thiolate to produce polymeric structures, the product texture was ranging from waxy or rubbery (polymeric thiolates) to a powdered crystalline material. The metal thiolates were separated by vacuum filtration and washed several time by acetone, and then they were purified by dissolution in hot chloroform (50°C) and re-precipitated by ethanol addition.

Owing to the slow dissolution kinetic and limited solubility of silver salts in alcohols, silver dodecyl mercaptide, $AgSC_{12}H_{25}$, has been prepared by adding drop-by-drop an acetone solution of dodecanethiol ($C_{12}H_{25}SH$, Aldrich) to a silver nitrate solution ($AgNO_3$, Aldrich, 99.9%) in acetonitrile at room temperature, under stirring [20]. Stoichiometric amounts of unpurified reactants were used. A white crystalline powder promptly precipitated. The Ag–mercaptide powder was separated by pump filtration and washed several times with acetone. The XRD pattern of silver dodecyl mercaptide showed a periodicity at low Bragg's angles, characteristic of a lamellar structure. According to the literature, such texture causes scarce solubility of this chemical compound in organic solvents [21]. The dry powder was stable for months at room temperature, but Ag-thiolate solution–dispersion in organic solvents showed a scarce stability and became brown in a few days.

7. NANOCOMPOSITE PREPARATION

Nanocomposite preparation will be illustrated considering the Ag–polystyrene systems. To prepare silver–polystyrene nanocomposite films, $AgSC_{12}H_{25}$ was dissolved/dispersed in chloroform, and this solution mixed with a chloroform solution of polystyrene (Aldrich, $M_w = 230{,}000$ amu). The obtained white slurry was cast onto a glass substrate (Petri dishes) and allowed to dry at room temperature. Dry $AgSC_{12}H_{25}$–polystyrene blends were homogeneous and transparent. This material was removed from the glass surface and annealed at ~150°C using a hot plate. During the thermal treatment, the transparent precursor films soften and slowly became light yellow. Little amounts of gas evolved from the viscous polymer phase during annealing with small bubble formation. The color of annealed film promptly changed from yellow to dark brown by cooling the material at room temperature. A prompt chromatic variation always was observed by heating the samples at temperatures higher than the polymer glass transition temperature (T_g), and the color transition was completely reversible. The intensity of this thermochromic effect depended on the quantity of the silver phase contained inside the polymer films (i.e., films changed from black to light brown at high silver contents, and from dark brown to light yellow at low silver content). However, this phenomenon was observed also in the presence of very small silver amounts. Excessively long annealing treatments of precursor film caused the formation of a silver mirror inside the polymeric material, probably because of sintering processes.

The nanocomposite films were purified by prolonged dipping in acetone. In particular, polystyrene became swollen by acetone absorption, and all organic by-product completely diffused outside. In addition, thermochromic layers of very uniform thickness were obtained by dissolving the nanocomposite material in adequate quantities of chloroform and spin-coating the obtained viscous solutions on polymeric substrates (e.g., PET films).

8. NANOCOMPOSITE FILM CHARACTERIZATION

The microstructure of obtained nanocomposite materials was imaged by transmission electron microscopy (TEM). In the preparation of TEM specimens, the nanocomposite material was dissolved in heptane, and drops of the solution were placed on a Formovar-covered copper grid. The dry films were then sputtered by graphite. Transmission electron micrographs were obtained by a Philips EM208S microscope, using an accelerating voltage of 100 kV. Figure 5.7a shows the inner structure of a nanocomposite sample obtained by annealing a $Pd(SC_{12}H_{25})_2$–PS blend for 5 min at 170°C. As visible, a uniform, contact-free distribution of metal particles resulted inside this film. Particles were sufficiently monodispersed with an average size of 10 nm and did not produce aggregates; in addition, they showed a spherical shape. However, as shown in Figure 5.7b, a special inner morphology characterized films of gold clusters embedded in polystyrene. In these nanocomposite films, the clusters were organized in two-dimensional superstructures of different extensions. Such a special topology of nanocomposite films was not observed with other types of metal particles. Figure 5.7c shows the special topology originated inside the

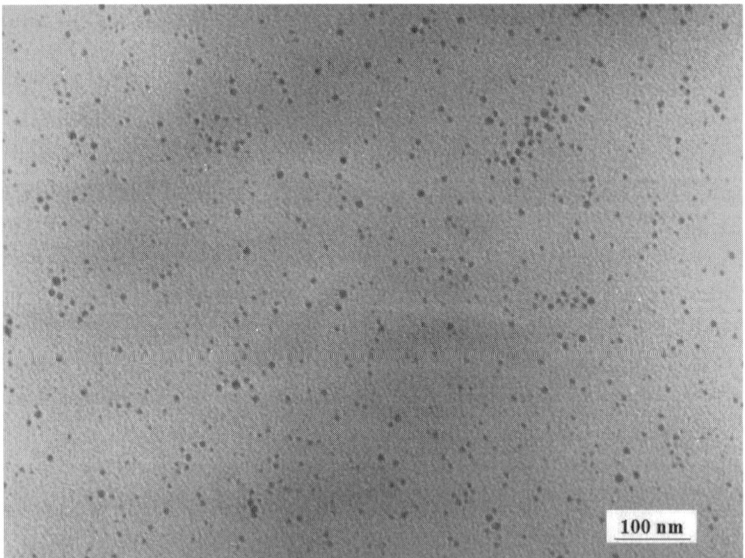

(a)

Figure 5.7. TEM micrographs of polystyrene-embedded metal clusters: **(a)** Palladium ($Pd(SC_{12}H_{25})_2$–polystyrene was annealed at 170°C for 5 min), **(b)** gold ($AuSC_{12}H_{25}$–polystyrene was annealed at 180°C for 5 min), and **(c)** silver ($AgSC_{12}H_{25}$–polystyrene was annealed at 150°C for 1 min).

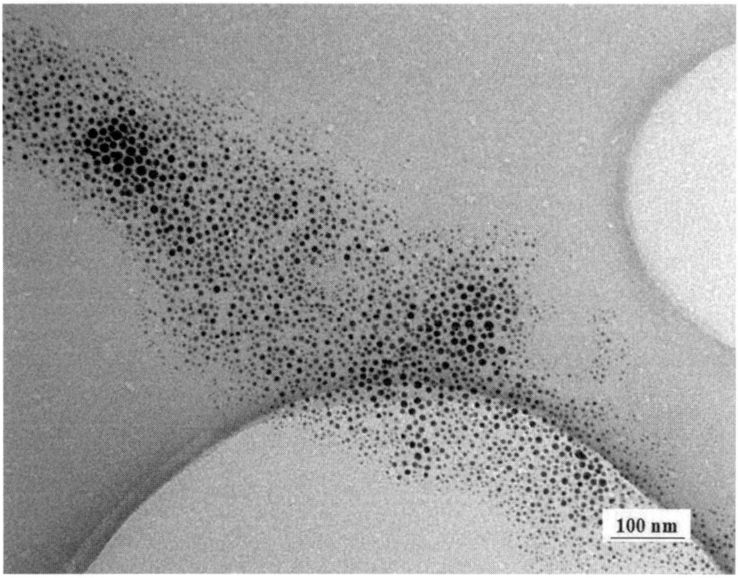

(b)

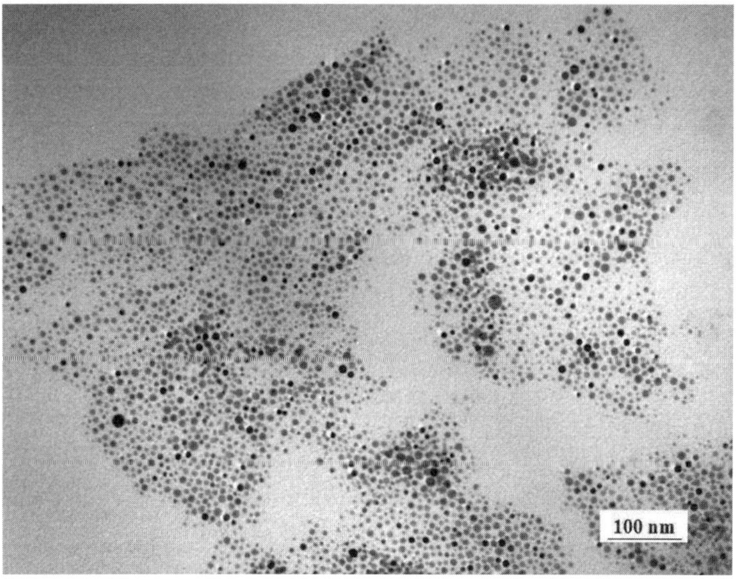

(c)

Figure 5.7. Continued

$AgSC_{12}H_{25}$–polystyrene films during thermal treatment. As visible, large two-dimensional domains of aggregated silver nanoparticles were originated by mercaptide decomposition in molten polymer. Silver aggregates had different shapes and size, and the included silver nanoparticles were not sintered together but were contact-free and uniformly distributed. Such a special microstructure was probably caused by the presence of a polystyrene phase among neighbor silver particles. In a typical preparation (i.e., 11.5% by weight of dodecyl mercaptide, annealed at 150°C for 5 min) the resulting silver clusters were polydispersed and had an average size of 8.0 nm, with a monomodal Gaussian size distribution.

Some mercaptide–polystyrene blends (e.g., $Pb(SC_{12}H_{25})_2$–PS, $AgSC_{12}H_{25}$–PS) were not stable at room temperature over a long time period. Figure 5.8a shows the inner structure of a $AgSC_{12}H_{25}$–polystyrene film aged for 8 weeks at 25°C. The silver clusters had a very small size (~1.8 nm) and were monodispersed and uniformly dispersed in the embedding polymeric matrix. Figure 5.8b shows the inner structure of a $Pb(SC_{12}H_{25})_2$–polystyrene aged for 12 weeks at 25°C.

When metals characterized by a surface plasmon absorption (e.g., gold and silver) were generated by this technique inside the polystyrene matrix, the nanocomposite films developed the characteristic color of the nano-sized metal. In particular, polystyrene films containing silver dodecyl thiolate developed a yellow color during the annealing treatment which changed to brown by cooling down the films, and polystyrene films containing gold thiolate developed a blue coloration. The presence of metal nanoparticles with plasmon absorption inside the polystyrene matrix can be sensibly detected by UV–Vis spectroscopy, because of the strong molar extinction characterizing these nano-sized metals. In particular, Figure 5.9a shows the optical spectrum of a silver–polystyrene nanocomposite sample dissolved in chloroform. As visible, in addition to the characteristic absorptions of polystyrene (from 200 nm to 260 nm), the spectrum also contains a broad absorption band located at ~430 nm corresponding to the silver surface plasmon absorption. Analogously, the presence of zero-valence gold nanoparticles inside the polystyrene matrix can be sensibly detected by UV–Vis spectroscopy, because of the characteristic surface plasmon absorption band of gold particles (see Figure 5.9b).

Figure 5.10 shows optical spectra of an Ag–PS nanocomposite film at two different temperatures (25°C and 110°C). These spectra were obtained by heating a nanocomposite film in the sampling chamber of the UV–Vis–NIR spectrophotometer. The spectrum at higher temperatures results from electronic absorptions of matrix and surface plasmon resonance of embedded silver nanoparticles (i.e., isolated Ag particles, absorbing at ~430 nm). At room temperature, a broad absorption band extending to most of visible spectral region was observed. This thermochromic effect characterizing the nanocomposite

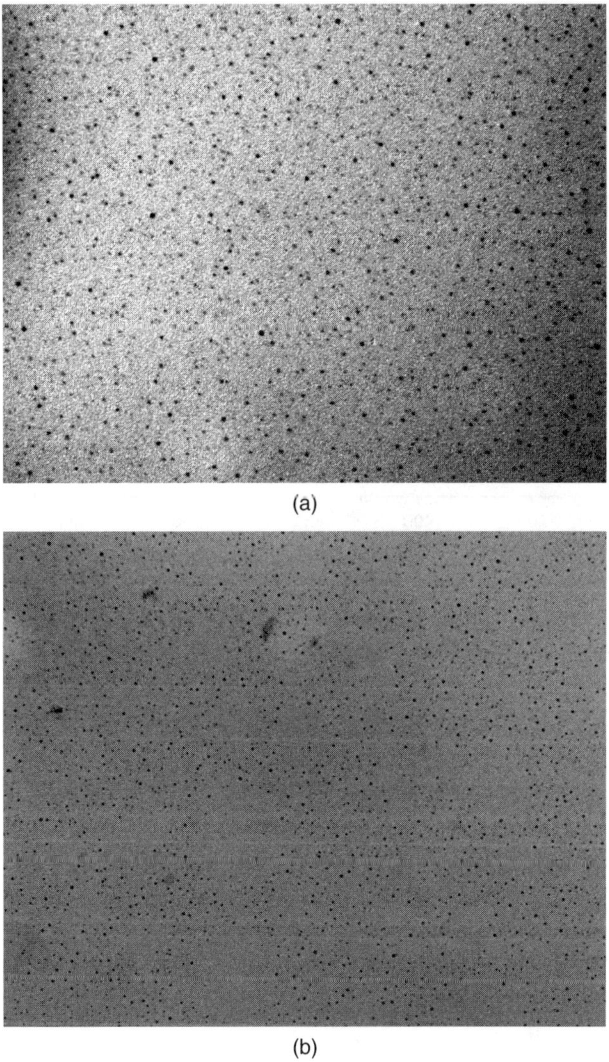

Figure 5.8. Mercaptide–polystyrene blends aged for a long time: (**a**) $AgSC_{12}H_{25}$–polystyrene aged for 8 weeks at 25°C, and (**b**) $Pb(SC_{12}H_{25})_2$–polystyrene aged for 12 weeks at 25°C.

films was probably related to the special nanocomposite microstructure, made of large nanoparticle aggregates. In fact, silver clusters are characterized by optical absorption in the visible region because of surface plasmon resonance. However, owing to dipole–dipole interactions, aggregated particles have optical absorption different from that of isolated particles. In particular, at short inter-particle distances, the interactions among particle dipoles cause a multiple

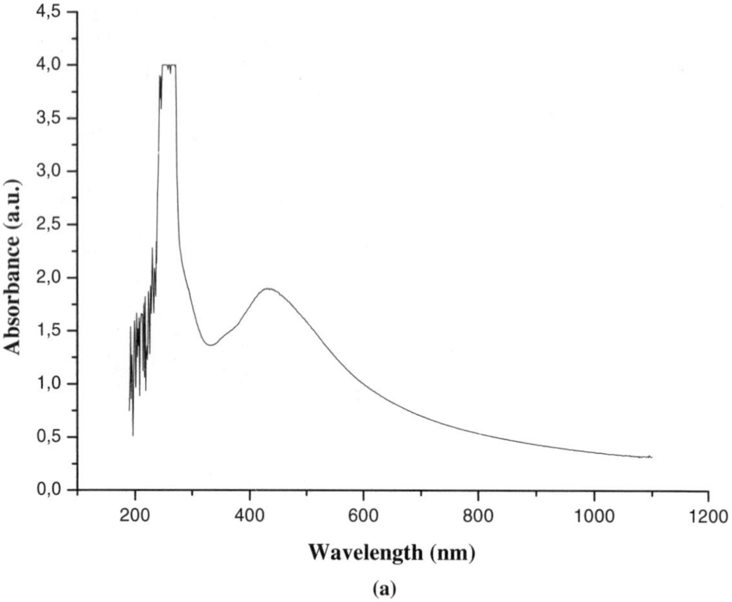

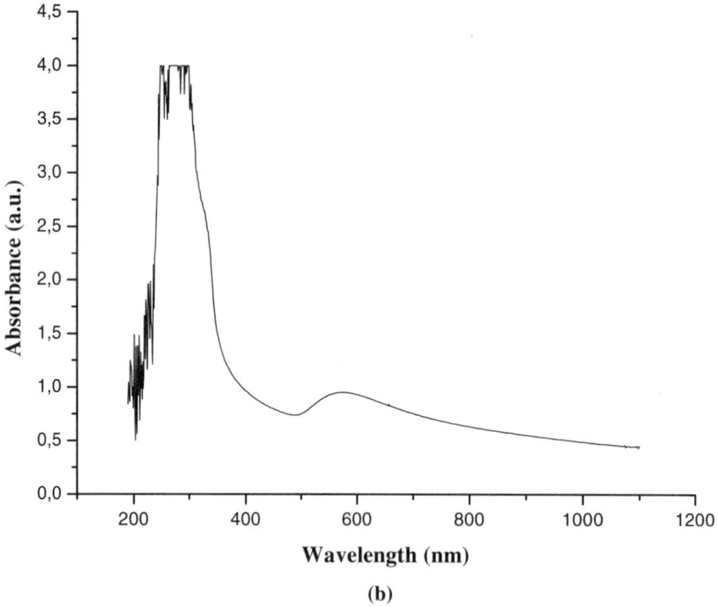

Figure 5.9. UV–Vis spectra of nanocomposite samples dissolved in chloroform: (**a**) Ag–PS and (**b**) Au–PS.

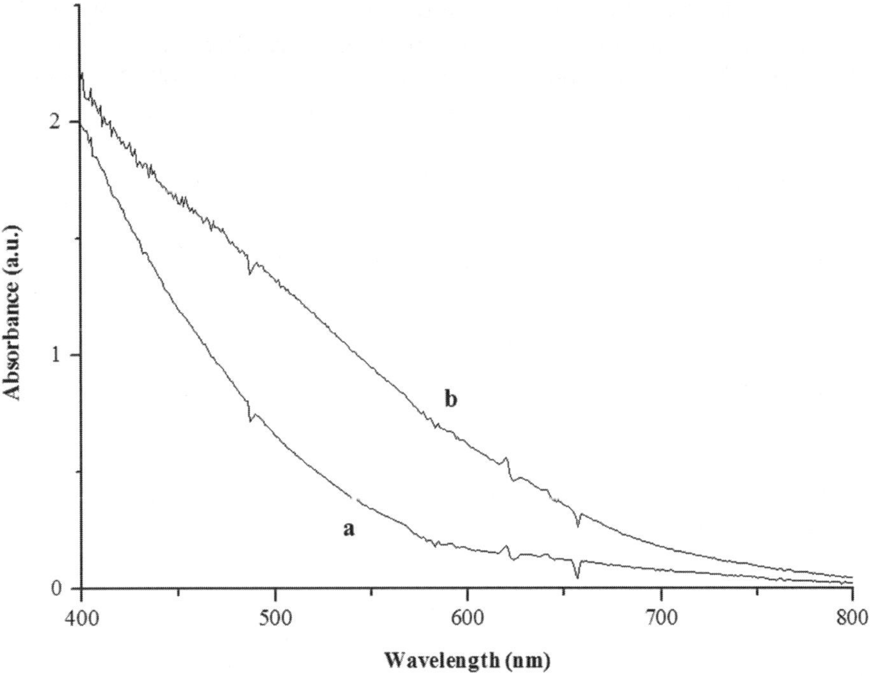

Figure 5.10. UV–Vis spectrum of Ag–PS at (a) 25°C and (b) 110°C.

splitting of silver surface plasmon absorption and consequently a broad optical absorption. Nanocomposite heating above the polystyrene glass transition temperature causes expansion of aggregates with significant increase in the interparticle distance and transition from "collective" surface plasmon absorptions to "individual" particle optical absorptions [22]. Such chromatic variation of polystyrene films filled by silver clusters can have important technological application, for example, in the area of thermal and chemical sensors. Gold-filled polystyrene films appeared dichroic; in particular, they appeared red or blue depending on the observation direction. Such a phenomenon was proba bly generated by the particular acicular shape of gold aggregates present in the films [1].

9. THERMOLYSIS MECHANISM

The mechanism involved in the marcaptide termolysis was investigated in the case of silver-based nanocomposite formation. Silver nanoparticles were completely removed from the polymeric matrix by prolonged centrifugation (30 min) of nanocomposite samples dissolved in chloroform at 13,000 rpm, and the isolated polystyrene was characterized by UV–Vis spectroscopy both as solid and chloroform solution. In addition to the characteristic absorptions of polystyrene phenyl side groups (i.e., $\pi \rightarrow \pi^*$ transitions at 200, 230, and 260 nm), the UV–Vis spectrum of solid film (see Figure 5.11a) revealed the presence of a broad absorption band whose maximum is located at ~350 nm. As visible in Figure 5.11b, when the optical spectrum was obtained from solution, the resulting peak split in two components located at 330 nm and 360 nm. Such absorptions indicated the presence of conjugated biphenylbutadiene groups along the polymer chains (in particular, sterically hindered *trans*-diphenylbutadiene groups should be present in the polymer chain [23]). Therefore, the polymer should participate in the thermolysis process and undergo a hydrogen abstraction reaction with formation of carbon–carbon double bonds in the chains. The UV–Vis characterization of silver particles isolated by centrifugation revealed the presence of polystyrene, which probably formed a network around silver particles in the aggregates (see Figure 11c). Heating in boiling solvent (decane) and/or prolonged sonication of separated silver nanoparticles did not allow the polystyrene removal from particles. Consequently, in addition to hydrogen abstraction, also polystyrene crosslinking processes should happen during the thermolysis reaction.

The cleavage of S–C bond in the silver mercaptide compound was calorimetrically studied. In particular, $AgSC_{12}H_{25}$–polystyrene blends were subjected to annealing treatments in a differential scanning calorimeter (DSC, TA Instruments, DSC-2920) at 10°C/min under the inert atmosphere of dry nitrogen. A typical DSC thermogram of nanocomposite samples is shown in Figure 5.12. As visible, the DSC thermogram included an endothermic peak at ~60°C, produced by the melting of silver thiolate phase, and a stronger endothermic peak at ~110°C, which according to the literature [19] was produced by the cleavage of S–C bond in the thiolate molecules. Above that temperature, the baseline of DSC termogram became very noisy. Such a phenomenon was probably due to the formation of volatile by-products, which gradually diffused through the film and evaporated at film surface, giving a number of very small endothermic peaks. The DSC thermogram included also an inflection in the baseline at 95°C; this signal was produced by the polystyrene glass-transition temperature.

According to UV–Vis and DSC observations, the mechanism for zero-valence silver formation inside polystyrene matrices should involve a reduction of silver ions by hydrogen atoms generated from polystyrene during thermolysis. In particular, according to the DSC thermogram, the Ag-mercaptide

→

Figure 5.11. UV–Vis spectra of **(a)** solid nanocomposite matrix (**insert**: pure polystyrene), **(b)** nanocomposite matrix in chloroform (**insert**: 1,4-dipheny-l,3-butadiene), and **(c)** nanocomposite filler (silver nanoparticles).

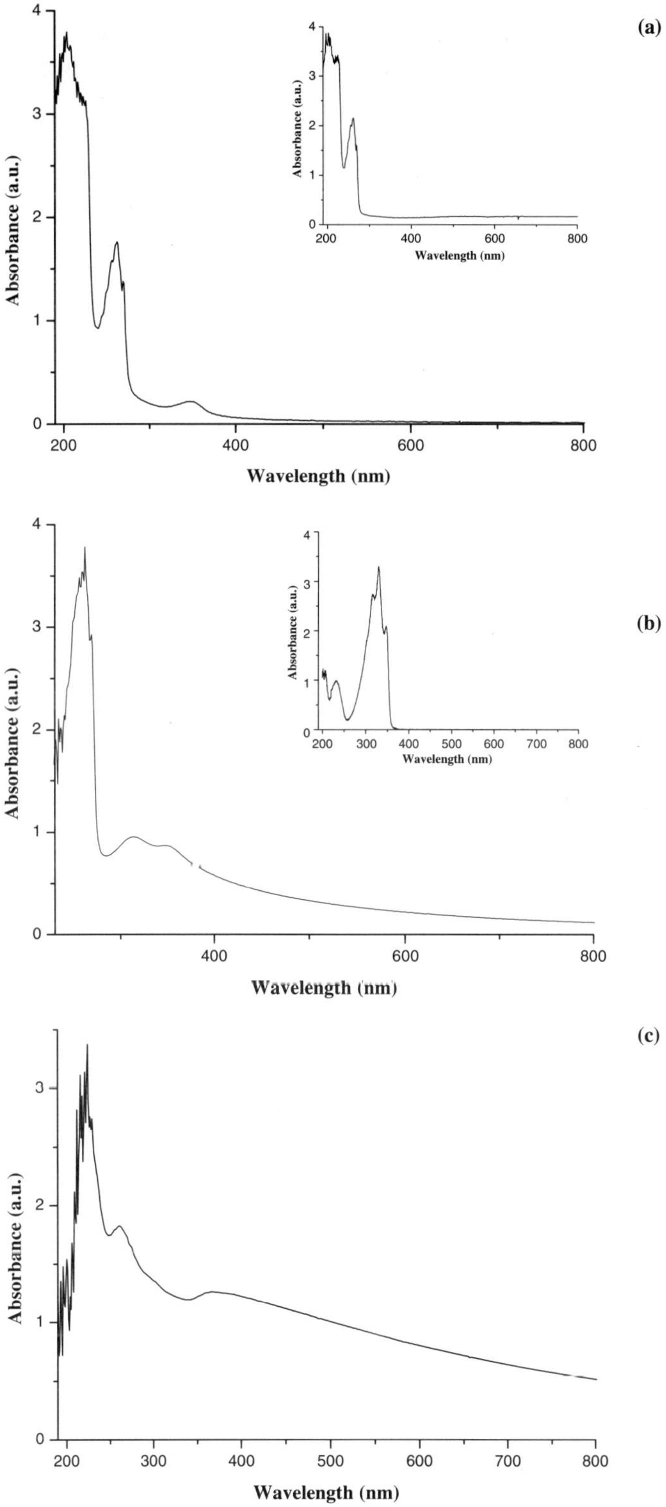
(a)
Absorbance (a.u.)
Wavelength (nm)
(b)
Absorbance (a.u.)
Wavelength (nm)
(c)
Absorbance (a.u.)
Wavelength (nm)

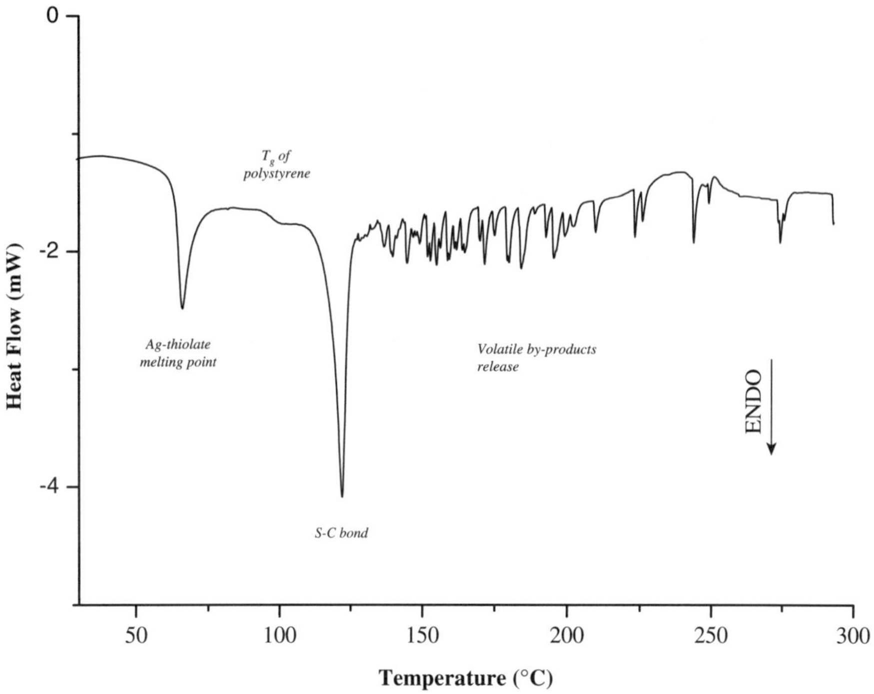

Figure 5.12. DSC thermogram of a $AgSC_{12}H_{25}$–polystyrene blend (10°C/min, under fluxing nitrogen).

molecules dissolved in the molten polystyrene matrix spontaneously decompose and generate alkyl, R·, and metal sulfide, AgS·, radicals by heating at 110°C through an S–C bond cleavage reaction:

$$\text{Ag–S–R} \rightarrow \text{Ag–S}\cdot + \text{R}\cdot \tag{3}$$

Since these molecular fragments are produced in a hydrogen-rich environment, they promptly stabilize by hydrogen abstraction from the surrounding hydrocarbon polymer. This process produces dodecane and acid metal sulfide species, in addition to polystyryl radicals. Hydrogen atoms should be principally removed from tertiary carbons in the polymer chains, because in this case resonant-stabilized carbon radicals result:

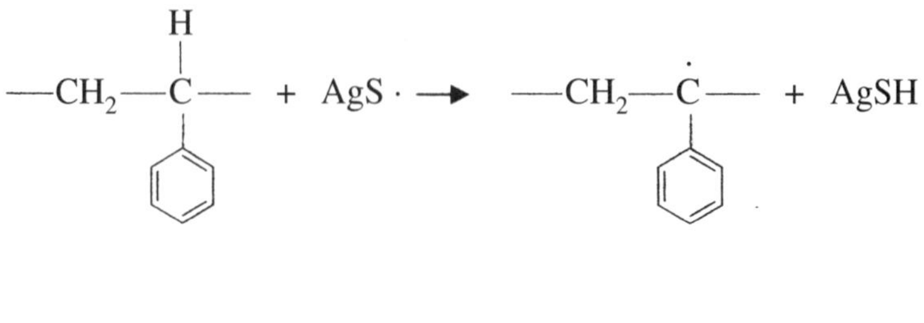

(4)

As known from studies on the mechanism of styrene radical polymerization, the generated carbon radicals may stabilize by β-elimination of hydrogen, producing conjugated double bonds in the polymer chain or bond together, creating a crosslinked three-dimensional network.

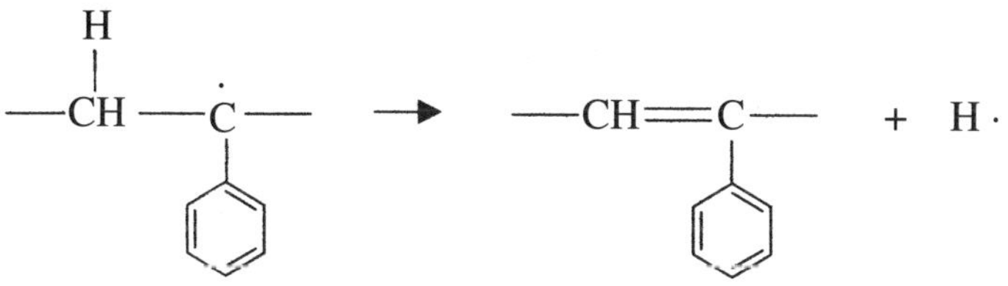

(5)

The free hydrogen atoms generated in the system may reduce silver ions to zero-valence state, giving H_2S as a by-product. Atomic hydrogen may also combine to produce molecular hydrogen and/or reduce directly silver mercaptide molecules to give silver atoms and alkyl–thiol that absorbs on the cluster surface. The presence of diphenylbutadiene groups in the polymer chain can be explained on the basis of the extremely easy abstraction of hydrogen atoms near the styryl groups.

According to this reaction mechanism, zero-valence metals are obtained with all those metal ions easily reducible by hydrogen (e.g., Au, Ag, Pd, Pt, Cu, etc.), whereas metal sulfides result when the thermal decomposition (calcinations) of the acid metal sulfide produced during thermolysis is the only possible chemical process (e.g., Cd, Zn, etc.):

$$Cd(SH)_2 \rightarrow CdS + H_2S\uparrow \quad (6)$$

10. CONCLUSION

In this chapter, two new approaches for the synthesis of metal–polymer nanocomposite materials have been described. The first method allows the preparation of contact free dispersions of passivated gold clusters in polystyrene, and it is based on a traditional technique for the colloidal gold synthesis—that is, the alcoholic reduction of tetrachloroauric acid in presence of poly(vinyl pyrrolidone) as polymeric stabilizer. The primary function of the stabilizer is to avoid cluster sintering, but it also allows us to isolate clusters by co-precipitation. It has been found that the obtained polymer-protected nanometric gold particles can be dissolved in alkane–thiol alcoholic solutions to yield thiol-derivatized gold clusters by thiol absorbtion on the metal surface. Differently from other approaches for thioaurite synthesis available in the literature, this method allows complete control over the passivated gold cluster structure since a number of thiol molecules can be equivalently used and the metal core size can be accurately varied by modifying the reactive mixture composition. The same technique can be utilized also for the preparation of cluster compounds of other metals. The second preparative approach describes the use of homoleptic mercaptides as an effective additive for plastic metallation. This method makes it possible to generate dispersions of monodisperse nanoparticles in the polymeric matrix with sizes and filling factors that can be tuned simply by varying the amount of dissolved thiolate precursor and annealing time or temperature. Alkanethiolates of transition metals are organic compounds characterized by a combination of chemical and physical properties really adequate for the generation of nano-sized metal or sulfide inclusions in polymers. In particular, these compounds can be easily synthesized and have a hydrophobic nature compatible with most polymers. In addition, the mercaptide thermolysis occurs at moderately low temperatures (150–200°C), but these compounds are sufficiently stable at room temperature, and consequently they can be handled and stored without special care.

REFERENCES

1. G. Carotenuto and L. Nicolais, Nanocomposites, Metal-Filled, in the *Encyclopedia of Polymer Science and Technology*, Wiley, New York (2003); W. Caseri, *Macromol. Rapid Commun.* **21**, 705–722 (2000).
2. A. G. deLeon, Y. Dirix, Y. Staedler, K. Feldman, G. Hahner, W. R. Caseri, and P. Smith, *Appl. Opt.* **39**(26), 4847–4851 (2001).
3. A. H. Lu, G. H. Lu, A. M. Kessinger, and C. A. Foss, *J. Phys. Chem B* **101**(45), 9139–9142 (1997).
4. Y. Dirix, C. Darribere, W. Heffels, C. Bastiansen, W. Caseri, and P. Smith, *Appl. Opt.* **38**(31), 6581–6586 (1999).
5. L. Zimmerman, M. Weibel. W. Caseri, U. W. Suter, and P. Walther, *Polym. Adv. Technol.* **4**, 1–7 (1993).
6. G. L. Fisher and R. W. Boyd, in *Nanostructured Materials Cluster, Composites, and Thin Films*, edited by V. M. Shalaev and M. Moskovits, Washington, DC (1998), p. 108.
7. R. D. Harris and J. S. Wilkinson, *Sensors and Actuators B* **29**, 261–267 (1995).
8. M. Brust, M. Walker, D. Bethell, D. J. Schiffrin, and R. Whyman, *J. Chem. Soc., Chem. Commun.* **15**, 801–802 (1994).
9. G. Carotenuto and L. Nicolais, *J. Mater. Chem.* **13**, 1038–1041 (2003).
10. J. Turkevich, P. C. Stevenson, and J. Hillier, *Discuss. Faraday Soc.* **11**, 55–75 (1951).
11. T. Sugimoto, *Adv. Colloid Interface Sci.* **28**, 65–108 (1987).
12. R. Zsigmondy and E. Hückel, *Z. Phys. Chem.* **116**, 291–305 (1925).

13. R. H. Terril, T. A. Postlethwaite, C. H. Chen, C. D. Poon, A. Terzis, A. Chen, J. E. Hutchinson, M. R. Clark, G. Wignall, J. D. Londono, R. Superfine, M. Falvo, C. S. Johnson, Jr., E. T. Samulski, and R. W. Murray, *J. Am. Chem. Soc.* **117**, 12537–12548 (1995); T. Prozorov and A. Gedanken, *Adv. Mater.* **10**(7), 532–535 (1998).

14. C. M. Friend and D. A. Chen, *Polyhedron* **16**(18), 3165–3175 (1997).

15. F. Fievet, in *Fine Particles Synthesis, Characterization, and Mechanism of Growth*, edited by T. Sugimoto, New York (2000), pp. 460–496.

16. A. B. R. Mayer, *Mater. Sci. Eng. C* **6**, 155–166 (1998).

17. G. Carotenuto, B. Martorana, P. Perlo, and L. Nicolais, *J. Mater. Chem.* **13**, 2927 (2003).

18. L. Qu, W. Tian, and W. Shu, *Polym. Degrad. Stab.* **76**, 185–189 (2002).

19. M. B. Sigman, A. Ghezelbash, T. Hanrath, A. E. Saunders, F. Lee, and B. A. Korgel, *J. Am. Chem. Soc.* **125**, 16050–16057 (2003).

20. I. G. Dance, K. J. Fisher, R. M. Herath Banda, and M. L. Scudder, *Inorg. Chem.* **30**, 183–187 (1991).

21. P. J. Thomas, A. Lavanda, V. Sabareesh, and G. U. Kulkarni, *Proc. Indian Acad. Sci. (Chem. Sci.)* **113**, 661–619 (2001).

22. U. Kreibig, M. Quinten, and D. Schoenauer, *Phys. Scripta* **T13**, 84–92 (1986).

23. R. L. Rucker, B. J. Schwartz, M. A. El-Bayoumi, and C. B. Harris, *Chem. Phys. Lett.* **235**, 471–478 (1995).

6

PLASMON ABSORPTION OF EMBEDDED NANOPARTICLES

A. Heilmann

Fraunhofer Institute for Mechanics of Materials, Halle (Saale), Germany

1. INTRODUCTION

The optical properties of metal nanoparticles embedded in an insulating host differ substantially from the optical properties of bulk metals. Under the influence of an electrical field, there is a plasmon excitation of the electrons at the particle surface. This resonance, which takes place at a certain energy of the incident light, results in an optical absorption, the so-called plasmon absorption or plasma resonance absorption [1, 2].

The excitation of the surface plasmon is found to be an extinction maximum or transmission minimum. The spectral position $\tilde{\nu}_r$, half-width (full width at half-maximum) Γ and relative intensity I_r depend on various physical parameters. First, the dielectric functions of the metal $\hat{\varepsilon}_{me}(\tilde{\nu})$ and of the polymer $\hat{\varepsilon}_{po}(\tilde{\nu})$ are involved. Second, the particle size and shape distribution play an important role. Third, the interfaces between particles and the surrounding medium, the particle particle interactions, and the distribution of the particles inside the insulating material have to be considered. For a description of the optical plasmon resonance of an insulating material with embedded particles, a detailed knowledge of the material constants of insulating host and of the nanoparticles

Metal-Polymer Nanocomposites, Edited by Luigi Nicolais and Gianfranco Carotenuto
ISBN 0-471-47131-3

as well as of the nanostructural properties is necessary. In addition to an appropriate theoretical model, an experimental material is also required which allows nanostructural investigations and optical measurements as well.

For the calculations of the optical properties of polymer films with embedded nanoparticles, two routes can be selected. In the exact route, the extinction cross sections $C_{ext}(\tilde{\nu})$ of single particles are calculated. The calculated extinction spectra for single particles—or, better, a summation of various excitation spectra for a particle assembly—can be compared with the experimental spectra of the embedded nanoparticles. In the statistic route, an effective dielectric function $\hat{\varepsilon}(\tilde{\nu})$ is calculated from the dielectric function of the metal $\hat{\varepsilon}_{me}(\tilde{\nu})$ and of the polymer material $\hat{\varepsilon}_{po}(\tilde{\nu})$ by using a mixing formula, the so-called effective medium theory. The optical extinction spectra calculated from the effective dielectric functions by using the Fresnel formulas can be compared with the experimental spectra.

The success of the calculations depends mainly on the introduction of the nanostructural information into the calculation. At both routes of modeling, the nanostructural information can be introduced in different levels. In the first level, only the metal particle filling factor or a mean particle size is introduced. In the next level, also the particle shape is introduced. The best representation of the nanostructure can be reached by introducing the particle size and shape distribution or the size and shape of each single particle into the optical theory.

At both routes, the determination of the size and shape of the nanoparticles is a prerequisite for description of the optical properties. Besides atomic force microscopy (AFM) and scanning electron microscopy (SEM), transmission electron microscopy (TEM) is the most powerful method to determine size and shape distributions of the nanoparticle assemblies. However, extensive sample preparation that is often required can cause preparation effects, and the TEM micrographs sometimes are not representative of the whole nanoparticle-containing insulating material. Therefore, an experimental material is required which can be investigated very easily without extensive TEM preparation.

As such exemplary experimental material, plasma polymer thin films with embedded silver particles are selected [3]. These films were made by simultaneous or alternating plasma polymerization and metal evaporation. The films can be deposited as multilayers consisting of two polymer thin films and a nanoparticle-containing film between these films. Because of the two plasma polymer layers on either side, the particles are completely embedded in a homogeneous media. The multilayer systems are very appropriate for determining particle size and investigating the interface between metal particles and plasma polymer matrix, because here metal nanoparticles are embedded in one plane. This allows a simple determination of the particle size and shape in the TEM.

Moreover, the size and shape of the nanoparticles can be artificially modified by thermally induced processes like reshaping and coalescence without destroying the polymer matrix. During these processes, the multilayer system remains intact and no material loss takes place. This has the advantage that the optical properties can be described by comparing two nanostructures with the same film thickness and metal content but with different particle size and shape distributions. The introduced two routes for optical computations will be demonstrated by comparing the optical properties of a sample before and after the thermal nanostructural changes.

2. EXPERIMENTAL

The multilayer system consisting of plasma polymer thin films with embedded silver particles were deposited by simultaneous plasma polymerization and metal evaporation in a special designed deposition reactor described in reference 3. The nanoparticle containing plasma polymer films were deposited as multilayer systems with embedded metal particles. The particles are embedded in one plane between two plasma polymer layers without nanoparticles. Since the multilayers have been deposited on glass substrates with a predeposited KCl thin film, the KCl film dissolves in distilled water and the detached film can be caught on common TEM grids. TEM micrographs of these multilayer systems show well-separated single particles. The vertical structure of the multilayers was investigated also by transmission electron microscopy in cross section [4] and XPS depth profiling [5] to confirm the embedding of the silver particles in one plane. Together with TEM micrographs in lateral direction, size and shape of the particles in all three dimensions could be determined.

Size and shape of the embedded particles can be varied during film deposition. They mainly depend on the amount of the evaporated metal, the pressure during the deposition, and the distance of the sample from the metal evaporation source. The particles have a broad size distribution that can be described in principle with a logarithmic-normal distribution [6].

At moderate thermal annealing much below the melting point of the metal particles, substantial changes of the size and shape of the particles were observed [3]: Atomic diffusion processes occur along the grain boundaries and the particle surface (reshaping). If two neighboring particles are closely connected, the particles coalesce and the number of the embedded particles decreases. Additionally, hints of diffusion of metal atoms through the polymer matrix (Ostwald-ripening) were found.

During thermal treatment inside the electron microscope, changes in the particle size and shape can be observed *in situ* [7]. The samples are placed on a sample heating unit and heated up to a certain temperature. However, if the electron beam is fixed during thermal annealing inside the electron microscope at one sample position, matrix changes and contamination were found. In the following, this sample part is called an irradiated part. This local contamination caused hardening of the polymer matrix. In this case, the thermally induced coalescence and reshaping of the metal particles takes place at higher temperatures compared to the nonirradiated parts of the samples. So, at one sample two nanostructural states can be observed: at the irradiated part the *frozen* deposited nanostructure and outside from the irradiated part the nanostructure after thermal treatment.

Figure 6.1 shows a TEM micrograph of a sample prepared as described above and treated with electron beam irradiation and *in situ* annealing in the electron microscope. Electron microscopy was carried out using a Philips electron microscope CM 20 FEG at 200-kV acceleration voltage. The electron beam was placed on the left part of the sample (left part of Figure 6.1) during annealing. The maximum temperature on the sample holder was 770 K; but due to the low thermal conductivity of the sample, we suppose that the real temperature at the investigated sample region was lower. Because of contamination and hardening of the polymer matrix by electron beam irradiation during the thermal annealing, the nanostructure of the irradiated sample part has remained as deposited, also after thermal annealing. The right part of Figure 6.1 shows the nanostructure of the sample part without electron beam irradiation during thermal annealing. At this sample part, thermally induced reshaping of the silver particles took place. So, Figure 6.1 shows the silver particle sizes and shapes before (*left*) and after (*right*) reshaping.

In order to obtain quantitative results about the size and shape distribution of the embedded particles, both sides of the TEM micrograph were analyzed

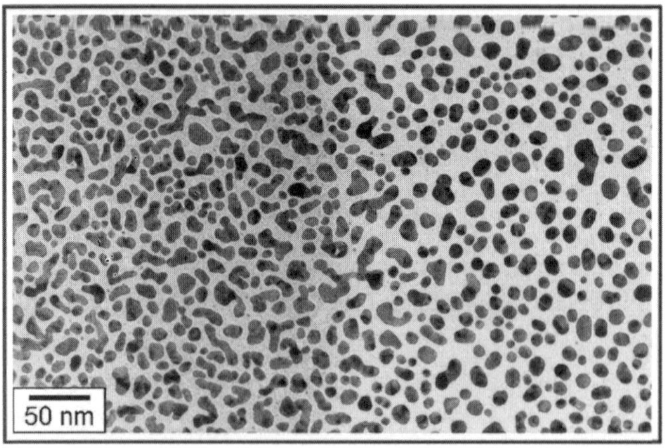

Figure 6.1. TEM micrograph of a plasma polymer multilayer with embedded silver nanoparticles depicting a nanostucture before (**left**) and after (**right**) reshaping and coalescence.

with an optical image processing system. Based on the pixel coordinates of the digital image, area v and perimeter u as well as the center of gravity of each particle were calculated. For approximately spherical particles, the particle size D is given as the mean value of 36 lines of intersections (ferets) $F_n(n)$ as particle projections at different angles $D = \frac{1}{36}\sum_{n=1}^{36} d_n F_n$. For particles with large deviations from the spherical shape, the shape factor $S = 4\pi Ar/U^2$ or the aspect ratio $R = F_{min}/F_{max}$, where F_{min} and F_{max} represent the minimal and maximal feret, have to be calculated additionally for particle form characterization [8]. With respect to the results of the cross-sectional TEM examination, the particles are depicted as elongated spheroids (rotational symmetric ellipsoids), which are described by the mean half-axis A (major axis) and the lower half-axis B (minor axis).

From Figure 6.1, 368 particles were analyzed both before and after annealing, and the results are given in two three-dimensional histograms (Figure 6.2). The x direction gives the minor axis B, the y direction gives the major axis A, and the z direction gives the number of particles found in the A–B interval (count). In particular in the left histogram, there is an important number of large particles with eccentricities $e = B/A < 1$. In the right histogram, the eccentricities of the particles are increased and the particle shapes are closer to the spherical shape ($A \approx B$). This is obviously a result of thermally induced reshaping. The optical computations in the following sections are demonstrated on two different particle assemblies. The two assemblies have different particle size and shape distributions, but the amount of silver in the multilayer system is exactly the same.

In Figure 6.3, the optical extinction spectra before and after thermal annealing are plotted for a comparable film sample that was deposited on quartz

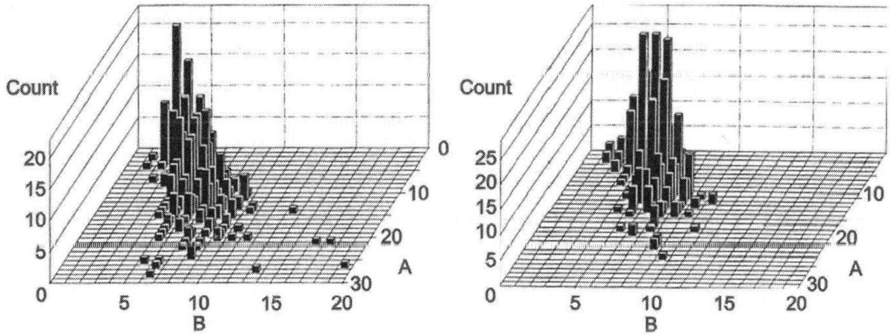

Figure 6.2. Histograms of the mean and lower half-axis for the silver particles from Figure 6.1: for the particles as deposited **(left)** and for the particles after coalescence and reshapening **(right)**.

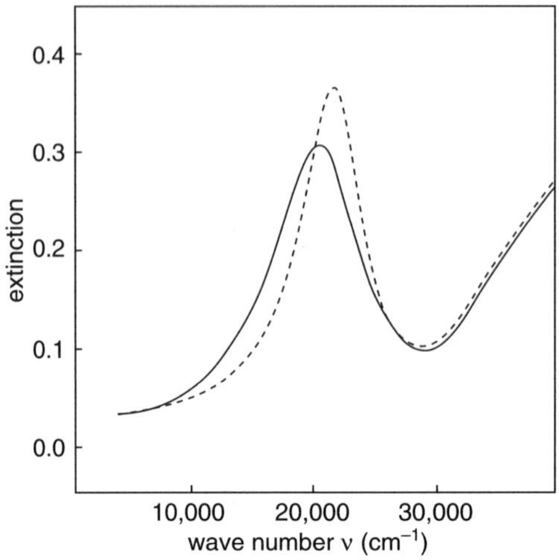

Figure 6.3. Measured extinction spectra before (solid line) and after (dashed line) thermal treatment.

substrates. The spectra were measured with a commonly used spectrophotometer in the spectral region between 4000 cm^{-1} and 50,000 cm^{-1}. Thermal treatment was performed under vacuum conditions (10^{-3} Pa) up to 480 K. If the particle size and shape change due to thermal annealing, changes of the plasma resonance absorption are expected. For silver particles embedded in a plasma polymer matrix, a shift of the plasmon position to higher wave numbers (blue shift) and a decrease of the half-width of the plasma resonance absorption were found. The plasma resonance absorption reaches a peak at $\tilde{\nu}_r$ = 20,410 cm^{-1} before annealing and at $\tilde{\nu}_r$ = 21,740 cm^{-1} after annealing. This means that the extinction peak shifts with $\Delta\tilde{\nu}_r$ = 1330 cm^{-1} to higher wave numbers. In addition, the half-width of the plasma resonance absorption which was determined by Lorentz peak fitting, decreases from $\Gamma = 5450$ cm^{-1} to $\Gamma = 4130$ cm^{-1}. Based on the experimental spectra from Figure 6.3, the results of the optical computations will be discussed in the next two sections.

3. OPTICAL COMPUTATIONS IN EXACT ROUTE

For the calculations of the optical properties of polymer films with embedded particles in the exact route, the extinction cross sections $C_{ext}(\tilde{\nu})$ of single particles were calculated based on the solution of the Maxwell equations for spherical particles which is the Mie theory [9, 10]. To reduce the effort for the calculations, often the quasi-static approximation (Rayleigh theory), which considers only dipole excitations, is used. The Rayleigh–Gans theory is a quasi-static approximation for nonspherical particles and includes the Rayleigh theory, which is only valid for spherical particles.

For only a few particle geometries, rigorous solutions for the extinction and scattering of particles are available (spheres: Mie theory [9], spheroids [11], infinite cylinder [12, 13]); the spectra of other particles must be described using approximations. In the Rayleigh approximation, Gans [14] developed a model for ellipsoidal particles, and Fuchs [15] developed a model for cubic particles. Both are based on the calculation of a dipole polarizability of the particle. In the following, the theory of Gans is introduced.

In the Rayleigh approximation, the particle is so small that an applied electric field $\vec{E}_0$ only induces a dipole with dipole moment $\vec{p}$ in the particle. In general, $\vec{E}_0$ and $\vec{p}$ must not be collinear, meaning that the polarizability $\hat{\alpha}$ of the particle is a tensor:

$$\vec{p} = \epsilon_0 \epsilon_m \hat{\alpha} \vec{E}_0 \tag{1}$$

In this notation, the elements of the polarizability tensor are only proportional to the particle volume, similar to the atomic polarizability. ϵ_m is the dielectric constant of the surrounding medium. A dipole with linear response oscillates with the same frequency as the applied electric field and, hence, emits an electric field $\vec{E}_s$ asymptotically given in the far field as

$$\vec{E}_s = \frac{e^{ikr}}{ikr} \frac{ik^3}{4\pi\epsilon_0\epsilon_m} \vec{e}_r \times (\vec{e}_r \times \vec{p}) \tag{2}$$

Using Poynting's theorem, finally, after some mathematics, the extinction cross section C_{ext} for the dipole is obtained as

$$C_{\text{ext}} = k \,\text{Im} \sum_{i=1}^{3} \sum_{j=1}^{3} \alpha_{ij} e_i e_j \tag{3}$$

Im means the imaginary part, and e_i and e_j are the normalized components of the incident field vector. For an ellipsoidal particle with three different elementary axes A, B, and C, the polarizability tensor only has diagonal elements α_{jj}, with

$$\alpha_{jj} = 4\pi ABC \frac{\epsilon(\omega) - \epsilon_m(\omega)}{\epsilon_m(\omega) + G_j(\epsilon(\omega) - \epsilon_m(\omega))} \tag{4}$$

G_j are factors taking into account the geometry of the ellipsoid. They satisfy the conditions

$$\sum G_j = 1, \qquad G_j \leq 1 \quad \forall j \tag{5}$$

For comparison with the experiments, the special case of spheroids has been found suitable. They are characterized by coincidence of two of the three elementary axes. One distinguishes elongated or prolate spheroids with $B = C$ and distinguishes flattened or oblate spheroids with $A = B$. In the following, A is always the major axis and B the minor axis of the spheroid. Accordingly, only two factors $G_1(e)$ and $G_2(e)$ are relevant, depending on the eccentricity e, defined for a spheroid as

$$e^2 = 1 - \left(\frac{B}{A}\right)^2 \tag{6}$$

G_1 differs for prolate and oblate spheroids:

$$\textit{prolate:} \quad G_1(e) = -g(e)^2 \left\{1 - \frac{1}{2e} \ln\left(\frac{1+e}{1-e}\right)\right\} \tag{7}$$

$$\textit{oblate:} \quad G_1(e) = \frac{g(e)}{2e^2} \left\{\frac{\pi}{2} - \tan(g(e))^{-1}\right\} - \frac{g(e)^2}{2} \tag{8}$$

with

$$g^2(e) = \frac{1 - e^2}{e^2} \tag{9}$$

In Figure 6.4, $G_1(e)$ is plotted versus the eccentricity e for prolate and oblate spheroids. It starts with $G_1(0) = \frac{1}{3}$, which is the case of a spherical particle (A

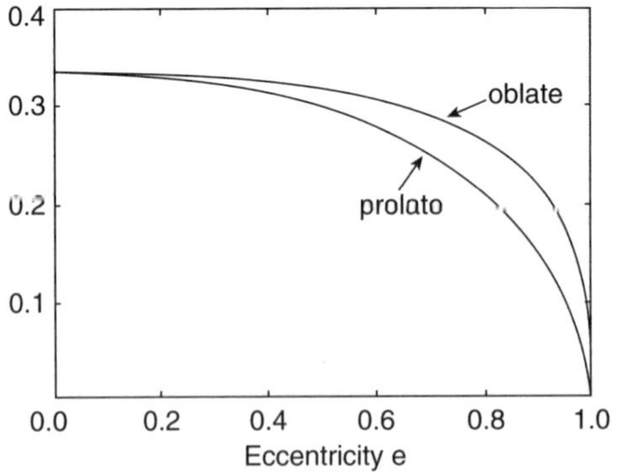

Figure 6.4. Geometry factor $G_1(e)$ as a function of the eccentricity e.

$= B = C$) and decreases with increasing eccentricity. Vice versa, following from equation (5), $G_2(e)$ also starts with $G_2(0) = \frac{1}{3}$, but it increases with increasing eccentricity. It follows now from equation (4) that for a spheroidal silver particle, two surface plasma resonances are induced, namely, when the optical constants of silver satisfy the conditions

$$\epsilon(\omega) = -\epsilon_m \frac{1 - G_j}{G_j} \tag{10}$$

For simplicity, consider the dielectric constant to be purely a Drude dielectric constant without damping, that is,

$$\epsilon(\omega) = 1 - \frac{\omega_p^2}{\omega^2} \tag{11}$$

where ω_p is the plasma frequency of the free electrons. Then, it becomes obvious that the plasma resonance for G_1 contributes to the spectrum at lower frequencies, while the plasma resonance for G_2 contributes at higher frequencies than the frequency at which the plasma resonance of the equivalent volume sphere with $G_1 = G_2 = \frac{1}{3}$ is peaked.

The optical extinction spectra for prolate spheroids were computed using optical constants of silver from references 16 and 17, taking into account additional damping of the plasma resonance absorption according to the model of the mean free path effect. Figure 6.5 gives a section enlargement of Figure 6.1, left, from the particles as deposited. Five particles are marked. The major axis A, the minor axis B, and eccentricities e are given as follows:

Particle 1:	$A = 9.7\,\text{nm}$,	$B = 7.3\,\text{nm}$,	$e = 0.66$
Particle 2:	$A = 12.1\,\text{nm}$,	$B = 11.0\,\text{nm}$,	$e = 0.41$
Particle 3:	$A = 21.0\,\text{nm}$,	$B = 7.0\,\text{nm}$,	$e = 0.94$
Particle 4:	$A = 23.6\,\text{nm}$,	$B = 13.1\,\text{nm}$,	$e = 0.83$
Particle 5:	$A = 33.9\,\text{nm}$,	$B = 6.8\,\text{nm}$,	$e = 0.98$

Figure 6.6 shows the calculated extinction spectra of these five selected particles. Peak position and magnitude of the G_1 mode obviously depend on the shape and size of the corresponding particle. The larger the eccentricity of the particle, the lower the wave number where the plasma resonance absorption of the G_1 mode is peaked. The peak magnitude approximately correlates with the particle volume, but for a quantitative comparison the dispersion of the optical constants in the polarizability must be taken into account. For that reason, the absorption peak magnitude of the larger particle 5 is smaller than the magnitude of particle 4, although its volume is larger than that of particle 4.

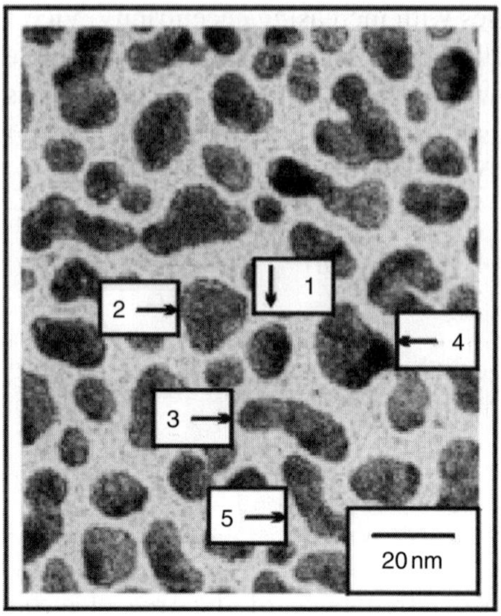

Figure 6.5. TEM micrograph of embedded silver nanoparticles (section enlargements from Figure 6.1).

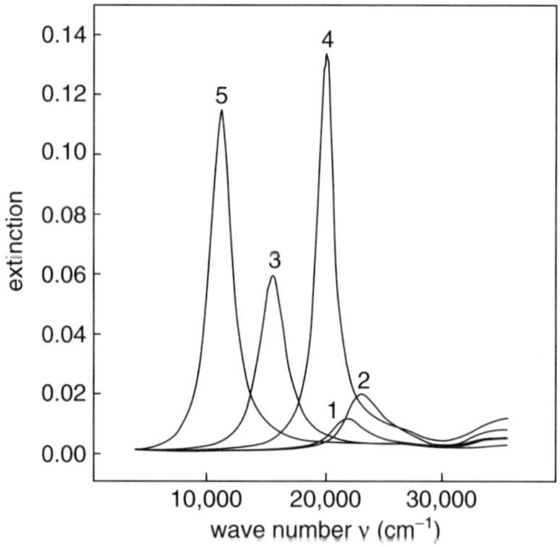

Figure 6.6. Extinction spectra of five single particles, as marked in Figure 6.5.

In Figure 6.7, computed extinction spectra for the particle assemblies before and after the nanostructural changes are presented. The results from image analysis are used to compute extinction spectra of 368 particles with varying size parameters; the total spectra of the sample are added up before (solid curve)

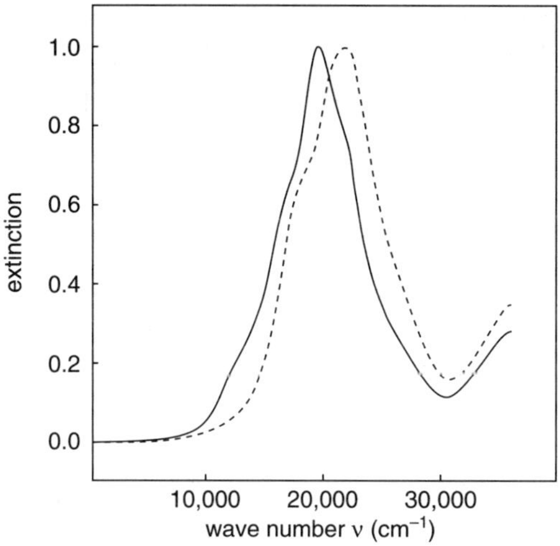

Figure 6.7. Calculated extinction spectra added up for 368 particles before (solid curve) and 368 particles after (dashed curve) thermal treatment, respectively.

and after thermal (dashed curve) treatment. For better comparison the spectra are normalized.

It is obvious that the calculated spectrum of the sample after thermal treatment is blue-shifted with a shift of $\Delta\tilde{\nu}_r = 1570\,\text{cm}^{-1}$. The peak positions before and after thermal treatment are $\tilde{\nu}_r = 19{,}420\,\text{cm}^{-1}$ and $\tilde{\nu}_r = 21{,}050\,\text{cm}^{-1}$, in good agreement with our experimental data at Figure 6.3. At the experimental spectra, the plasma resonance absorption is peaked at $\tilde{\nu}_r = 20{,}410\,\text{cm}^{-1}$ before and at $\tilde{\nu}_r = 21{,}740\,\text{cm}^{-1}$ after thermal treatment. This means that the experimental extinction peak shifts with $\Delta\tilde{\nu}_r = 1330\,\text{cm}^{-1}$ to higher wave numbers. The half-widths with $\Delta\tilde{\nu}_r = 7200\,\text{cm}^{-1}$ and $\Delta\tilde{\nu}_r = 6730\,\text{cm}^{-1}$ are larger than the measured half-widths of $\Gamma = 5450\,\text{cm}^{-1}$ and $\Gamma = 4130\,\text{cm}^{-1}$, respectively. One possible reason for this divergence is that the particles that have a more complicated shape are projected to ellipsoids. Another reason may be that the computed spectra exhibit resonances at lower wave numbers which are missing in the measured spectra in Figure 6.3. They are caused by the restricted validity of the used model in the computations. For silver particles with sizes exceeding about 20 nm, the Rayleigh approximation is not more valid and more precise computations are needed.

The agreement between the experimental and calculated spectra is rather good. The calculations give a good explanation of the blue shift of the plasma resonance absorption of the silver nanoparticles due to their nanostructural changes during the thermal treatment. In the computations with the

Rayleigh–Gans theory, the values for major and minor particle half-axes were used for each single particle. The nanostructural information given from the TEM micrographs was very well considered for the optical calculations.

4. OPTICAL COMPUTATIONS IN THE STATISTICAL ROUTE

In the statistic route, an effective dielectric function $\hat{\varepsilon}(\tilde{\nu})$ is calculated from the dielectric function of the metal $\hat{\varepsilon}_{me}(\tilde{\nu})$ and the of polymer material $\hat{\varepsilon}_{po}(\tilde{\nu})$ by using a formula, the effective medium theory. The most general effective medium theory is the Bergman theory in which the nanostructure of the composite material can be considered by a spectral density function. The Bergman theory includes the solutions from the Bruggeman theory and the Maxwell Garnett theory for spherical, parallel-oriented, and random-oriented ellipsoidal particles.

In addition to the volume filling factor, some theories include parameters that describe nanostructure—for example, the depolarization factor L or the spectral density $g(x)$. If the nanostructure is unknown, these parameters and the filling factor act only as fit parameters. If the calculations fit well to the experimental data, it is possible to obtain information about an unknown nanostructures, although there are situations where this can be misleading.

The most popular effective medium theories are the Maxwell Garnett theory [18], which was derived from the classical scattering theory, and the Bruggeman theory [19]. With these theories, an effective dielectric function is calculated from the dielectric functions of both basic materials by using the volume filling factor. At some extensions of these theories, a unique particle shape for all particles is assumed. There is also an other concept based on borders for the effective dielectric functions. The borders are valid for a special nanostructure. Between these borders, the effective dielectric function varies depending on the nanostructure of the material. The Bergman theory includes a spectral density function $g(x)$ that is used as fit function and correlates with the nanostructure of the material [20].

In 1906, J. C. Maxwell Garnett used the Maxwell Garnett theory, equation (12), for the first time to descibe the color of metal colloids glasses and of thin metal films. Equation (12) can be deviated from the Rayleigh scattering theory for spherical particles [21], or from the Lorentz–Lorenz assumption for the electrical field of a sphere and the Clausius–Mossotti Equation by using the polarizability of an metal particle if only dipole polarization is considered [22].

The Maxwell Garnett theory is expressed as

$$\frac{\hat{\varepsilon}(\tilde{\nu})-\hat{\varepsilon}_{po}(\tilde{\nu})}{\hat{\varepsilon}(\tilde{\nu})+2\hat{\varepsilon}_{po}(\tilde{\nu})}=f\frac{\hat{\varepsilon}_{me}(\tilde{\nu})-\hat{\varepsilon}_{po}(\tilde{\nu})}{\hat{\varepsilon}_{me}(\tilde{\nu})+2\hat{\varepsilon}_{po}(\tilde{\nu})} \tag{12}$$

where $\hat{\varepsilon}_{me}(\tilde{\nu})$ is the dielectric function of the metal, $\hat{\varepsilon}_{po}(\tilde{\nu})$ is the dielectric function of the polymer, $\hat{\varepsilon}(\tilde{\nu})$ is the effective dielectric function, and f is the volume filling factor; this theory was developed for spherical particles only and without interactions of neighbored particles. All particles are completely encapsulated in an insulation matrix material. A further prerequisite was also that the size of the particles is small in comparison with the wavelength of the incident light. At higher filling factors, the particle–particle interactions can no longer be neglected.

For the case where the particles do not have a spherical shape, various extensions of the Maxwell Garnett theory for nonspherical particles were introduced. The particles are spheroidal with the same shape (ratio of major-axis A and minor axis B), but with different sizes still in the wavelength limit. It remains only to choose between a parallel or a random orientation of the mean axis of the ellipsoids.

$$\frac{\hat{\varepsilon}(\tilde{\nu})-\hat{\varepsilon}_{po}(\tilde{\nu})}{\hat{\varepsilon}_{po}(\tilde{\nu})+L[\hat{\varepsilon}(\tilde{\nu})-\hat{\varepsilon}_{po}(\tilde{\nu})]}=f\frac{\hat{\varepsilon}_{me}(\tilde{\nu})-\hat{\varepsilon}_{po}(\tilde{\nu})}{\hat{\varepsilon}_{po}(\tilde{\nu})+L[\hat{\varepsilon}_{me}(\tilde{\nu})-\hat{\varepsilon}_{po}(\tilde{\nu})]} \tag{13}$$

At the extensions of the Maxwell Garnett theory for parallel-oriented ellipsoids, equation (13) [23, 24], all particles are ellipsoids with parallel the mean axis. Only one depolarization factor L is necessary. L describes the ratio between the axes of the ellipsoids, and values for L between $0 \le L \le \frac{1}{3}$ (rod), $L = \frac{1}{3}$ (sphere) and $\frac{1}{3} \le L \le 1$ (disk) are possible. For the encapsulation of spherical particles ($L=\frac{1}{3}$), equation (13) gives the same result as the Maxwell Garnett theory, equation (12). For particles embedded in one plane, the orientation of the ellipsoids in ratio to the substrate is parallel or perpendicular to the plane of the incident light. An orientation of the ellipsoidal particles diagonally to the substrate cannot be considered.

$$(1-f)\frac{\hat{\varepsilon}(\tilde{\nu})-\hat{\varepsilon}_{po}(\tilde{\nu})}{\hat{\varepsilon}_{me}(\tilde{\nu})-\hat{\varepsilon}(\tilde{\nu})}=\frac{f}{3}\sum_{i=1}^{3}\frac{\hat{\varepsilon}_{po}(\tilde{\nu})}{\hat{\varepsilon}_{po}(\tilde{\nu})+[L_i(\hat{\varepsilon}_{me}(\tilde{\nu})-\hat{\varepsilon}_{po}(\tilde{\nu})]} \tag{14}$$

The extensions of the Maxwell Garnett theory for random-oriented ellipsoids, equation (14) [25], needs three depolarization factors L_1, L_2, L_3 with $\Sigma L_i = 1$ to describe the embedded ellipsoids. Frequently, ellipsoids with a symmetrical axis of rotation are assumed with $L_2 = L_3$. Extreme geometries are rods with $L_1 \gg L_2 = L_3$ and disks with $L_1 \ll L_2 = L_3$. For $L_i = \frac{1}{3}$, the extensions of the Maxwell Garnett theory for random-oriented ellipsoids, equation (14), give the same result as the Maxwell Garnett theory, equation (12).

There is also another extension of the Maxwell Garnett theory for random-oriented ellipsoids [29] which is different from equation (14). Other extensions of the Maxwell Garnett theory are described for chiral aggregates of spheres [26], for thin films with columnar structures [27], and for embedded spherical particles of several metals [28].

For the following calculation, experimentally determined dielectric functions for silver [30] and for a plasma polymer [31] were taken. The effective dielectric functions $\hat{\varepsilon}(\tilde{\nu})$ were calculated with the Maxwell Garnett theory for parallel-oriented particles, equation (13). From the effective dielectric function, transmission or extinction spectra can be calculated by using the Fresnel formulas [10] for the optical system air–composite media–quartz substrate. As a further parameter, the thickness of the film with embedded particles and the thickness of other present layers that do not contain metal nanoparticles have to be included. The calculated extinction spectra can be compared with the experimental spectra.

For the calculation of the transmission or extinction spectra, the thickness of each layer d_{p1}, d_c, and d_{p2} must be considered in the right way. In the following, the results from the particle size and shape analysis from the TEM micrograph of Figure 6.1 were transfered to the parameters that are necessary for the effective medium theories. These are the volume filling factor f, the depolarization factor L, and the composite film thickness d_c.

From the image analysis of the lateral TEM micrographs, the area filling factor F_{fl} and the mean values for the particles size $\bar{D}$ and the particle shape $\bar{S}$ are given. The maximum mean particle diameter $\bar{D}_{max}$ is defined with $\bar{D}_{max} = \bar{D} + \sigma_D$. The area filling factor F_{fl} only agrees with the volume filling factor f if the particles have prism shape with ellipsoidal basic area. In the case $f = F_{fl}$, the height of all prisms is equal to the composite film thickness d_c.

The cross-sectional TEM investigations have depicted the fact that vertical projection of the particles are also ellipsoids with the diameters D_A, D_B, and D_C. It is useful to assume that two lower diameters are identical ($D_B = D_C$, rotational symmetry). The diameters $\bar{D}_A$ and $\bar{D}_B$ can be calculated from $\bar{D}$ and $\bar{S}$ using $\bar{D}_A = \bar{D}(1+\sqrt{1-\bar{S}})$ and $\bar{D}_B = \bar{D}(1-\sqrt{1-\bar{S}})$.

By comparison of the ratio of prism volume to the ellipsoid volume, a maximal volume filling factor f_{max} follows from the area filling factor using $f_{max} = \frac{2}{3}F_{fl}$. This upper bound of the volume filling factor is valid if all particles have an identical size of $\bar{D}_{max}$. In this case, the minimum composite film thickness corresponds with the mean vertical diameter of the particles $d_{c_{min}} = \bar{D}_B$ if all particles lie exactly in one plane. Because there is a statistical particle size distribution, the composite film thickness must be defined from the maximal vertical diameter of the largest particle $d_{c_{max}} = D_{B_{max}} \cdot D_{B_{max}}$ can be obtained from the image analysis data. It is better to calculate $D_{B_{max}}$ with $D_{B_{max}} = D_{max}(1-\sqrt{1-S_{max}})$. The filling factor f is decreased further by the ratio between the mean particle diameter $\overline{D}_B$ and maximum vertical particle diameter $D_{B_{max}}$ with $f_{min} = (\overline{D}_B/D_{B_{max}})\frac{2}{3}F_{fl}$. Based on this geometrical assumptions, the lower and the upper bound $f_{min} \leq f \leq f_{max}$ act a criteria for the success of the calculations with the effective medium theories.

The depolarization factor L in equation (13) as well as the depolarization factors L_i in equation (14) can be calculated from the mean aspect ratio $\overline{Q}$ of the particles as well as from the mean shape factor $\overline{S}$. For ellipsoids with three doubled half-axes $\overline{D}_A$, $\overline{D}_B$, and $\overline{D}_C$, L_i is given by equation (15) [23].

$$L_i = \frac{\overline{D}_A, \overline{D}_B, \overline{D}_C}{16} \int \frac{du}{(\overline{D}_i + 4u)\sqrt{(\overline{D}_A^2 + 4u)(\overline{D}_B^2 + 4u)(\overline{D}_C^2 + 4u)}} \tag{15}$$

With equation (15) the depolarization factors L and L_i can be calculated from the mean shape factor $\overline{S}$. In the left part of Figure 6.8, L as a function from S is given. The lower curve is valid for rods ($L < \frac{1}{3}$), and the upper curve is valid for disks ($L > \frac{1}{3}$). The depolarization factors L_i for rods ($L_1 > L_2 = L_3$) and for plates ($L_1 < L_2 = L_3$) are also given in Figure 6.8. For rods, the depolarization factor L_1 is given from the upper curve in the left part of Figure 6.8, while $L_2 = L_3$ is given from the lower curve. For disks, the right part of Figure 6.8 is valid. L_1 is given from the lower curve, and $L_2 = L_3$ is given from the upper curve.

Figure 6.9 presents two calculated optical extinction spectra. The spectra can be compared to the spectra from Figure 6.3. Based on the results from the image analysis and the considerations concerning the filling factor and the depolarization factors, the following parameters were used. Spectra before reshaping: $d_{p1} = 40$ nm, $d_c = 10$ nm, $d_{p2} = 40$ nm, $f = 0.2$, and $L = 0.21$. Spectra after reshaping: $d_{p1} = 38$ nm, $d_c = 13$ nm, $d_{p2} = 39$ nm, $f = 0.15$, and $L = 0.25$. The decrease of the area filling factor in the coalesced part in Figure 6.1 was compensated by an increase of the composite film thickness. The spectral positions

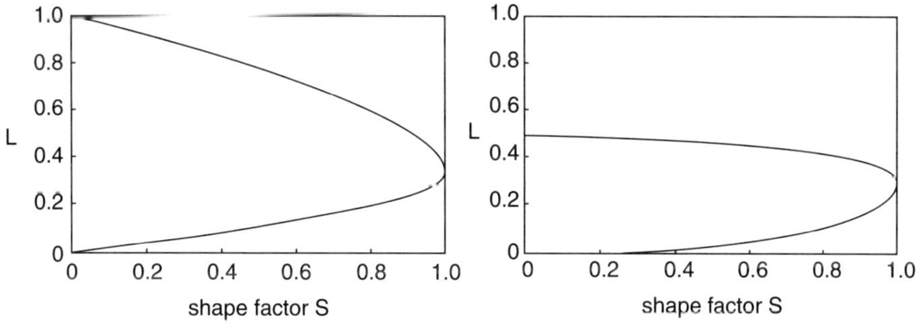

Figure 6.8. *L* and *L_i* as a function of the particle shape factor *S*.

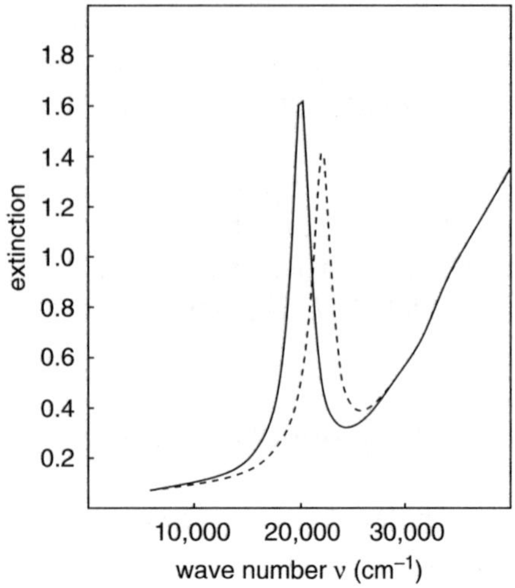

Figure 6.9. Calculated extinction spectra (Maxwell Garnett theory) for parallel-oriented ellipsoids before (solid curve) and after (dashed curve) thermal treatment.

at $\tilde{\nu}_r = 20{,}200\,\text{cm}^{-1}$ before and at $\tilde{\nu}_r = 21{,}900\,\text{cm}^{-1}$ of the plasmon resonance are comparable to the spectral positions in the experimental spectra, the plasma extinction peak shifts with $\Delta\tilde{\nu}_r = 1700\,\text{cm}^{-1}$ to higher wave numbers. This is in good agreement with the experimental spectra of Figure 6.3, but obviously the half-width is too low. The blue shift that occurs at reshaping was reproduced successfully under consideration of different depolarization factors. However, in the effective medium theory the depolarizcation factor that represents the nanostructure can only be used as mean value for all particles.

5. SUMMARY

In this chapter, two different routes for modeling of the optical properties of nanoparticles embedded in thin polymer films were presented. Both routes can give comparable results. The only difference between the Rayleigh theory and the Maxwell Garnett theory for spherical particles, respectively, is the different mathematical formulation. The calculated spectra are comparable. In principle, this must be noted also for the Rayleigh Gans theory and the Maxwell Garnett theory for parallel-oriented ellipsoidal particles. The difference is only in which way the nanostructural information is introduced. This is done in the Rayleigh Gans theory as the size and shape of each single particle and in the Maxwell Garnett theory as statistical particle size and shape distribution for the particle assembly. However, in the effective medium theory the depolarization factor can only be used as mean value for all particles. Therefore, the computation of optical spectra of single particles that are characterized by image processing and the summation of all contributions to a complete spectrum can be preferred.

The presented computations demonstrated passable ways to describe the optical properties of embedded nanoparticles. If reliable nanostructural information is given, computation can be successful without extensive effort. Commercial programs for the presented calculations are available. For a more precise description of nanoparticle-containing insulating materials, the implementation of the nanostructural information given by electron microscopy has to be introduced in a more detailed manner into the optical theory, the computational effort must increase enormously, and it cannot be ensured that qualitatively new results will be reached. Furthermore, in most cases, computational results for one experimental system cannot be copied easily to an other experimental system. The presented computations show a recommended shortcut for optical modeling that can be used for a large number of nanoparticle-containing insulating materials. Depending on the kind of material and the given nanostructural information, the computations will result in a more or less good attempt to determine the relationship between nanostructure and optical plasmon resonance for embedded nanoparticles.

REFERENCES

1. H. Raether, *Surface Plasmons*, Springer Tracts in Modern Physics Vol. 111, Springer, Berlin (1988).
2. U. Kreibig, and M. Vollmer, *Optical Properties of Metal Clusters*, Springer Series in Material Science, Vol. 25, Springer, Berlin (1995).
3. A. Heilmann, *Polymer Films with Embedded Metal Nanoparticles*, Springers Series in Materials Science, Vol. 52, Springer, Heidelberg (2002).
4. W. Grünewald, A. Heilmann, and C. Reinhardt, *Appl. Surf. Sci.* **93**, 157 (1996).
5 A. Heilmann, J. Werner, M. Kelly, B. Holloway, and E. Kay, *Appl. Surf. Sci.* **115**, 365 (1997).
6. C. G. Granqvist and R. A. Buhrman, *J. Appl. Phys.* **47**, 2200 (1976).
7. A. Heilmann and J. Werner, *Thin Solid Films* **317**, 21 (1998).
8. A. Heilmann, J. Werner, D. Schwarzenberg, S. Henkel, P. Grosse and W. Theiss, *Thin Solid Films*, **270**, 103 (1995).
9. G. Mie, *Ann. Phys.* **25**, 377 (1908).
10. C. F. Bohren, and D. R. Huffman, *Absorption and Scattering by Small Particles*, Wiley, New York (1983).
11. S. Asano, and G. Yamamoto, *Appl. Opt.* **14**, 29 (1975).
12. W. Seitz, *Ann. Phys.* **21**, 1013 (1906).

13. W. von Ignatowski, *Ann. Phys.* **18**, 495 (1905).
14. R. Gans, *Ann. Phys.* **37**, 881 (1912).
15. R. Fuchs, *Phys. Rev. B* **11**, 1732 (1975).
16. M. Quinten and U. Kreibig, *Appl. Opt.* **32**, 6173 (1993).
17. M. Quinten, *Z. Phys. B* **101**, 211 (1996).
18. J. C. M. Garnett, *Philos. Trans. Royal Society London* **203**, 385 (1904); **205**, 237 (1906).
19. D. A. G. Bruggeman, *Ann. Phys.* **24**, 636 (1935).
20. D. J. Bergman, *Phys. Rev. B* **23**, 3058 (1981); *Phys. Rep. (Phys. Lett. C)* **43**, 377 (1978).
21. Lord J. W. Rayleigh, *Philos. Mag.* **34**, 481 (1892).
22. C. Kittel, *Introduction to Solid State Physics*, Wiley, New York (1995).
23. D. Polder and J. H. van Santen, *Physica* **XII**, 257 (1946).
24. R. W. Cohen, G. D. Cody, M. D. Coutts, and B. Abeles, *Phys. Rev. B* **8**, 3689 (1973).
25. H. Fricke, *Phys. Rev.* **24**, 575 (1924).
26. A. H. Shivola and I. V. Lindell, *Electr. Lett.* **26**, 119 (1990).
27. G. B. Smith, *Opt. Commun.* **71**, 279 (1989).
28. D. E. Aspnes, *Thin Solid Films* **89**, 249 (1982).
29. C. G. Granqvist and O. Hunderi, *Phys. Rev. B* **18**, 2897 (1978).
30. P. B. Johnson and R. W. Christy: *Phys. Rev. B* **6**, 4370 (1972).
31. A. Heilmann, G. Kampfrath, and V. Hopfe, *J. Phys. D* **21**, 986 (1988).

7

MAGNETOOPTICS OF GRANULAR MATERIALS AND NEW OPTICAL METHODS OF MAGNETIC NANOPARTICLES AND NANOSTRUCTURES IMAGING

V. I. Belotelov

Institute of General Physics, RAS, Moscow, Russia; and M. V. Lomonosov Moscow State University, Faculty of Physics, Physics of Oscillations Department, Moscow, Russia

P. Perlo

Fiat Research Center, Orbassano, Italy

A. K. Zvezdin

Institute of General Physics, RAS, Moscow, Russia; and Fiat Research Center, Orbassano, Italy

1. INTRODUCTION

Among different methods of investigation of magnetic objects, optical methods take leading positions because of their relative simplicity and self-descriptiveness. Magnetooptics of the uniform magnetic media and magnetic structures with the size greater than the wavelength of electromagnetic radia-

Metal-Polymer Nanocomposites, Edited by Luigi Nicolais and Gianfranco Carotenuto
ISBN 0-471-47131-3

tion has been studied in detail for the last several decades. Magnetooptical (MO) techniques of such objects allowed investigation of their magnetic structure both in statics and in dynamics. By means of these methods, images of the magnetization distribution can be obtained in polarization microscopy on the bases of the Faraday, Kerr, or Voigt effects.

At the same time, nowadays, there is great interest in the study of the magnetic properties of low-dimension systems, nanostructures, and granular materials. Magnetic nanostructures are viewed as a promising material for ultrahigh density (up to 1 Tbit/cm^2) data recording. Note that such a giant record density approaches the superparamagnetic limit (10 Tbit/cm^2) [1], which seems to be the ultimate value of the surface record density (today's value is 30 Gbit/cm^2 [2]). The typical size of a one-domain magnetic particle in structured nanomedia is estimated at 5–100 nm. MO investigations of nanostructures are also being conducted and yield fruit. Conventional methods like MO Faraday and Kerr observation along with new optical methods, such as scanning near-field optical microscopy and polarized microscopy, can give a great amount of information about these structures.

Usually, MO methods in their adaptation to the nanoscale world preserve simplicity in realization and cheapness. Optical imaging of magnetic structures is nondestructive, provides high sensitivity, and is not limited by protective nonmagnetic layers. Most importantly, magnetooptical imaging does not require magnetic probes, in contrast to magnetic force microscopy, and consequently does not influence the magnetic structure under investigation and allows measurements in applied magnetic field. Spatial resolution that can be achieved in MO imaging [in scanning near-field optical microscopy (SNOM)] is of the same order (10 nm) as the best resolution obtained by other (e.g., magnetic force microscopy) methods. At the same time some MO methods (e.g., MO Kerr magnetometry) also perform high temporal resolution.

Nevertheless, some difficulties in the experimental setup arise: organization of high precision sample positioning, smooth displacements, production of miniature probes, and so on. At the same time, new features of interaction between light and nanoscaled magnetic objects have not been fully studied, and theory techniques for nanoscale magnetooptics have been poorly developed. MO properties of nano-sized objects strongly depend on their transport properties (conductivity). They are also greatly influenced by quantum effects; consequently, dielectric tensor for bulk media is no longer appropriate for their description.

Taking, the aforesaid, into account, elaboration of consistent theory for the determination of transport and optical properties of magnetic nanostructures is quite urgent. Along with that, some improvements and innovations in experimental technique are also needed in this area.

Magnetooptics of granular materials and other nanoscaled magnetic objects is a thriving part of modern optics. At present it is only in the initial stages, and further investigation in this direction are to be performed.

In this chapter we give a brief outline of the conventional magnetooptics of granular materials, and then we discuss some theoretical and experimental aspects of MO observations of nanoscale magnetic structures.

2. OPTICS OF THE COMPOSITE MEDIA

In this section we shall consider propagation of the electromagnetic radiation through the media with inhomogeneities—that is, composite media. For example, it can be randomly spaced metallic particles embedded into dielectric polymer matrix. In general, when a wave propagates in the same medium, it docs not rcsolvc thc individual particlcs, and the composite appears as a homogeneous material. Let us consider this question in more detail. When a heterogeneous material is probed by an electromagnetic wave, the resolution limit of the probing wave is set by $\lambda/2$. In addition, the Rayleigh scattering strength varies as λ^{-4}. It follows that in the so-called long wavelength limit the heterogeneous material would appear homogeneous to the probing wave, with the material properties characterized by effective parameters. In the long wavelength approximation it is possible to regard the material as being homogeneous and possessing an appropriate effective value of the permittivity, permeability, and conductivity. These effective parameters can be found in terms of the properties of the composite constituent phases by means of a homogenization procedure or an effective-medium theory. Since for the optical frequencies the range permeability constant is close to unity, we in what follows shall consider only dielectric constant [3].

Two main composite topologies should be marked out: cermet and aggregate (Figure 7.1) [4]. In the cermet topology, each inclusion is completely surrounded by host material; and for not very large filling factors of the metallic phase composite, it performs as a dielectric material. For the aggregate topology the metallic inclusions are allowed to touch, making the composite conductive. The transition between these two topologies happens when the filling factor reaches the percolation threshold f_{pc} at which the metallic network first becomes connected. In the vicinity of f_{pc}, the dielectric constant and therefore the optical properties exhibit dramatic changes. In general, the percolation threshold occurs at $0.2 < f_{pc} < 0.75$. Thus for cubic lattices the maximum possible f are 0.52, 0.68, and 0.74, for sc, bcc, and fcc lattices, respectively [5]. In $Co–SiO_2$ for instance, f_{pc} is about 0.55, while it equals to 0.25 for $Co–Al_2O_3$ [6].

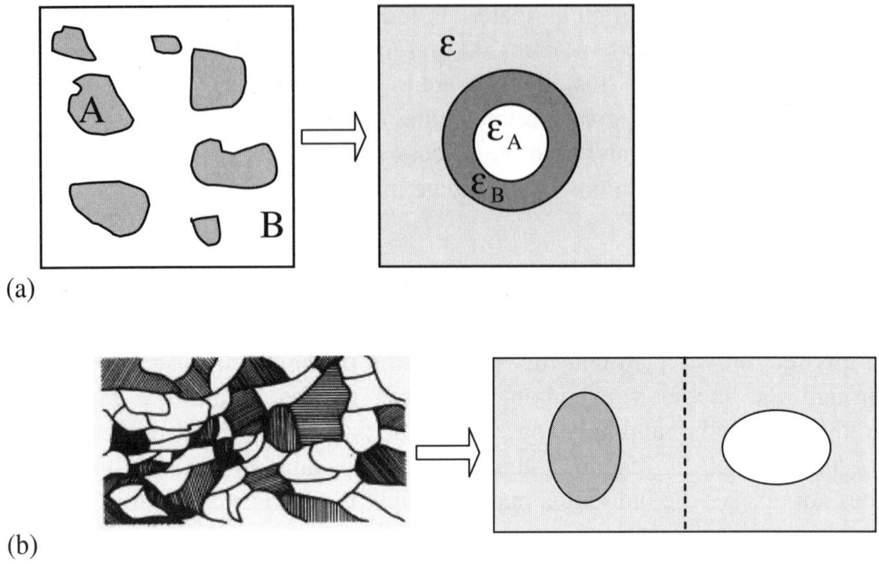

Figure 7.1. Effective medium theory approach for cermet (**a**) and aggregate (**b**) topologies.

In regular arrays the percolation transition is sharp, because all of the spheres approach contact at once. Practically no ordered structure of the inclusions can take place in the composite. That is why when f is not very small, disordered composites generally contain (a) local metallized regions (particle clusters) where the critical density has been reached and (b) other nonmetallized (dielectric) regions with lower local particle densities. With increasing volume filling factor f, the metallized clusters grow at the expense of the dielectric regions. Thus, in contrast with the behavior of regular arrays, the metallization transition in disordered composites occurs at different times in different regions of the sample. Nevertheless, at $f = f_{pc}$ the entire suspension becomes conductive.

There are a number of formulas for the determination of the effective dielectric constant: Maxwell Garnett (MG) formula, Bruggeman (BG) equation, Hanai–Bruggeman (HBG) formulas, empirical Lichtenecker mixture equations, and so on.

The empirical equations are obtained by the intuitive suggestions with the use of experimental data. Here we cite an instance of two Lichtenecker equations [7–9] for two-phase and multiphase composites:

$$|\varepsilon_{\text{eff}}| = (1-f)|\varepsilon_1| + f|\varepsilon_2|, \qquad \frac{\varepsilon_{\text{eff}}'}{\varepsilon_{\text{eff}}'} = \frac{\varepsilon_1''}{\varepsilon_1'} + \frac{\varepsilon_2''}{\varepsilon_2'}, \qquad \ln\varepsilon_{\text{eff}} = \sum_i f_i \ln\varepsilon_i$$

where f_i is the volume fraction of ith constituent, and ε_i' and ε_i'' are real and imaginary parts of the dielectric constant, respectively. For other empirical formulas, one can address the following references.

The MG, BG, and HBG formulas are based on the more comprehensive approaches, one of which is presented further.

2.1. Homogenization Technique, the Effective-Medium Formulas

The approach is based on considering the composite as made up of elementary structural units (ESU)—that is, in fact, the main principle of the effective-medium theory [4, 10, 11]. For example, in a composite with the cermet topology, the basic unit may be taken as a coated grain (Figure 7.1a). If the inclusions are allowed to touch—that is, if the aggregate topology is presented—then the two phases should be considered on an equal basis, and a grain of constituent A and a grain of constituent B are two basic units (Figure 7.1b). Once the ESU are chosen, the next step is the embedding of each individual unit in a homogeneous effective medium characterized by a yet-undetermined effective dielectric constant ε_{eff} or effective conductivity σ_{eff}.

The basic definition of an effective medium is that the ESU, when embedded in the effective medium, should not be detectable in an experiment using electromagnetic measurement. In other words, the extinction of the ESU should be the same as if it were replaced with a material characterized by ε_{eff}. This criterion makes it fruitful to use a recently derived [12] "optical theorem" for absorbing media; it relates the extinction of the spherical cell compared to that of the surrounding medium with the scattering amplitude in the direction of the impinging beam $S(0)$ (forward scattering amplitude) by

$$C_{\text{ext}} = 4\pi \operatorname{Re}\left[S(0)/k^2\right]$$

where k is the wave vector of the light in the effective medium. From the definition of an effective medium, it follows that $C_{\text{ext}} = 0$, that is $S(0) = 0$, which expresses the fundamental property of an effective medium. Consequently, the effective-medium general condition is then [13]

$$\sum_i f_i S_i(0) = 0 \qquad (1)$$

where f_i is the volume fraction of the ith unit. Since $S_i(0)$ is an implicit function of ε_{eff}, equation (1) represents a condition for its determination.

If ESU is a coated sphere (Figure 7.1a), then the Lorenz–Mie [14] theory gives $S(0)$ as

$$S(0)=i(kb)^3\frac{(\varepsilon_B-\varepsilon)(\varepsilon_B-\varepsilon)+f_A(2\varepsilon_B+\varepsilon)(\varepsilon_A-\varepsilon_B)}{(\varepsilon_B+2\varepsilon)(\varepsilon_A+2\varepsilon_B)+f_A(2\varepsilon_B-2\varepsilon)(\varepsilon_A-\varepsilon_B)}+O\left((kb)^5\right)$$

The filling factor is

$$f=\frac{a^3}{b^3}$$

where $a(b)$ is the radius of the inner (outer) sphere in Figure 7.1a. This yields the Maxwell Garnett formula

$$\varepsilon_{eff}=\varepsilon_B\frac{\varepsilon_A(1+2f)+2\varepsilon_B(1-f)}{\varepsilon_A(1-f)+\varepsilon_B(2+f)} \tag{2}$$

which is often written in a different form:

$$\frac{\varepsilon_{\text{eff}}-\varepsilon_B}{\varepsilon_{\text{eff}}+2\varepsilon_B}=f\frac{\varepsilon_A-\varepsilon_B}{\varepsilon_A+2\varepsilon_B}$$

The MG approach is applicable for so-called cermet topology [15, 16]. This formula has been accepted as satisfactory when the exact interparticle interactions are not important—for example, in the case of dilute dispersions or components of low polarizabilities.

For small volume fractions, by expanding equation (2) to the first order in the volume fraction, one can find the absorption coefficient of the composite α that is related to the imaginary part of ε_{eff}:

$$\alpha=9f\frac{\omega\varepsilon_B^{3/2}}{c}\frac{\varepsilon_2}{(\varepsilon_1+2\varepsilon_B)^2+\varepsilon_2^2}$$

where ε_1 and ε_2 are real and imaginary parts of ε_A [17]. This absorption coefficient has a maximum at ω for which the condition $\varepsilon_1 + 2\varepsilon_B = 0$ takes place, which is known as the surface plasmon resonance.

In the aggregate topology the ESU are two spheres (Figure 7.1b). These two spheres are submerged into the effective medium, and their relative concentrations are f and $1 - f$, where f is a volume fraction of the constituent A. Using the dipolar approximation of $S(\theta)$ for a sphere [12] leads to the Bruggeman equation:

$$f\frac{\varepsilon_B-\varepsilon_{\text{eff}}}{\varepsilon_B+2\varepsilon_{\text{eff}}}+(1-f)\frac{\varepsilon_A-\varepsilon_{\text{eff}}}{\varepsilon_A+2\varepsilon_{\text{eff}}}=0 \quad (3)$$

For small concentrations of inclusions (f close to 1 or to 0), it appears that a number of composite films are better described by the separated grain structure—that is, ruled by the MG equation. For intermediate values of f, the BG approach gives better results. Nevertheless, both approaches are not very precise. That is why symmetrized MG approximations have been proposed by introducing an average over the two following random units: an inclusion of A coated by a shell of B and an inclusion of B coated by a shell of A. Using this picture, the film or plate can be modeled as a mixture of two types of coated spheres or ellipsoids. Dielectric-coated metal ellipsoids are denoted as type 1 units, and metal-coated insulator ellipsoids are denoted as type 2 units (Figure 7.2). The problem is to evaluate the proportion of these two random units for a given value of the filling factor f. Sheng [4] has introduced a probabilistic growth model where the probability of occurrence of these two units as a function of f is given by the number of possible configurations of each type of units. This number is given by the volume available to the internal ellipsoid in the external ellipsoid. For details see reference 4.

2.2. The Hanai–Bruggeman Formula

The HBG formulas, also known as the asymmetric Bruggeman formulas, are obtained assuming that the MG model is exact at low filling factors and then, following an iterative procedure, adding a small fraction of particles at each step [18, 19]. This model is recognized as valid at least for $f < 0.8$.

The first HBG formula is for spherical particles:

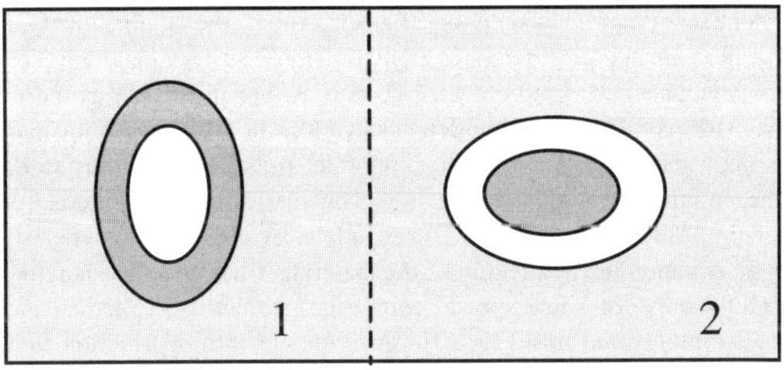

Figure 7.2. Random units 1 and 2 in the effective medium (symmetrized Maxwell Garnett approximation).

$$\frac{\varepsilon_2 - \varepsilon_{\text{eff}}}{\varepsilon_2 - \varepsilon_1}\left(\frac{\varepsilon_1}{\varepsilon_{\text{eff}}}\right)^{1/3} = 1 - f$$

and the second one is for lamellar particles:

$$\varepsilon_{\text{eff}} = \varepsilon_2 \frac{3\varepsilon_1 + 2f(\varepsilon_2 - \varepsilon_1)}{\varepsilon_2 - f(\varepsilon_2 - \varepsilon_1)}$$

Some authors have proposed a generalized HBG equation that includes all possible types of inclusion geometry [20]:

$$\frac{\varepsilon_2 - \varepsilon_{\text{eff}}}{\varepsilon_2 - \varepsilon_1}\left(\frac{\varepsilon_1}{\varepsilon_{\text{eff}}}\right)^{L} = 1 - f$$

The generalization is accomplished by letting the depolarization factor L vary between 0 and 1.

Thus, if the wavelength of the light is much greater than the grain size, the long wavelength approximation and effective medium theory can be applied to determine the effective value of the composite dielectric constant and, consequently, describe composites optical properties. However, if the size of the structure is of the order of tens and even units of nanometers, then the effective medium approach is not applicable. Indeed, within this approximation, the effective permittivity of a composite is determined as a function of the permittivity for each composite component and, in turn, the nanocomposite components are characterized by the same tensor of permittivity as those used for bulk media.[1]

[1] However, in the case of granular metallic structures in which the grain sizes do not exceed several nanometers, the conditions of applicability of this approach are violated, because the mean free path of electrons in these materials is considerably greater than the mean grain size, and spin-dependent scattering and tunneling of electrons become significant. Thus, it is evident that the effective-medium approximation, as applied to these systems, is rather contradictory and does not offer an adequate description of their properties. For the optical properties of the composites with magnetic constituents, spin-dependent electron scattering both inside conducting regions and at rough interfaces plays a very important role. The only way to describe optical properties of the composite in such a case is to consider directly an interaction of light and electrons in the composite. It can be done, for example, by means of the consistent solution of Boltzmann kinetic equation in each region of the material. This procedure is rather complex and can be done exactly only for some special composite geometries: spherical inclusions in the uniform matrix or multilayered film [21, 22]. Since the electron distribution function is found, the current distribution can be readily calculated and after using the averaging procedure one can determine dielectric constant for the material. Some other theoretical approaches can be found in references 23–25.

If we are interested not only in the total optical response of the medium but also in its response at submicron scales and nanoscales (i.e., imaging of the nanosized structures), then it is vital to have a possibility of calculation of fine electromagnetic field distribution at the fixed distances above the sample. One of the powerful methods that can be utilized for that is the Green function technique, which will be described in Section 7.

2.3. Effective Medium Approach for Magnetic Composites

Basic equations of the homogenization theory are applicable not only for scalar values of the constituents dielectric constants ε_i, but also for the tensorial quantities. Indeed, evaluation of these equations does not demand any special requirements regarding the character of the electric displacement D and electric intensity E relation: $D = \hat{\varepsilon}\varepsilon_0 E$ in which dielectric constant $\hat{\varepsilon}$ can be tensorial. This fact enables us to calculate the effective dielectric constant $\varepsilon_{\mathrm{eff}}$ for the composites with anisotropic constituents. It can be, for example, polymer composites with magnetic granular, the permittivity of which is tensorial quantity. The explicit form of the $\hat{\varepsilon}$ tensor for magnetic media will be given in the next section.

Here we consider effective-medium equations for a composite consisting of the anisotropic granules characterized by permittivity $\hat{\varepsilon}_A$ placed into an isotropic matrix with permittivity $\varepsilon_B\hat{I}$, where $\hat{I}$ is the unit tensor.

We assume that the anisotropic part of $\hat{\varepsilon}_A$ tensor, which we denote as $\delta\hat{\varepsilon}_A$, is much smaller than the diagonal one. Thus, $\hat{\varepsilon}_A$ tensor can be presented as

$$\ddot{\varepsilon}_A = \varepsilon_A \hat{I} + \delta\ddot{\varepsilon}_A$$

where $|\delta\hat{\varepsilon}_A| \ll \varepsilon_A$. Effective permittivity $\hat{\varepsilon}_{\mathrm{eff}}$ can also be written in the form $\hat{\varepsilon}_{\mathrm{eff}} = \varepsilon_{\mathrm{eff}}\hat{I} + \delta\hat{\varepsilon}_{\mathrm{eff}}$. The anisotropic part of the $\hat{\varepsilon}_{\mathrm{eff}}$ tensor is to be found from the conventional mixture equations. Substitution of the tensorial form of the phase permittivities into the MG equation (2) gives

$$\delta\hat{\varepsilon}_{\mathrm{eff}} = \left(\frac{3\varepsilon_B\sqrt{f}}{\varepsilon_A(1-f)+\varepsilon_B(2+f)}\right)^2 \delta\hat{\varepsilon}_A \tag{4}$$

If the $\delta\hat{\varepsilon}_A$ tensor has only off diagonal nonzero components, then the Bruggeman equation leads to the following expression for $\delta\hat{\varepsilon}_{\mathrm{eff}}$:

$$\delta\hat{\varepsilon}_{\mathrm{eff}} = \frac{f\varepsilon_{\mathrm{eff}}\delta\varepsilon_A}{(\varepsilon_A+2\varepsilon_{\mathrm{eff}})^2} \bigg/ \left[f\frac{\varepsilon_A}{(\varepsilon_A+2\varepsilon_{\mathrm{eff}})^2} + (1-f)\frac{\varepsilon_B}{(\varepsilon_B+2\varepsilon_{\mathrm{eff}})^2} \right] \tag{5}$$

ε_{eff} is to be found from the conventional BG equation (3). The BG approach also works for the case when some diagonal components of the $\delta\hat{\varepsilon}_A$ tensor are nonzero, but this time expression for $\delta\hat{\varepsilon}_{eff}$ is more cumbersome.

Since equation (4) was derived from the MG equation, it is applicable for small concentrations of inclusions. On the contrary, equation (5), which was derived from the BG equation, is more appropriate for the intermediate values of f.

Determination of the $\delta\hat{\varepsilon}_{eff}$ tensor allows us to describe anisotropic optical properties of the composites. In particular, for magnetic composites the effective value of the MO parameter Q, which plays an important role in the magnetooptics, can be evaluated. The definition of the parameter Q is given in the next section.

3. MAGNETOOPTICS OF UNIFORM MEDIA

Particular properties of a medium in the macroscopic theory of the MO phenomena are defined by the form of the $\hat{\varepsilon}$ and $\hat{\mu}$ tensors. It is enough to consider just the $\hat{\varepsilon}$ tensor, because the properties that we describe below are similar for $\hat{\varepsilon}$ and $\hat{\mu}$ tensors. Besides, for visible and near-infrared light the $\hat{\mu}$ tensor is approximately equal to the unit tensor. In the magnetically ordered state the $\hat{\varepsilon}$ tensor depends on the order parameter. In ferromagnets the order parameter is magnetization M, in antiferromagnets it is the sublattice magnetization, and so on. We shall confine our discussion to ferromagnets.

Let us consider the simplest case of an optically isotropic ferromagnet. The presence of magnetization reduces the symmetry to the single-axis one. The $\hat{\varepsilon}$ tensor can be represented as a sum of symmetric and asymmetric tensors, which is given by [26]

$$\hat{\varepsilon} = \begin{pmatrix} \varepsilon_1 & 0 & 0 \\ 0 & \varepsilon_1 & 0 \\ 0 & 0 & \varepsilon_0 \end{pmatrix} + ig \begin{pmatrix} 0 & -m_z & m_y \\ m_z & 0 & -m_x \\ -m_y & m_x & 0 \end{pmatrix} \tag{6}$$

The D vector for ferromagnetic medium can be written as

$$D = \varepsilon_0 E + i[g \times E] + b(E - m(m \cdot E)), \tag{7}$$

where $m = M/M$, $b(M) = \varepsilon_1 - \varepsilon_0$, ε_0 is the dielectric permittivity of the medium at $M = 0$, and g is the gyration vector. In an isotropic medium, normally $g = aM$. If there is an absorption, then

$$\varepsilon_0 = \varepsilon_0{}' + i\varepsilon_0{}'', \qquad g = g' + ig'', \qquad b = b' + ib''$$

are complex functions of the frequency. The second terms in the formulas (6) and (7) define the gyrotropic effects: magnetic gyrotropic birefringence and magnetic circular dichroism. The last terms define optical magnetic anisotropy: magnetic linear birefringence and magnetic linear dichroism.

The constants g and b become zero when M approaches zero.

Usually for the characterization of the MO effects the MO (Voigt) parameter Q is used. It is defined as

$$Q = Q' + iQ'' = g/\varepsilon_1$$

Normally, $|Q| \ll 1$. If the $\hat{\mu}$ tensor can no longer be considered as a unit tensor, then both Q and Q_M need to taken into account and the MO medium is called bi-gyrotropic.

In crystals the dependence of the $\hat{\varepsilon}$ tensor on M is more complicated, namely,

$$\varepsilon_{ik} = \varepsilon_{ik}^0 - ie_{ikl}g_l + \delta_{iklm}M_lM_m$$

where $g_l = a_{lq}M_q$ and e_{ikl} is the antisymmetric three-dimensional-order pseudotensor (the Levi–Civita tensor). The polar tensors ε_{ik}^0, a_{lq}, and δ_{iklm} are defined by the crystallographic symmetry.

4. MAGNETOOPTICAL EFFECTS IN TRANSMISSION

The MO Faraday effect manifests itself in a rotation of polarization plane of a linearly polarized light at an angle Φ when the light propagates along the medium magnetization M (Figure 7.3).

In electromagnetic theory the Faraday effect can be explained as follows. When the medium magnetization has non zero projection on the wave vector k_0 of the incident radiation, two independent fundamental Maxwell equations solutions are circular polarized waves with different refractive indexes n_+и n_-, respectively. At the output of the magnetic medium these waves gain phase shift and when added give linearly polarized wave with rotated polarization plane. That is why Faraday effect is also called magnetic circular birefringence [26, 27].

The angle of the light polarization plane rotation is given by

$$\theta = \theta_F h = -\frac{k_0 g h}{2n_0}$$

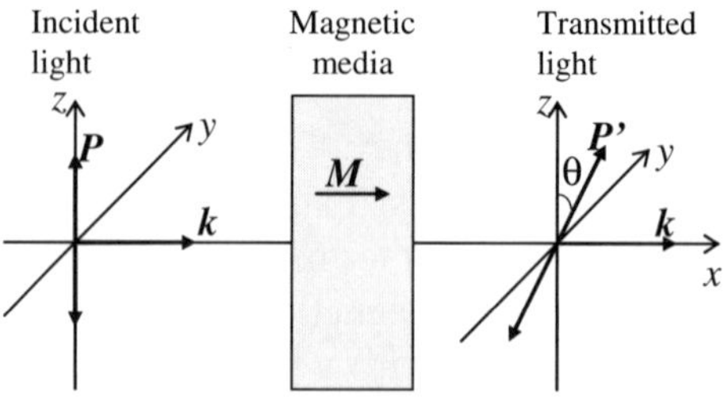

Figure 7.3. The MO effect resulting from the interaction of optical radiation with the medium for the geometry when radiation passes along the medium magnetization—the Faraday effect.

where θ_F is a specific Faraday rotation (rotation of polarization plane of the wave per unit length of the sample, typical values of θ_F are 1000–30,000°/sm), $n_0 = \frac{1}{2}(n_+ + n_-)$, h is the sample's thickness, k_0 is the wave vector module of the incident light, and g is the module of the medium gyration vector [26].

If a medium has absorption, the absorption coefficients of the right- and left-handed circular polarized light are different. This phenomenon is called *magnetic circular dichroism.* After transmission through a medium that exhibits such properties, the light changes its polarization from linear to elliptical. The elliptical polarization is characterized by the orientation angle θ, which is analogous to the Faraday angle, and ellipticity ψ. These quantities are calculated by [26]

$$\theta = -\frac{k_0 g' h}{2n_0}, \qquad \psi = -\frac{k_0 g'' h}{2n_0}$$

Specific magnetic circular birefringence and magnetic circular dichroism can be unified into one general concept—the complex specific Faraday rotation:

$$\tilde{\theta}_F = \theta_F + i\psi_F = -\frac{k_0 g h}{2n_0}$$

The Faraday effect is widely used for the magnetic structure visualization in transparent samples with an easy direction axis not parallel to the sample's surface [27]. In this case, electromagnetic radiation passing through the domains of opposite magnetization gains slightly different rotation of the polarization plane. This enables us to visualize domains and domain walls by means of analyzer. Depending on the mutual orientation of polarizer and analyzer, one can obtain images with different types of the contrast (Figure 7.4).

If the magnetic sample is transparent but magnetized in its plane, then for the visualization of the magnetic structure the Cotton–Mouton or Voigt effect can be used. It arises when electromagnetic radiation propagates in the direction that is perpendicular to the medium's magnetization. A linearly polarized light that has its polarization plane oriented at an angle to the magnetization direction becomes elliptically polarized after propagation through the medium, with the longer axes of the ellipse being approximately parallel to the incident polarization. This effect results from the difference of refractive indexes of two components of light radiation, which are linearly polarized parallel and perpendicular to the direction of magnetization. Consequently, magnetic linear birefringence takes place.

The Voigt effect is often revealed in experiment as a relative phase shift of the two polarization components per unit length of a sample:

$$B_{C-M} = (\omega/c)|\mathrm{Re}(n_{\parallel} - n_{\perp})|$$

where $n_{\parallel} = (\varepsilon_0)^{1/2}$ and $n_{\perp} = (\varepsilon_1 - \varepsilon_0^{-1}g^2)^{1/2}$ are the refractive indexes for the light polarized parallel and perpendicular to the gyration vector.

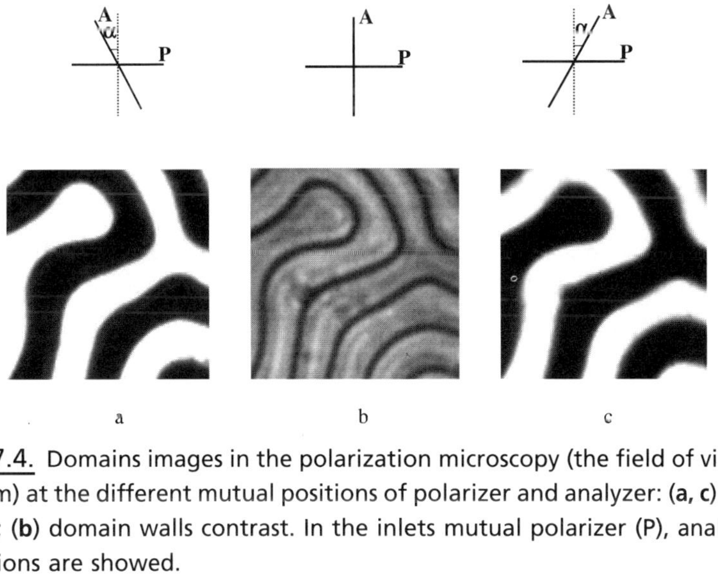

Figure 7.4. Domains images in the polarization microscopy (the field of vision size is 20 mkm) at the different mutual positions of polarizer and analyzer: **(a, c)** Domain contrast; **(b)** domain walls contrast. In the inlets mutual polarizer (P), analyzer (A) orientations are showed.

a b c
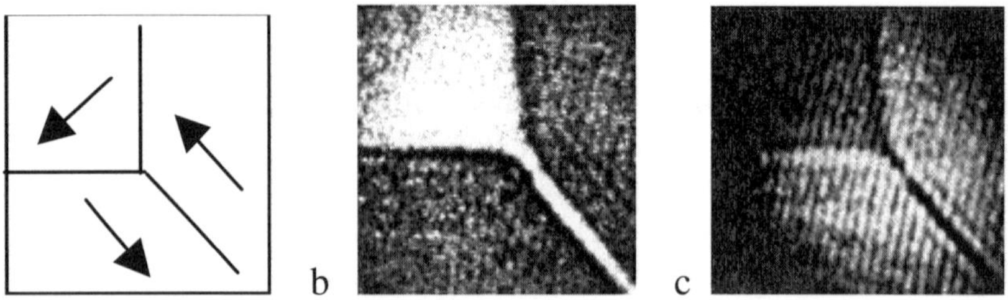

Figure 7.5. Domain structure observation by means of the Cotton–Mouton effect in the iron–garnet film with crystallographic orientation (100). (**a**) Magnetization distribution in the sample; (**b, c**) images obtained at different polarizer and analyzer orientations [28].

The Voigt effect is quadratic in magnetization (i.e., the second-order effect), in contrast to the Faraday effect, which is linear in magnetization.

Analogous to circular magnetic dichroism, linear magnetic dichroism is also possible in absorbing medium. This effect originates from the difference in absorption coefficients of the two fundamental linearly polarized waves. The presence of the magnetic linear dichroism results in a rotation of the orientational angle of the ellipse during the wave propagation.

When the Voigt effect is utilized for the sample's magnetic pattern observation, the analyzer half-wave plate is set in order to convert elliptical polarization into either linear polarization or new elliptical polarization, with the smaller eccentricity depending on the sign of rotation in the incident ellipse [28]. The polarization transformation courses differ in terms of the images of differently magnetized domains (Figure 7.5).

5. THE MAGNETOOPTICAL KERR EFFECT

Along with MO effects that take place during transmission of light through a magnetized substance, there are a number of effects that manifest themselves when the light is reflected from the surface of a magnetized material. These phenomena are conventionally designated MO Kerr effects (MOKE). There are three types of the Kerr effect, which are differentiated depending on a mutual orientation of the magnetization, with respect to both the wave propagation direction and the normal to the surface [26, 29].

The complex polar Kerr effect consists of both the rotation of polarization plane and the appearance of the ellipticity if a linearly polarized light reflects from a sample surface and the sample is magnetized normally to this surface (Figure 7.6a). If the light falls from the vacuum perpendicularly on the sample

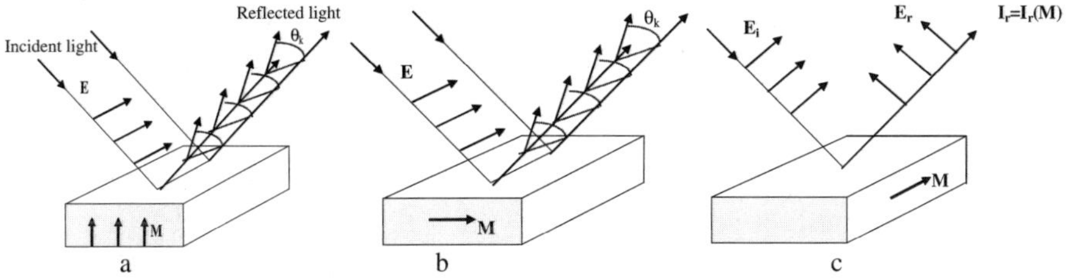

Figure 7.6. MO Kerr effects taking place when light is reflected from the surface of a magnetized material: polar (a), longitudinal (b), and transverse (c) effects.

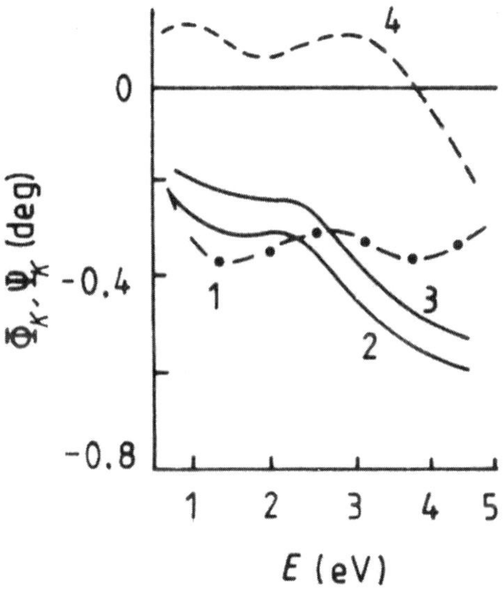

Figure 7.7. The polar Kerr rotation (curve 1) and the ellipticity (curve 2) for EuO single crystal at 10 K and $H = 40$ kOe [31].

surface, the expression for the complex polar Kerr effect takes the following form:

$$\tilde{\Phi}_K = \Phi_K + i\Psi_K = \frac{inQ}{n^2 - 1}$$

where Φ_K is the Kerr rotation angle and Ψ_K is the ellipticity. The polar Kerr effect versus photon energy is presented in Figure 7.7.

The longitudinal (meridional) Kerr effect means both the rotation of polarization plane and appearance of the ellipticity when a linearly polarized light reflects from a sample surface provided that the magnetization vector belongs to both the sample plane and the light incidence plane (Figure 7.6b).

The polar and longitudinal Kerr effects constitute the group of longitudinal MO effects. Under certain conditions, variations of the intensity of linearly polarized reflected light are observed in the configuration of either the polar or longitudinal Kerr effect geometry [30].

Like the above-mentioned effects, the transversal (equatorial) Kerr effect is linear on magnetization. The transversal effect may be observed only in absorbing materials. It is manifested as the intensity variations and the phase shift of a linearly polarized light reflected from a magnetized material, if the magnetization lies in the sample plane but is perpendicular to the light incidence plane (Figure 7.6c). For p-polarization of the incident light (radiation polarization is perpendicular to the incident plane) the relative change in the reflected light intensity is given by

$$\delta_p = \frac{I - I_0}{I_0} = 2\,\mathrm{Im}\,\rho_{12}^{\mathrm{p}}$$

where I and I_0 are the intensities of the reflected light in the magnetized and nonmagnetized states, respectively,

$$\rho_{12}^{p} = \frac{r_{12}^{p} Q \sin\varphi}{2\left(n^2 - \sin^2\varphi\right)^{1/2}}$$

r_{12}^{p} is the conventional Fresnel coefficient for reflection, and n is the refraction coefficient for the magnetic medium.

MO Kerr effects are used to observe the domain structure of opaque materials. If the magnetization vector is perpendicular to the sample plane, then the polar Kerr effect is utilized. If the magnetization lies in the sample plane, then structure images can be obtained either in the longitudinal geometry with polarizers or in the transverse geometry without polarizers.

6. NONLINEAR MAGNETOOPTICAL EFFECTS

Nonlinear MO effects and, in particular, the second harmonic generation effect are becoming increasingly important because they are nondestructive and can be remotely sensed *in situ* with high spatial and temporal resolution at any interface accessible by light.

In centrosymmetric media, the electric field contribution to the second-order optical polararizability is forbidden by symmetry [32]. At the same time, the majority of widely spread magnetic materials (Fe, Co, Ni, FeNi, etc.) possess inversion symmetry, so second harmonic generation appears only in ultrathin surface layer where inversion symmetry is broken. The magnetization of a material does not usually break this symmetry, but it can modify substantially the form of the nonlinear susceptibility for surface second harmonics generation [33, 34]. This explains the fact that nonlinear MO effects usually exceed corresponding linear effects by several orders of magnitude [35–42]. Thus for granular Co–Cu magnetic films (thickness is 200 nm) in polar geometry the nonlinear Kerr effect is 8°, whereas the linear effect is no more than several tenths of a degree [38]. In the transversal configuration the nonlinear effect surpasses the linear effect more than 40-fold. While investigating ultrathin films (thickness of 1–10 nm) it was revealed that the nonlinear effect, in contrast to the linear one, practically does not depend on the sample thickness and even for nanometer scales is about 10°–20° [39] (Figure 7.8). Thus, the technique of nonlinear optical signal observation is very convenient and sensitive for investigation of the magnetic materials surfaces.

The surface nonlinear optical polarization of second order can be written in the form [34]

$$P_s^{2\omega} = \chi_{ijk}^{(2)} E_j E_k \tag{8}$$

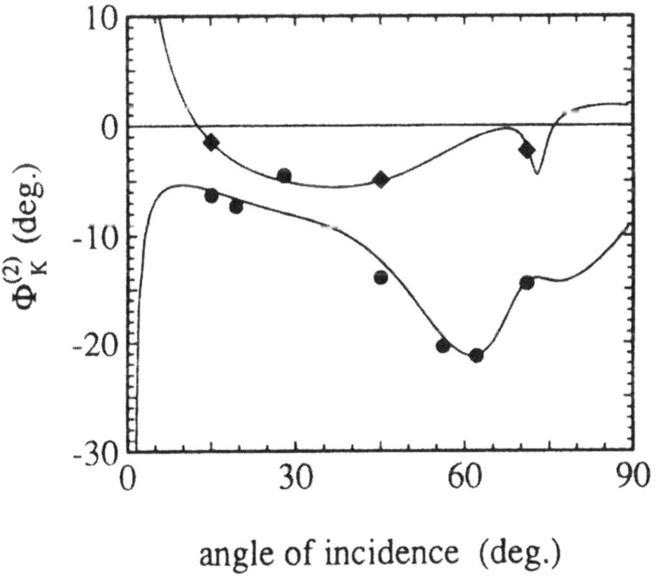

Figure 7.8. Nonlinear Kerr rotation $\Phi_k^{(2)}$ for an Fe–Cr multilayer as a function of the angle of incidence. Filled circles: s-input polarization; filled diamonds: p-input polarization. The curves are theoretical fits [39].

where the surface nonlinear susceptibility tensor $\chi_{ijk}^{(2)}$ is a function of magnetization M, and E is the electric field of the light wave. The symmetry of $\chi_{ijk}^{(2)}$ is defined by the time-reversal symmetry and by the symmetry of the particular surface under consideration. The time-reversal properties, neglecting dissipation, requires that the real part of $\chi_{ijk}^{(2)}$ is an even function of M, while the imaginary part is an odd function of M. The latter feature can be particularly useful for probing of the magnets surfaces.

From a symmetry viewpoint the second-order surface polarization in the linear (-in-M) approach can be written as

$$P_s^{2\omega} = P_{s0}^{2\omega} + P_{sm}^{2\omega} \tag{9}$$

where

$$P_{s0}^{2\omega} = \chi_1 E(EN) + \chi_2 E^2 N \tag{10}$$

and

$$P_{sm}^{2\omega} = \chi_3 E(E(mN)) + \chi_4 E^2[mN] + \chi_5[Em](EN) + \chi_6[EN](Em) \tag{11}$$

are contributions that are independent of the magnetization $P_{s0}^{2\omega}$ and are linear in the magnetization $P_{sm}^{2\omega}$, where χ_i, $i = 1, 2$, are nonlinear optical parameters, and χ_i, $i = 3, 4, 5, 6$, are nonlinear MO parameters; and N is the vector normal to the surface. It is clear that only these two independent combinations that are second order in E and have the symmetry of the polar vector can be composed from the polar vectors N and E to get $P_{s0}^{2\omega}$; and only four independent combinations can be composed from the polar vectors N and E and the axial vector m to get $P_{sm}^{2\omega}$.

The relation (11) should be regarded as an expansion of $P(E, N, M)$ in E, N, and M. We restrict the analysis to the terms that are quadratic in E and linear in N and M. The ratio of the light-wave field E to the magnitude of the intraatomic field E^*, the ratio $\zeta = E_{surf}/E^*$ of the surface electric field E_{surf} (which breaks the even symmetry at the surface) to E^* (for N), and the magnitude of the magnetooptical gyrotropy (M), which is determined by the MO parameter Q and usually satisfies the condition $Q \ll 1$, are small parameters. Here expansion in N actually means expansion in NE_{surf}, as was shown in reference 35.

The use of the expansions (10) and (11) reduces the number of parameters needed to describe nonlinear MO phenomena (in comparison to the general formula (8). We can show this by comparing equation (8) with equations (9) and (10). Formula (10) can naturally be represented in a matrix form, just as equation (8), where the third-rank tensor has the following form in Voigt's notation:

$$\begin{bmatrix} 0 & 0 & 0 & 0 & e_{15} & 0 \\ 0 & 0 & 0 & e_{15} & 0 & 0 \\ e_{31} & e_{31} & e_{33} & 0 & 0 & 0 \end{bmatrix} \qquad (12)$$

It is symmetric relative to interchange of the indices j and k. Such a form for the tensor $\chi_{ijk}^{(2)}$ corresponds to the limiting ∞m symmetry group (the Curie group). A uniform electric field, for example, has this symmetry. It follows from equation (10) that $e_{33} = e_{31} + 2e_{15}$; that is, the tensor $\chi_{ijk}^{(2)}$ $(M = 0)$ is specified in our case by two independent parameters, rather than three, as required by ∞m symmetry. However, there is no contradiction between formulas (10) and (11), since formula (10) corresponds to the linear approximation with respect to z. Taking into account the next term with respect to z in the expansion in (10) in, for example, the form $N(NE)^2$, we obtain $e_{33} = e_{31} + 2e_{15} + O(\zeta^2)$.

7. GREEN FUNCTIONS APPROACH. OPTICS OF SEPARATE NANOOBJECTS

Averaged signal from inhomogeneous medium with nanoscaled objects generally is not very informative, and it is vital in some cases to detect signal from separate nanostructures. This type of situation arises, for example, in read–write processes at high storage densities, when the sizes of the written information bits and distances between them do not exceed several tens of nanometers. At this stage, characterization of the optical properties of the sample by the effective permittivity is no longer appropriate, and one has to use methods more rigorous than the homogenization technique. Among these methods, the dyadic Green functions approach deserves particular consideration.

As is known from the differential equations theory, the Green functions are used for solution of the linear inhomogeneous differential equations:

$$\hat{L}_{\psi(r)} = f(r) \qquad (13)$$

where $\hat{L}$ is the linear differential operator. The Green function for this equation is given by

$$\hat{L}G(r, r') - \delta(r - r')$$

With the use of the Green function, the solution of equation (13) can be expressed as

$$\psi(r) = \psi_0(r) + \int G(r, r') f(r') dr',$$

where $\psi_0(r)$ is the solution of corresponding homogeneous equation. The standard Helmholtz equation with nonzero right side

$$\Delta E(r) + k_0^2 E(r) = F(r)$$

where $k_0 = \omega/c$, can be solved in terms of the function

$$G(r, r') = \frac{e^{ik_0|r-r'|}}{|r-r'|}$$

which is called the *scalar free-space Green function.*

However, while dealing with the inhomogeneous medium, one faces more complex equation. The medium inhomogeneities can be described by a relatively small additional term $P_1(r,t)$ in the medium polarization:

$$P(r,t) = \varepsilon_0 \chi(z,\omega) E(r,t) + P_1(r,t)$$

The fact that susceptibility $\chi(z,\omega)$ depends only on the Z-spatial coordinate means that we examine here a medium with a stratified structure. The polarization vector $P_1(r,t)$ is determined by the kind of inhomogeneities. If it is related to the magnetization distribution, then in linear magnetooptics we have

$$P_1(r,t) = i\varepsilon_0 \varepsilon Q \cdot m \times E(r,t)$$

where $m = M/M$ is the unit magnetization vector and Q is the MO parameter (see Section 3). If one is to investigate the MO response at the second harmonics, then $P_1(r,t)$ is determined by equation (8).

In the Maxwell equation

$$\nabla \times \nabla \times E(r,t) = -\mu_0 \frac{\partial^2}{\partial t^2} D(r,t)$$

let us substitute $E(r,t) = E(r,\omega)e^{-i\omega t}$ and $D(r,t) = D(r,\omega)e^{-i\omega t}$ and use the relation

$$D(r,\omega) = \varepsilon_0 E(r,\omega) + P(r,\omega) = \varepsilon_0 \varepsilon(z,\omega) E(r,\omega) + P_1(r,\omega)$$

where $\varepsilon(z,\omega) = 1 + \chi(z,\omega)$. Then the equation for the Fourier coefficient of the electric field $E(r,\omega)$ is given by

$$\nabla \times \nabla \times E(r,\omega) - \varepsilon(z,\omega) k_0^2 E(r,\omega) = \frac{k_0^2}{\varepsilon_0} P_1(r,\omega) \quad (14)$$

where $k_0 = \omega/c$. This equation is inhomogeneous, and thus the apparatus of the Green function can be used to solve it. But the Green function this time is dyadic [43]. It is introduced by the equations

$$\left(\frac{\partial^2}{\partial x_\lambda \partial x_\mu} - \delta_{\lambda\mu}\frac{\partial^2}{\partial x_\mu^2} - \varepsilon(r)\frac{\omega_0^2}{c^2}\delta_{\lambda\mu}\right)G_{\mu\nu}(r, r', \omega) = -\delta_{\lambda\nu}\delta(r - r') \qquad (15)$$

which are solved with ordinary Maxwell boundary conditions. Therefore, the expressions for the fields derived using the Green function *a priori* meet these boundary conditions. The method of Green functions makes it possible to derive analytical formulas for the Fourier transform of the electric field of a diffracted wave. This approach proved to be valid in solving a great variety of scattering problems: light reflection and transmission in rough surface media [43, 44], plasmon polariton propagation [45], investigation of nonlinear surface MO effects [36, 40], and calculation of the field distribution in the near-field region [46–49].

With the use of this function, one can convert the partial differential equation (14) into an integral equation

$$E_\mu(r, \omega) = E_\mu^{(0)}(r, \omega) - \frac{k_0^2}{\varepsilon_0}\int d^3r' G_{\mu\nu}(r, r', \omega)P_{1\nu}(r, \omega), \qquad (16)$$

where $E_\mu^{(0)}(r,\omega)$ is a solution of the corresponding homogeneous equation

$$\nabla \times \nabla \times E(r, \omega) - \varepsilon(z, \omega)k_0^2 E(r, \omega) = 0$$

$E_\mu^{(0)}(r,\omega)$ describes electric field without taking into account inhomogeneities. To solve the integral equation (16), the first Born approximation can be utilized. In the first Born approximation, additional medium polarization $P_1(r,\omega)$ depends only on the unperturbed electric field $E_\mu^{(0)}(r,\omega)$, and the total electric field is

$$E_\mu(r, \omega) = E_\mu^{(0)}(r, \omega) + E_\mu^{(1)}(r, \omega)$$

where $E_\mu^{(1)}(r, \omega) = -\frac{k_0^2}{\varepsilon_0}\int d^3r' G_{\mu\nu}(r, r', \omega)P_{1\nu}^{(0)}(r', \omega)$

To proceed further, it is appropriate to introduce the Fourier representations of $G_{\mu\nu}(r,r',\omega)$ and $P_{1\nu}^{(0)}(r,\omega)$:

$$G_{\mu\nu}(r, r', \omega) = \frac{1}{4\pi^2}\int d^2k_\parallel \exp(ik_\parallel(r_\parallel - r_\parallel'))g_{\mu\nu}(z, z', \omega, k_\parallel) \qquad (17)$$

$$P_{1\nu}^{(0)}(r,\omega) = \frac{1}{4\pi^2}\int d^2k_{\|}\exp(ik_{\|}r_{\|})\hat{P}_{1\nu}^{(0)}(z,\omega,k_{\|}) \tag{18}$$

where $k_{\|} = (k_x, k_y, 0)$ and $r_{\|} = (x, y, 0)$. The form of these representations is dictated by the dielectric constant $\varepsilon(z,\omega)$. The Fourier transforms of the Green functions are determined by solving ordinary differential equations. This procedure is outlined, for example, in references 43–45. Examples of the Green function Fourier transforms for the half-space or plate can be found in references 43 and 45.

Using equations (17) and (18), we can express Fourier transform of the scattered electric field $\hat{E}_{\mu}^{(1)}(z,\omega,k_{\|})$ in terms of the Fourier transform $\hat{P}_{1\nu}^{(0)}(z,\omega,k_{\|})$ of the medium inhomogeneities polarization:

$$\hat{E}_{\mu}^{(1)}(z,\omega,k_{\|}) = -\frac{k_0^2}{16\pi^4}\int dz' g_{\mu\nu}(z,z',\omega,k_{\|})\hat{P}_{1\nu}(z',\omega,k_{\|}) \tag{19}$$

Thus, to pass to the electric field distribution in the real space, one needs to make inverse Fourier transformation.

With the electric field $E(r,\omega)$ found, one can calculate optical properties of the material—transmission, reflection, and absorption—and simulate an image of its structure. For the letter, it is important that relation (19) describes in fact the angular spatial spectrum of the field of the wave diffracted from the grains or other inhomogeneities of the sample (e.g., magnetic structure). The scattered wave field can be written as a superposition of electromagnetic plane waves with various wave vectors $k_{\|}$ and complex amplitudes defined by equation (19). Based on equation (19), one can theoretically investigate optical properties of the materials and calculate images of their dielectric and magnetic structures obtained in the various types of the microscopy. Two examples of such calculations for the scanning near-field and dark-field microscopies are presented in Section 10. The microscopic image is formed by those spatial Fourier components of the wave field that arrive at the microscope objective. A set of harmonics passed through the system is controlled by the objective numerical aperture $NA = n\sin u$, where n is the refractive index of the medium surrounding the objective and u is the aperture angle of the objective. Thus, the features of the image formed in the optical system are controlled by the inverse Fourier transform that takes into account only the field harmonics passed through the system:

$$E_{\mu}^{(1)} = \frac{1}{4\pi^2}\int_{\{k_{\|}\}} dk_{\|}\hat{E}_{\mu}^{(1)}e^{ik_{\|}r} \tag{20}$$

Once electric field at the image plane is found, one can calculate the intensity distribution at this plane: $I(r_{\|}) = |E(r_{\|})|^2$; that is, one can simulate an image of the investigated structure.

8. SCANNING NEAR-FIELD OPTICAL MICROSCOPY

Detailed study of the magnetization distributions on submicrometer spatial scales and the elaboration of the corresponding high-resolution techniques is presently one of the most important problems in applied and fundamental micromagnetism. In the past decade, significant progress has been made in the development of methods for studying micro- and nanostructures with the advent and onrush of magnetic force microscopy, which is a version of scanning probe microscopy [50]. Currently, magnetic force microscopy is characterized by the best spatial resolution in studying magnetic microstructures (tens of nanometers [51]). However, its application offers a number of problems: complicated interpretation and identification of images, the uncontrollable influence of the microprobe on the magnetization distribution in samples and vice versa, the problem of positioning the force microscope tip above a surface point under study, and the low scanning velocity, which makes real time observation of fast processes impossible [52]. In this connection, some other alternative methods of magnetic imaging should be investigated. In this chapter we shall analyze scope and limitations of the promising approach to optical microscopy of magnetic structures, namely, scanning near-field optical microscopy (SNOM). It combines the rich capabilities of optical methods with the high resolution of scanning probe microscopy [53, 54]. In the SNOM, a probe scans the specimen, with the distance between them being smaller than the wavelength from its surface.

The SNOM can be utilized in a variety of different imaging modes including illumination mode, collection mode and dual mode (Figure 7.9) [55, 56]. In all modes, the resolution depends on the aperture size and the probe-specimen spacing rather than on the wavelength. In the illumination mode, the probe acts as the optical near-field generator, which illuminates the specimen. The signal (transmitted, reflected, or emitted light) is collected in the far-field using conventional optics. In the collection mode of the SNOM, the sample is illuminated using far-field optics and the signal is collected by the SNOM tip. Both the illumination mode and the collection mode of the SNOM have been utilized and have achieved high-resolution imaging [53–58]. The motivation in choosing one imaging mode over another is usually dictated by the particular experiment. The tip can also be used to both illuminate the sample and to collect the signal, a configuration known as dual SNOM mode [59]. This arrangement is attractive because it is easy to implement (the need for far-field optics—that is, the conventional optical microscope—is eliminated). In this mode, however, both the illumination light and the collected signal must pass through the aperture, causing significant reduction in the signal-to-noise ratio.

Like conventional optical microscopy, the SNOM can be performed in transmission or in reflection. The most common method is the transmission SNOM in which a thin, transparent sample is excited by the tip (i.e., illumina-

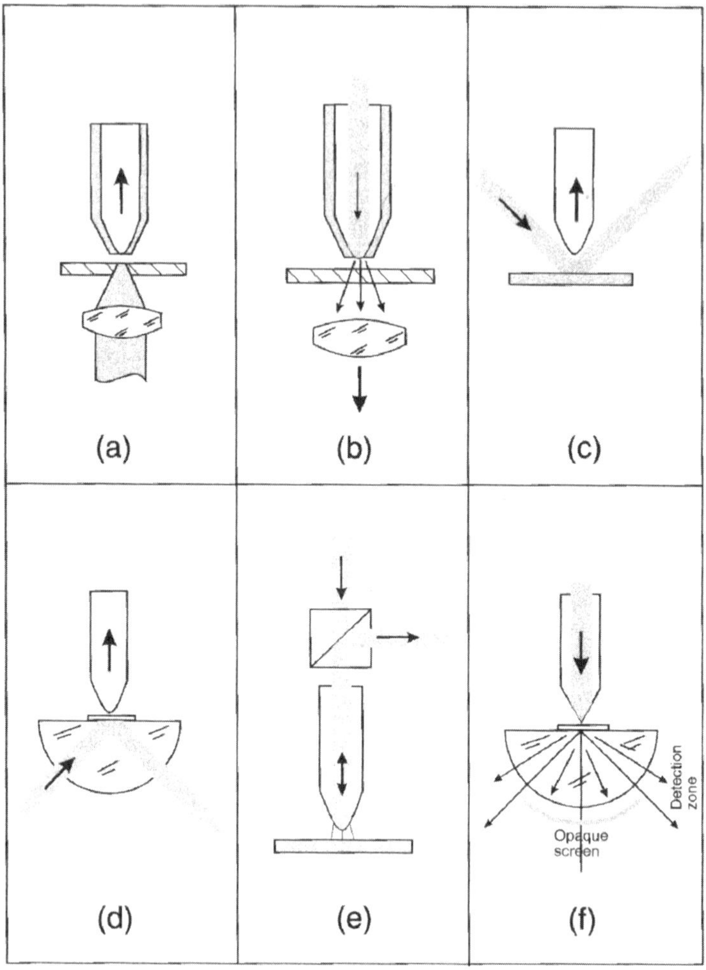

Figure 7.9. Types of the SNOM configurations. (**a**) Transmission collection mode. The tip is generally metallized except for its nano-sized end. (**b**) Transmission illumination mode. (**c**) Reflection collection mode. (**d**) Photon scanning tunnel mode. The illumination beam is totally reflected inside a substrate. (**e**) Dual illumination collection mode. It is a combination of (**a**) and (**b**). (**f**) Reflection illumination mode. It is an inverted photon tunnel mode (**d**) [53].

tion mode) and the signal light is collected in the far-field on the opposite side of the sample. For opaque samples, however, the reflection geometry is required. In the reflection SNOM, the tip and collection optics must be placed on the same side of the sample, which presents several important problems in comparison with transmission SNOM [60, 61].

Moreover, among SNOM modifications operating in the collection mode, there is a so-called photon scanning tunnel mode, where light is incident at the angle of total internal reflection (Figure 7.9d). The optical near field localized in the neighborhood of the specimen surface is detected with a near-field probe.

The first application of the SNOM for the MO studies happened in 1992 [62], when it was demonstrated that near-field MO observation can be obtained in the same manner as conventional far-field observation—that is, by using two cross-polarizers. Betzig et al. [62] visualized 100-nm magnetic domains and claimed spatial resolution of 30–50 nm. The possibility of MO domain imaging was confirmed in both the transmission regime (Faraday geometry) [63, 64] and the reflection regime (Kerr microscopy) [65–67].

By now the MO SNOM development is only at the initial stage. Published experimental results are not very numerous and mainly addressed the investigation of the simple magnetic structures in Co–Pt multilayers [62] and iron–garnet films [63–70]. Apart from that, some problems related to the MO images interpretation arise. At the same time, the MO SNOM has indisputable potential for the imaging of magnetic structures due to its high resolution unlimited by the light diffraction. That is why further experimental and theoretical work is essential. In particular, utilization of the computer simulation of the near-field images with some *a priori* model of the object with the subsequent comparison of simulated and observed images looks quite natural for overcoming the problem of images interpretation. In such simulation the dyadic Green function technique can be very fruitful. Let us consider its application for the calculation of magnetic nanoparticle images in SNOM.

We examine here the passive probe model [71], which ignores the effect of the probe on the SNOM image and assumes that the signal detected is proportional to the near-field intensity at the nanostructure surface in the absence of the probe. This hypothesis may be valid either if the field scattered by the tip is very small or if it is not reflected back by the sample. Thus, from this qualitative analysis, we may expect the probe to be passive either if the tip is very small or if the sample has a low reflectivity. Therefore, a metallic tip close to a metallic sample may not satisfy the assumption of a passive probe, whereas a tiny metallic tip above a dielectric (or magnetic) might be considered as a passive probe.

A passive probe model simplifies calculation substantially. Indeed, such an approach enables us to work in the first Born approximation, while for calculation of the near field we only need to calculate the scattered field $\hat{E}^{(1)}_{\mu}(z,\omega,k_{\parallel})$ by equation (19) with the Fourier transform $\hat{P}^{(0)}_{1\nu}(z,\omega,k_{\parallel})$ chosen to be compatible with the kind of nanoscaled structure. The specificity of work in the near-field reveals information on the domain of integration $\{k_{\parallel}\}$ in equation (20). Since the main idea of the SNOM lies in dealing with nonradiation evanescent waves that correspond to the high spatial frequencies, domain $\{k_{\parallel}\}$ must include a wide range of $k_{\parallel}$, including $k_{\parallel}$ for which $|k_{\parallel}| > k_0$. The fulfillment of the last inequality is necessary to obtain a high spatial resolution. For example, a resolution of 10 nm, which is reachable by the SNOM, can be achieved if all spatial harmonics with $k_{x,y} \in (-25k_0, 25k_0)$ are integrated. We should note here that the scanning altitude [which is z in equation (19)] must not be larger than several tens of nanometers. A more detailed description of the Green function method application for the calculations in the near field can be found in reference 46. Some results of the images modeling are presented in Section 10.

9. POLARIZED ANISOTROPIC DARK-FIELD MICROSCOPY

Dark-field optical microscopy techniques feature significant potential for detecting and studying such magnetic formations as domain boundaries and substructures (Bloch lines, Bloch points) characterized by an intricate magnetization distribution and a characteristic size of tenths and hundredths of a micrometer. Dark-field optical microscopy is widely used in the studies of phase objects in physics, mineralogy, biology, and so on [72]. In the simplest version of this method, a sample is illuminated through a condenser with a diaphragmed center, so that the light flux incident onto the sample represents a hollow cone (Figure 7.10). If the objective aperture is smaller than the condenser aperture,

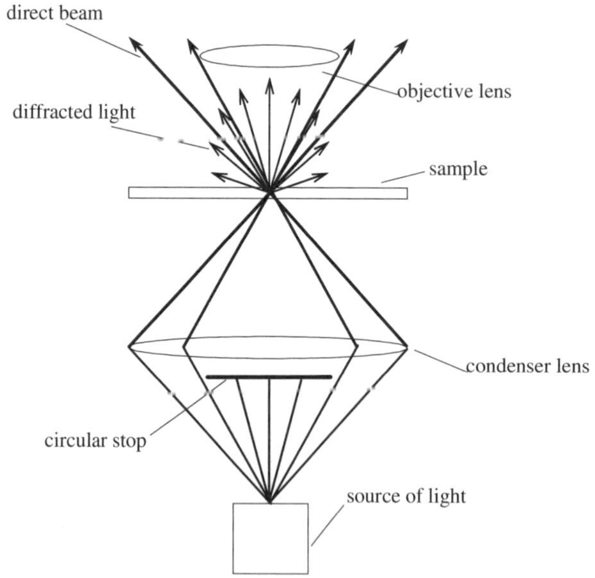

Figure 7.10. The schematic diagram of the phase-object observation using the dark-field method.

only light scattered at the sample phase structure hits the objective, forming a structure image.

In terms of the spatial resolution, the dark-field method is virtually equivalent to conventional optical microscopy [73]. At the same time, the detection sensitivity of the method under consideration is much higher than that of the conventional transmission microscopy, since the image is not subject to background illumination. This makes it possible to apply the dark field to observe and detect phase inhomogeneities smaller than or of the order of fractions of a micrometer (ultramicroscopy). Submicrometer magnetic structures may also be related to such phase inhomogeneities, because one can use the MO (Faraday, Kerr, Cotton–Mouton) effects. Currently, such magnetic structures also attract significant interest in the context of searches for new methods for data recording, storing, and reading.

The dark-field method was first applied to study magnetic media in reference 73, where the domain structure was visualized in ferrite–garnet films. A significant contribution to the development of MO dark-field microscopy was made by Thiaville and co-workers [74, 75], who optically detected the vertical Bloch lines (VBLs) in iron–garnet films and described the visualization mechanism.

The VBLs represent stable magnetic vortices and separate domain wall regions of opposite polarity [76]. The VBL sizes in many typical ferrite–garnet films do not exceed tenths of a micrometer.

In references 21 and 74–77, experiments were carried out using the anisotropic dark-field microscopy (ADFM) method, which is a modification of the conventional dark-field method wherein a narrow beam is cut out from the dark-field illumination cone. In references 74 and 75, the light incidence plane was normal to the domain wall planes (Figure 7.11).

In the ADFM images obtained in references 74 and 75 there is no illumination from domains, while domain walls are seen as contrast lines. Some portions of these lines exhibit lighter or darker regions, which are the VBLs of different magnetic topology (Figure 7.11). Further studies showed that the possibility of observing the Bloch lines in this dark-field configuration is related to the microscopic domain wall tilt in the VBL localization regions [21] rather than to light scattering immediately on the magnetization distribution in the Bloch line. Furthermore, another dark-field configuration making it possible to observe the Bloch lines is when the light incidence plane is parallel to the domain walls [21, 76, 77]. There is no illumination from either domains or domain walls in the images obtained in this configuration, while the VBLs are visualized as bright symmetric objects (Figure 7.11) regardless of magnetic topology.

The theoretical study reported in reference 78 allowed one to conclude that the VBL visualization mechanism in the last case is associated with light scattering on the VBL magnetic structure.

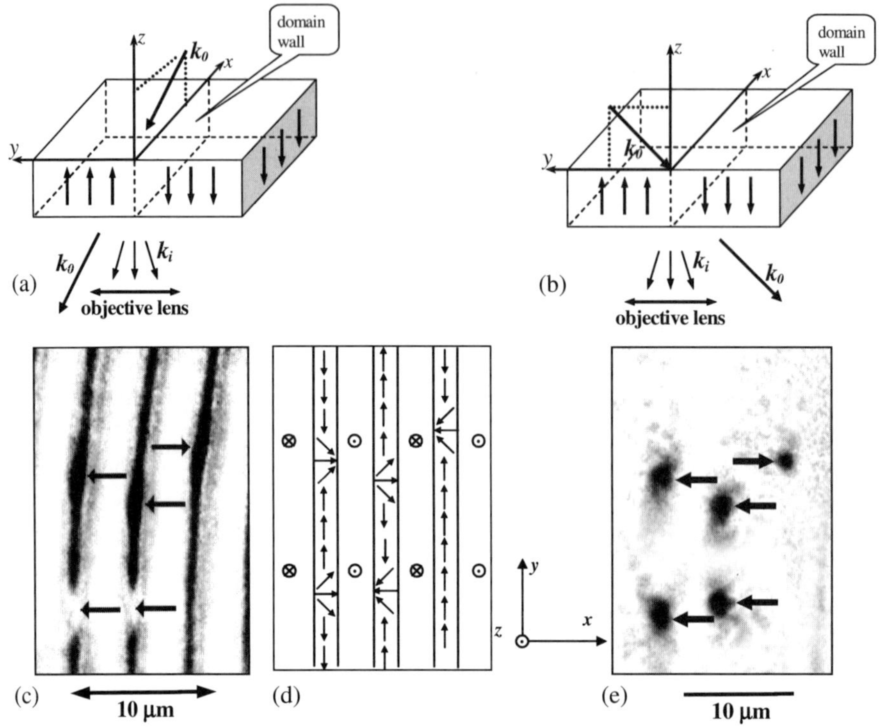

Figure 7.11. Observation of the VBLs in the ADFM. (**a, b**) Two main observation configurations. (**c, e**) Typical images of domain walls and Bloch lines (marked by arrows) obtained in observation configurations corresponding to (**a**) and (**b**), respectively (contrast is inverted). (**d**) Magnetic structure of the sample. The Bloch lines are indicated by arrows.

Thus, the results obtained in references 74–79 showed that the dark-field method can be used to detect magnetic structures of size ~0.1 mkm. Varying the specific observation geometry (e.g., the angle of light incidence and the angle formed by the light incidence plane and the domain wall plane), it becomes possible to obtain various images of magnetic microstructures. In this case, the possibility of imaging through various mechanisms is not improbable.

The high sensitivity of the dark-field anisotropic observation method suggests its possible application for magnetic nanoparticles imaging that we discuss in brief in Section 10.

10. NANOSCALE MAGNETIC ELEMENTS, MAGNETIC DOTS IMAGING

In the bulk, ferromagnetic materials usually form domain structures to reduce their magnetostatic energy. However, in very small ferromagnetic systems—that is, magnetic dots with the diameter not exceeding several hundred nanometers, magnetic nanobridges, and nanowires—the formation of domain walls is not always energetically favored. Thus, in magnetic dots, some other configurations such as curling spin or collinear-spin configurations arise instead of the domain structure. If the dot thickness is much smaller than its diameter, then, as a rule, the following magnetization configuration takes place. Magnetization directions change gradually in-plane so as not to lose too much exchange energy, but to cancel the total dipole energy. In the vicinity of the dot center, the angle between adjacent spins then becomes increasingly larger when the spin directions remain confined in-plane. Therefore, at the core of the vortex structure, the magnetization turns out-of-plane and parallel to the plane normal [80, 81]. Calculations based on the discrete Monte Carlo method [82] or on the solving of the Landau–Lifshitz equation with the full-scale magneto-dipole interaction [83, 84] show that no out-of-plane component of the magnetization occurs if the dot thickness becomes too small. When the thickness exceeds a certain limit, the top and bottom spin layers will tend to cancel each other, and again no perpendicular magnetization should be observed. For the dots with diameters smaller than 100 nm, a collinear-type single-domain phase is energy-wise preferable. So with respect to the size and shape of the magnetic dots, different magnetic structures are realized.

Furthermore, many interesting and not always predictable spin configurations happen in the dynamics when the external magnetic field is applied [83].

Though there are some other methods for the investigation of magnetization structures at the nanometer scales (magnetic force microscopy, spin tunnel microscopy), the MO observation can also give much information.

One of the MO methods that can be suitable here is the MOKE (Section 5)-based measurement of the hysteresis loops by plotting the signal (rotation or ellipticity) as a function of the applied magnetic field [85–87]. Cowburn et al. [87] reported such measurements at the longitudinal MOKE on nanoscale supermalloy ($Ni_{80}Fe_{14}$–Mo_5) dot arrays. From their experiments it was possible to conclude that a collinear-type single-domain phase is established in the particles with diameters smaller than a critical value (about 100 nm) and that a vortex phase likely occurs in dots with larger diameters. Aign et al. [88] used polar MO Kerr microscopy to show the influence of dipolar interaction on the magnetization reversal in the individual Co dots and on the collective behavior of the dots array. From a comprehensive MO microscopy study, coupled to numerical simulations, they arrived at a detailed understanding of the parameters that drive the single dot and collective magnetization reversal behaviors, as, for instance, the formation of magnetostatically frustrated zones in the arrays.

Investigation of magnetic nanoobjects structure on ultrafast timescales is possible with the use of stroboscopic effects [89–91]. Domain walls propagation in magnetic nanowires was also observed by the MOKE microscopy [92]. Thus among the variety of MO methods the MOKE approach remains at the forefront of magnetic nanostructure research performing high spatial and temporal resolution.

10.1. Imaging in SNOM

At the same time, as suggested by theoretical calculations, the size of the spin configuration peculiarities can be fairly small, and, consequently, conventional magnetization measurements fail to distinguish them from the surrounding magnetic structure. Moreover, observation of individual properties of the closely packed magnetic dots also demands for super-resolution. At this stage the SNOM microscopy—an optical technique with high spatial resolution—must be utilized. In spite of some technically removable difficulties in the extraction of the MO signal from the near-field images, this method allows a detailed investigation of the nanoscaled magnetization distribution.

To prove this statement, let us examine properties of the SNOM nonmagnetic and magnetic images. We consider scanning tunneling optical microscopy configuration of the SNOM (see Section 8). In this configuration a nanostructure is supposed to be located on the surface of a transparent substrate illuminated by linearly polarized light (wavelength $\lambda = 0.5\,\mu m$) at an incident angle beyond the limit of the total internal reflection (Figure 7.12). The light diffracted

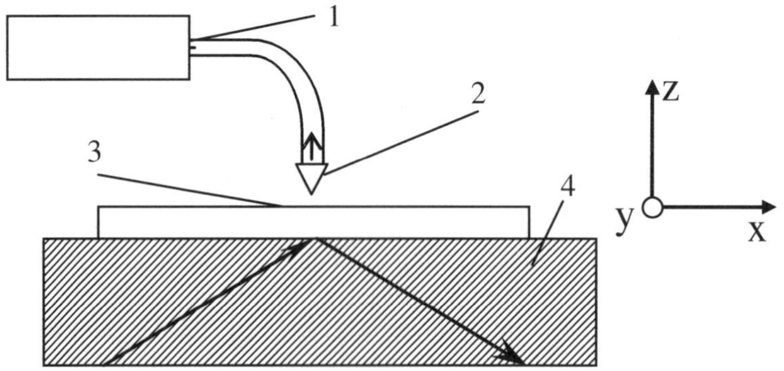

Figure 7.12. Schematic for scanning tunneling optical microscopy configuration. 1, detector; 2, optical probe; 3, sample; 4, substrate.

on the nanostructure is collected close to the sample surface by the tip of an optical fiber.

While performing calculations the following parameters were assumed. The refractive index of the semi-infinite substrate is $n = 1.5$, the incident angle is $\theta = 70°$ (angle of internal reflection for $n = 1.5$ equals to 42°). Polarization of the sample was supposed to be $P_{1i}^{(0)} = \varepsilon_0 \chi_{i,j} E_j^{(0)}$ (see Section 7), where the tensor of susceptibility is

$$\chi_{i,j} = \begin{pmatrix} n_1^2-1 & 0 & 0 \\ 0 & n_1^2-1 & 0 \\ 0 & 0 & n_1^2-1 \end{pmatrix} + \xi n_1^2 \begin{pmatrix} 0 & -im_zQ & im_yQ \\ im_zQ & 0 & -im_xQ \\ -im_yQ & im_xQ & 0 \end{pmatrix} \quad (21)$$

$m = (m_x, m_y, m_z) = M/M$, the unit magnetization vector. The first term in $\chi_{i,j}$ is a nonmagnetic part that can be called a crystalline one, and the second term in $\chi_{i,j}$ is a magnetic part (parameter $\xi = 0$ for nonmagnetic particle and $\xi \neq 0$ for the magnetic one).

For the images calculation the Green function technique described in Section 7 was utilized. Two main configuration of illumination were considered: s-polarization (incident polarization is perpendicular to the plain of incidence) and p-polarization (incident polarization is parallel to the plain of incidence). The electromagnetic field was calculated at a definite distance from the sample surface ("constant" height of scanning).

Simulations of the near-field images of nonmagnetic nanoparticles mounted on the surface of the dielectric substrate reveal one important feature (Figure 7.13): The plane of polarization rotates at the particle edges. Nonmagnetic rotation of polarization in near field is attributed to the fulfilling of boundary conditions at the boundaries between two conductors.

This fact should be taken into account in the interpretation of MO images, since MO observation exploits the property of a magnetized material to rotate the plane of wave polarization.

Bearing in mind that Q for magnetic materials is on the order of several hundredths, one can infer that a magnetic particle is difficult to be observed because of the unavoidable nongyrotropic rotation of the plane of polarization at its edges.

The MO contribution can be directly detected only at sites where the nonmagnetic contribution to the image approaches zero. Specifically, for a circular particle magnetized along the surface parallel to the plane of light incidence, the MO contribution becomes noticeable in crossed polarizers (the exponentially decaying background field is eliminated) at the edges parallel to the incident light (compare Figures 7.13 and 7.14).

Note that the nonmagnetic contribution is several hundred times higher than that of the MO component.

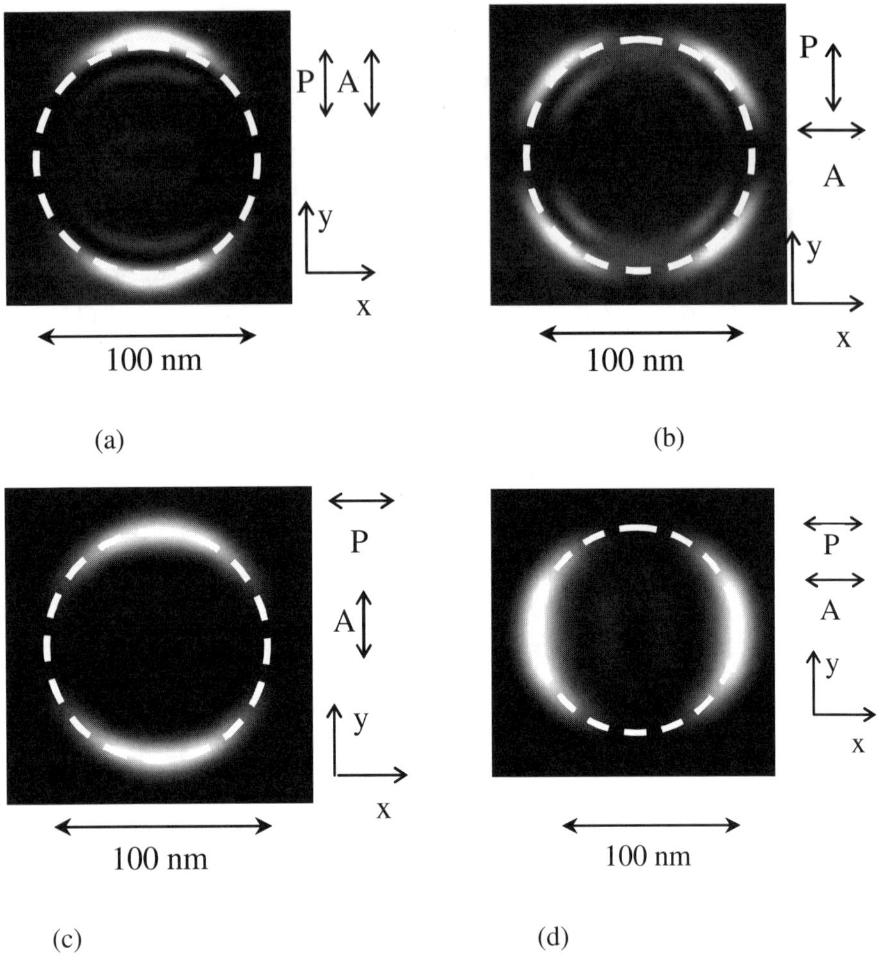

Figure 7.13. Simulated images of a nonmagnetic circle Ag particle (refractive index $n = 0.14 + i4.0$, diameter is 100 nm, thickness is 10 nm) under various illumination and observation conditions. The spatial resolution is 10 nm. A, analyzer; P, polarizer [46].

From the aforesaid, it follows that nonmagnetic inhomogeneities of nanoobjects (nanoparticle edges, lattice defects, etc.) make MO observations with SNOM difficult. Therefore, some special techniques are needed to extract the MO signal: modulation of the light polarization or working in the apertureless SNOM [68]. Another possibility is to detect the SNOM signal at the second optical harmonic that is generated on the surface of the sample [36, 40].

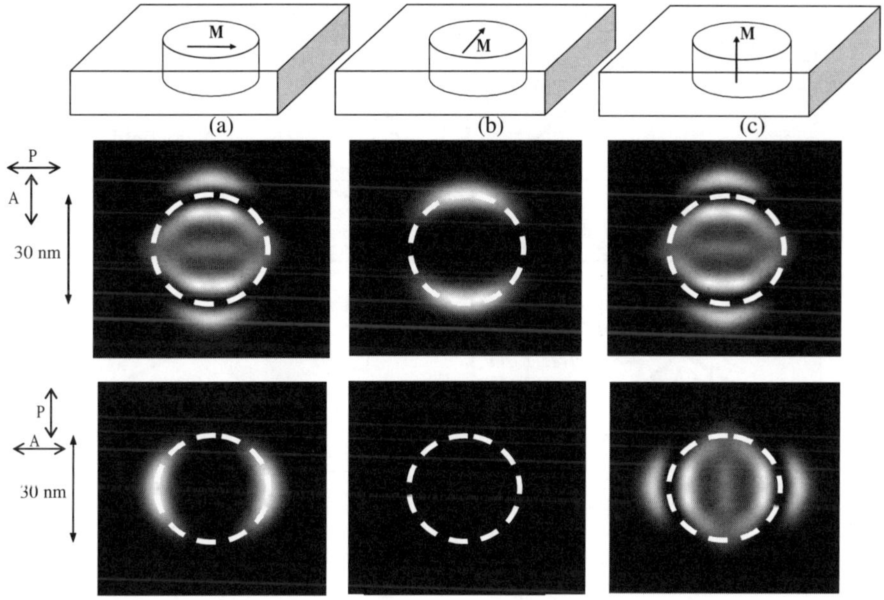

Figure 7.14. Simulated images of a 30-nm magnetic region embedded in a 10-nm-thick film under various illumination and observation conditions. The images of the nanoregion magnetized along the (**a**) *x*, (**b**) *y*, and (**c**) *z* axis are shown. The scan height is 11 nm, and the spatial resolution is 5 nm. A, analyzer; P, polarizer.

The other problem that arises during MO observation is that the SNOM resolution operating in transmission mode is limited not only by the size of the fiber-tip aperture but also by the thickness of the magnetic film [93–96].

The qualitative explanation of the decrease of resolution in SNOM gives Figure 7.15 [93]. Super-resolution beyond relay criterion is achievable by the presence of evanescent spatial harmonics in detected signal. With the increase of the sample's thickness, the contribution of these evanescent harmonics is reduced sharply: The contribution from the regions of the film that are located further than several hundred nanometers (for $\lambda \sim 0.5$ mkm) is almost entirely constituted by the far-field radiating components that are responsible for low-frequency space harmonics and low resolution. So we can conclude that a magnetic medium lying below this level does not contribute to the high resolution but only makes it worse. Consequently, the best results with transmission SNOM can be achieved only for thin (thickness is less than 300 nm).

Such a limitation for resolution does not exist for reflection measurements because this time only thin surface layers contribute to the image.

The other structure to be considered is a rectangular permalloy magnetic film of thickness 1 nm and size 200×500 nm. Simulation of the magnetization

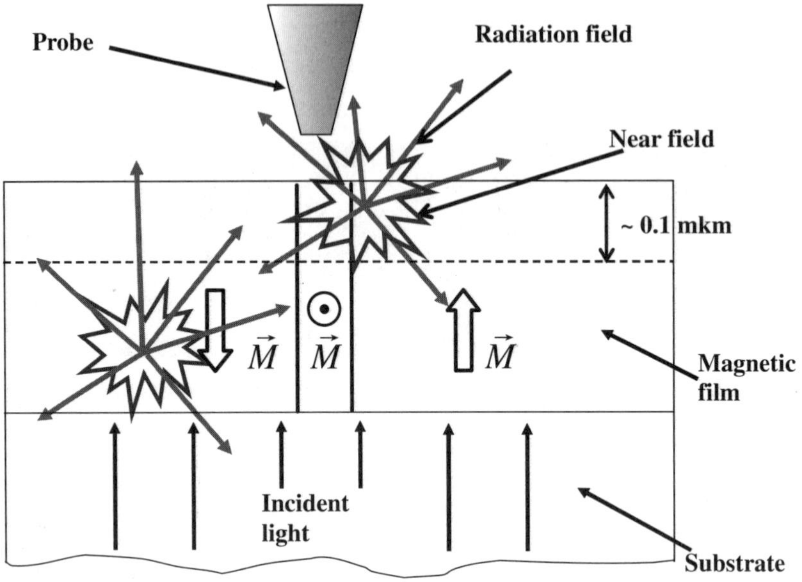

Figure 7.15. Radiation and nonradiation (near-field) fields arising in the magnetic film with the domain structure. Only near-field spatial harmonics radiated from the surface layer of several hundredths of a micrometer can reach the probe.

distribution in the sample with given geometrical and magnetic properties can be performed using the Landau–Lifshitz equation taking into account the full-scale magneto-dipole interaction [83] (Figure 7.16a).

In Figure 7.16b–e the calculated near-field images of the entire part of the sample for different orientation of cross-polarizers and plane of incidence are presented. Here again we consider the SNOM operating in the photon tunneling configuration. The brightest parts of the images correspond to the boundaries between the regions with uniform magnetization—that is, the domain walls. That is why it is possible to determine the shape and the size of magnetic nanometer-sized structures in the SNOM. For example, from Figure 7.16 one can estimate the width of the domain wall as 50 nm. This value is close to the "real" domain wall's width determined from the magnetization distribution in Figure 7.16a.

Finally we look at a SNOM image of the magnetic nanocontact region [83, 84, 97] (Figure 7.17a).

In this case the crystalline part of $\chi_{i,j}$ plays the main role in the image formation. This is because the banks of the nanocontact are not only magnetic edges but also crystalline ones. As we have already discussed, this brings about a very serious problem for the MO SNOM investigation of nanocontacts: In

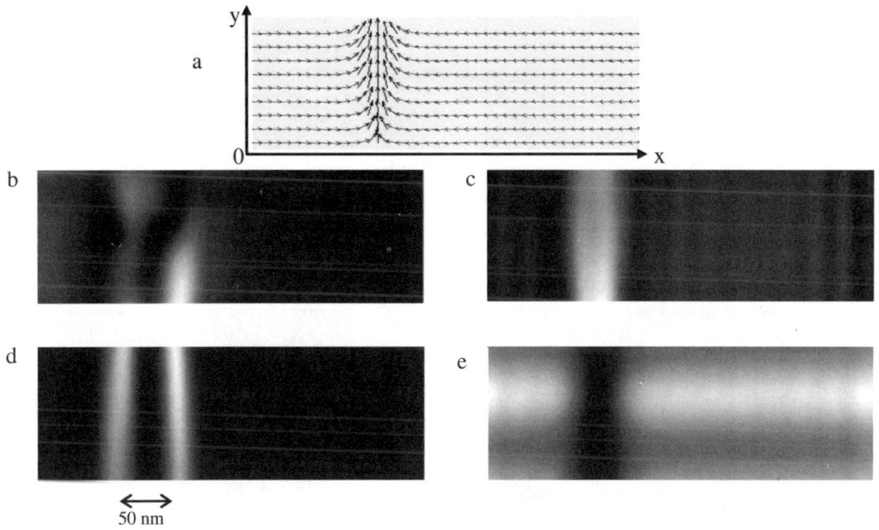

Figure 7.16. Magnetization distribution in rectangular permalloy magnetic film of thickness 1 nm (**a**) and simulated near-field images obtained in crossed polarizers for different orientations of incident polarization and plane of incidence: p-polarization, plane of light incidence is XZ plane (**b**); s-polarization, plane of light incidence is the *XZ* plane (**c**); p-polarization, plane of light incidence is *YZ* plane (**d**); s-polarization, plane of light incidence is the *YZ* plane. Height of scanning is 10 nm. Spatial resolution is about 15 nm. Intensities in (**c**) and (**e**) are two orders of magnitude smaller than that in (**b**) and (**d**).

near field the nonmagnetic rotation of polarization takes place on the crystalline edges of the structure, leading to extremely high intensities in the area of edges even during observation in cross-polarizers (Figure 7.17b). If crystalline contribution is subtracted from the image, then the magnetic structure becomes visible and one can observe the domain wall located at the center of the nanocontact. It is worth noting that the presence or absence of the domain wall influences substantially the transport properties of the nanocontact leading to the giant magnetoresistance effect [98].

10.2. Imaging in the ADFM

As we can see in Section 9, the ADFM technique is sensitive to the orientation of the nanoparticle magnetization vector and, being simple, may be useful and handy for studying the magnetization reversal. Combination of the ADFM technique with a high-speed photography setup presents a means for studying this

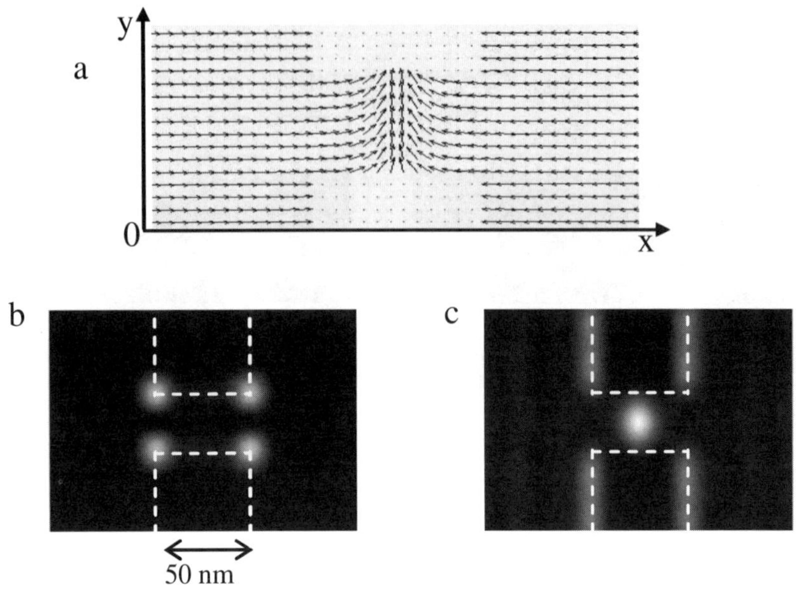

Figure 7.17. Magnetization distribution in magnetic nanocontact (**a**) and simulated near-field images obtained in crossed polarizers (s-polarization is incident, p-polarization is detected) for the case of light incidence parallel to the *XZ* plane: with (**b**) and without (**c**) taking into account the crystalline part of $\chi_{i,j}$ in equation (21). Height of scanning is 10 nm. Spatial resolution is about 15 nm.

process *in situ* with at least a nanosecond temporal resolution [99]—that is, during exposure to magnetic field pulses. We note that, according to theoretical and experimental estimations, the time of nanoparticle magnetization reversal is 10^{-12}–10^{-9} sec [100, 101].

The ADFM technique can also yield data on the magnetization state of an array of particles in the static case [21]. Indeed, by varying the azimuth angle of light incidence, one can construct the dependence of the array image intensity on the angle. The shape of the curves will depend on the scatter in the nanoparticle magnetization directions. Notwithstanding the fact that separate ADFM observation of nanoparticles is impossible at distances between particles shorter than $\lambda/2$, the change in the dark-field image intensity during the array magnetization reversal allows one to determine the average rate of remagnetization and to detect the presence of particles exhibiting dynamic properties differing from those of the majority of particles. Furthermore, the dependence of the image intensity on the azimuth angle of light incidence allows one to estimate the scatter in the particle magnetization directions in the static case.

References

1. K. O'Grady and H. Laidler, *J. Magn. Magn. Mater.* **200**, 616–633 (1999).
2. D. Weller, Plenary lecture at the Joint European Symposium on Magnetism, 28 August–1 September 2001, Grenoble, France (JEMS-2001).
3. L. D. Landau and E. M. Lifshitz, *Electrodynamics of Continuous Media*, Pergamon Press, New York, 1960.
4. P. Sheng, *Philos. Mag. B* **65**, 357–384 (1991).
5. W. T. Doyle and I. S. Jacobs, *Phys. Rev. B* **42**, 9319–9327 (1990).
6. F. Brouers, *J. Phys. C: Solid State Phys.* **19**, 7183–7193 (1986).
7. H. Ragossnig and A. Feltz, *J. Eur. Ceramic Soc.* **18**, 429–444 (1998).
8. M. Wu, H. Zhang, Xi Yao, and L. Zhang, *J. Phys. D: Appl. Phys.* **34**, 889–895 (2001).
9. L. Tsang, J. A. Kong, and R. T. Shin, *Theory of Microwave Remote Sensing*, John Wiley & Sons, New York, 1985, pp. 425–475.
10. Y. Rao, C. P. Wong, and J. Qu, *IEEE Trans. CPMT* **23**, 680 (2000).
11. D. Stroud, *Superlattices Microstruct.* **23**, 567 (1998).
12. H. C. Van de Hulst, *Light Scattering by Small Particles*, Dover, New York, 1969 or 1981.
13. M. Lax, *Rev. Mod. Phys.* **23**, 287 (1951).
14. M. Kerker, *The Scattering of Light and Other Electromagnetic Radiation*, Academic, New York, 1969.
15. J. C. Maxwell Garnett, *Philos. Trans. R. Soc. A* **203**, 385 (1904).
16. R. Ruppin. *Opt. Commun.* **182**, 273–279 (2000).
17. D. M. Wood and N. W. Ashcroft, *Phys. Rev. B* **25**(10), 6255 (1982).
18. H. Ragossnig and A. Feltz, *J. Eur. Ceram. Soc.* **18**, 429–444 (1998).
19. D. A. G. Bruggeman, I. *Ann. Phys. Leipzig* **24**, 636–664 (1935).
20. R. Pelster and U. Simon, *Colloid Polym. Sci.* **277**, 2 (1999).
21. V. I. Belotelov, A. K. Zvezdin, V. A. Kotov, et al., *Phys. Solid State* **45**, 1957 (2003).
22. N. F. Kubrakov, A. K. Zvezdin, K. A. Zvezdin, et al., *JETP* **87**, 600 (1998).
23. R. Atkinson, N. F. Kubrakov, A. K. Zvezdin, and K. A. Zvezdin. *J. Magn. Magn. Mater.* **156**, 169 (1996).
24. R. Q. Hood and L. M. Falikov. *Phys. Rev.* **B46**, 8287 (1992).
25. D. Bozec, V. G. Kravets, J. A. D. Matthew, and S. M. Thompson, *J. Appl. Phys.* **91**, 8795 (2002).
26. A. K. Zvezdin, V. A. Kotov, *Modern Magneto-Optics and Magneto-Optical Materials*, IOP Publishing, Philadelphia (1997), 363 pages.
27. R. V. Pisarev, *Physics of Magnetic Dielectrics*, Moscow, Nauka (1974).
28. B. E. Argile and E. Terrenzio, *J. Appl. Phys.* **55**, 2569–2579 (1984).

29. G. S. Krinchik, *Physics of Magnetic Phenomena*, Moscow State Universty Press, Moscow (1985).
30. G. S. Krinchik and E. E. Chepurova, *Proceedings ICM-73*, Moscow State University Press, Moscow (1973), p. 139.
31. J. Shoenes, *J. Magn. Soc. Jpn. Suppl.* **11**, 99 (1987).
32. P. Guyot-Sionnest and Y. R. Shen, *Phys. Rev. B* **35**, 4420 (1987).
33. U. Pustogowa, W. Hubner, and K. H. Benneman, *Phys. Rev. B* **48**, 8607 (1993).
34. Ru-Pi Pan, P. D. Wel, and Y. R. Shen, *Phys. Rev. B* **39**, 1229 (1989).
35. A. K. Zvezdin and N. F. Kubrakov, *JETP* **116**, 141 (1999).
36. V. I. Belotelov, A. P. Pyatakov, S. A. Eremin et al., *Physics of the Solid State* **42**, 1873 (2000).
37. U. Pustogowa, W. Hubner, and K. H. Benneman, *Phys. Rev. B* **49**, 1031 (1994).
38. O. A. Aktsipetrov, E. A. Gan'shina, V. S. Guschin et al., *J. Magn. Magn. Mater.* **196**, 80 (1999).
39. B. Koopmans, M. J. K. Koerkamp, and T. Rasing, *Phys. Rev. Lett.* **74**, 3692 (1995).
40. V. I. Belotelov, A. P. Pyatakov, G. G. Musaev et al., *Opt. Spectrosc.* **91**, 626 (2001).
41. O. A. Aktsipetrov, V. A. Aleshkevich, A. V. Melnikov et al., *J. Magn. Magn. Mater.* **165**, 421 (1997).
42. T. M. Crawford, C. T. Rogers, and T. J. Silva et al., *J. Appl. Phys.* **81**, 354 (1997).
43. A. A. Maradudin and D. L. Mills, *Phys. Rev. B* **11**, 1392 (1975).
44. D. L. Mills and A. A. Maradudin, *Phys. Rev. B* **12**, 2943 (1975).
45. V. A. Kosobukin, *Phys. Solid State* **35**, 884 (1993).
46. V. I. Belotelov, A. P. Pyatakov, A. K. Zvezdin, et al., *Tech. Phys.* **48**, 1–6 (2003).
47. O. J. F. Martin, C. Girard, and A. Dereux, *Phys. Rev. Lett.* **74**, 526 (1995).
48. V. A. Kosobukin, *JTP* **43**, 824 (1998).
49. M. Paulus, P. Gay-Balmaz, and O. J. F. Martin, *Phys. Rev. E* **62**, 5797 (2000).
50. Y. Martin and H. K. Wickramasinghe, *Appl. Phys. Lett.* **50**, 1455 (1987).
51. S. Hosaka, A. Kikukawa, and Y. Honda, *Appl. Phys. Lett.* **65**, 3407 (1994).
52. I. V. Yaminskioe and A. M. Tishin, *Usp. Khim.* **68**, 187 (1999).
53. D. Courjon and C. Bainier, *Rep. Prog. Phys.* **57**, 989 (1994).
54. D. W. Pohl, *Thin Solid Films* **264**, 250–254 (1995).
55. M. Ohtsu *Near Field Nano/Atom Optics and Technology*, Springer, Tokyo (1998), 297 pages.
56. G. S. Zhdanov, M. N. Libenson, and G. Marcinovskiy, *Phys. Uspekhi* **41**, 719 (1998).
57. E. Betzig and J. K. Trautman, *Science* **257**,189–195 (1992).
58. M. A. Paesler and P. J. Moyer, *Near-Field Optics Theory, Instrumentation and Applications*, John Wiley & Sons, New York (1996).

59. D. Courjon, J. M. Vigoureux, M. Spajer et al., *Appl. Opt.* **29**, 3734 (1990).
60. K. D. Weston, J. A. DeAro, and S. K. Buratto, *Rev. Sci. Inst.* **67**, 2924–2929 (1996).
61. E. Betzig, P. L. Finn, and J. S. Weiner, *Appl. Phys. Lett.* **60**, 2484–2486 (1992).
62. E. Betzig, J. K. Trautman, R. Wolfe et al., *Appl. Phys. Lett.* **61**, 142 (1992).
63. U. Hartmann, *J. Magn. Magn. Mater.* **157/158**, 545 (1996).
64. E. Betzig, J. K. Trautman, J. S. Weiner et al., *Appl. Opt.* **31**, 4563 (1992).
65. T. J. Silva, S. Schultz, *Rev. Sci. Instrum.* **67**, 715 (1996).
66. P. Fumagalli, A. Rosenberger, G. Eggers et al., *Appl. Phys. Lett.* **72**, 2803 (1998).
67. G. Eggers, A. Rosenberger, N. Held, et al., *Ultramicroscopy* **71**, 249 (1998).
68. H. Wioland, O. Bergossi, S. Hudlet, et al., *Eur. Phys. J. AP* **5**, 289 (1999).
69. T. Lacoste, T. Huser, and H. Heizelmann, *Z. Phys. B.* **104**, 183 (1997).
70. A. A. Ejov, A. S. Logginov, D. A. Muzychenko, A. V. Nikolaev, and V. I. Panov. *Poverkhn. Yavieniya Polim.* **11**, 56 (2000).
71. M. Nieto-Vesperinas and J. C. Dainty, *Scattering in Volumes and Surfaces*, Madrid (1990).
72. R. W. Ditchburn, *Light*, 2nd ed., Blackie, London (1963).
73. B. Kuhlow and M. Lambeck, *Physica B & C (Amsterdam)* **80**, 374 (1975).
74. A. Thiaville and F. Boileau, J. Miltat et al., *J. Appl. Phys.* **63**, 3153 (1988).
75. A. Thiaville and J. Miltat, *IEEE Trans. Magn.* **26**, 1530 (1990).
76. A. S. Logginov, A. V. Nikolaev, and V. V. Dobrovitski, *IEEE Trans. Magn.* **29**, 2590 (1993).
77. A. S. Logginov, A. V. Nikolaev, and V. N. Onishchuk, in *Proceedings of Conference on Physics of Condensed States*, Vol. 2, Sterlitamak. Gos. Pedagog. Institute, Sterlitamak (1997), p. 50.
78. V. I. Belotelov, A. S. Logginov, and A. V. Nikolaev, *Radiotekh. Élektron. (Moscow)* **46**, 870 (2001).
79. A. P. Malozemoff and J. C. Slonczewski, *Magnetic Domain Walls in Bubble Materials*, Academic, New York (1979).
80. A. Hubert and R. Schafer, *Magnetic Domains*, Springer, Berlin (1998).
81. T. Shinjo, T. Okuno, R. Hassdorf et al., *Science* **289**, 930 932 (2000)
82. J. Sasaki and F. Matsubara, *J. Phys. Soc. Jpn.* **66**, 2138 (1997).
83. K. A. Zvezdin, *Solid State Phys.* **42**, 1 (2000).
84. A. K. Zvezdin and K. A. Zvezdin. *JETP Lett.* **95**, 762–767 (2002).
85. D. A. Allwood, G. Xiong, M. D. Cooke et al., *J. Phys. D: Appl. Phys.* **36**, 2175–2182 (2003).
86. R. P. Cowburn, *Phys. Rev. B* **65**, 092409 (2002).
87. R. P. Cowburn, D. K. Koltsov, A. O. Adeyeye et al., *Phys. Rev. Lett.* **83**, 1042 (1999).
88. T. Aign, P. Meyer, S. Lemerle et al., *Phys. Rev. Lett.* **81**, 5656 (1998).

89. V. I. Belotelov, A. S. Logginov, and A. V. Nikolaev, *Physics of the Solid State*. **45**, 519 (2003).
90. D. P. E. Dickson, N. M. K. Reid, and C. A. Hunt, *J. Magn. Magn. Mater.* **125**, 345 (1994).
91. W. Wernsdorfer, E. Bonet Orozco, and K. Hasselbach et al., *Phys. Rev. Lett.* **78**, 1791 (1997).
92. J. P. Jamet, J. Ferre, P. Meyer et al., *IEEE Trans. Magn.* **37**, 2120 (2001).
93. V. I. Belotelov, A. S. Logginov, A. P. Pyatakov et al., *Technical Digest IQEC 2002*, Moscow, June, p. 368.
94. A. A. Ezhov, A. S. Logginov, D. A. Muzychenko, A. V. Nikolaev, and V. I. Panov, *Phys. Metals Metallogr.* (Supplementary Issue), **92**(1), 277 (2001).
95. P. Bertrand, L. Colin, V. I. Safarov. *J. Appl. Phys.* **83**, 6834 (1998).
96. A. A. Ejov, A. S. Logginov, D. A. Muzychenko, A. V. Nikolaev, and V. I. Panov, *Poverchn. Yavieniya Polim.* **11**, 56 (2000).
97. A. K. Zvezdin and A. F. Popkov, *JETP Lett.* **71**, 209 (2000).
98. N. Garsia, M. Munoz, and Y. W. Zhao, *Phys. Rev. Lett.* **82**, 2923 (1999).
99. L. P. Ivanov, A. S. Logginov, and G. A. Nepokooechitskiy, *Sov. Phys. JETP* **57**, 583 (1983).
100. D. P. E. Dickson, N. M. K. Reid, and C. A. Hunt, *J. Magn. Magn. Mater.* **125**, 345 (1994).
101. W. Wernsdorfer, E. Bonet Orozco, K. Hasselbach et al., *Phys. Rev. Lett.* **78**, 1791 (1997).

8

OPTICAL EXTINCTION OF METAL NANOPARTICLES SYNTHESIZED IN POLYMER BY ION IMPLANTATION

A. L. Stepanov

Kazan Physical–Technical Institute of the Russian Academy of Sciences, Kazan, Russian Federation; and Institute for Experimental Physics and Erwin Schrödinger Institute for Nanoscale Research, Karl-Franzens-University Graz, Graz, Austria

1. INTRODUCTION

The problem of designing new polymer-based composite materials containing metal nanoparticles (MNPs) is of current interest, particularly in the fabrication of magnetooptic data storages, picosecond optical switches, directional connectors, and so on. The nonlinear optical properties of these composites stem from the dependence of their refractive index on incident light intensity. This effect is associated with MNPs, which exhibit a high nonlinear susceptibility of the third order when exposed to ultrashort (picosecond or femtosecond) laser pulses [1].

Light-induced electron excitation in MNPs (so-called surface plasmon resonance, SPR) [2], which shows up most vividly in the range of linear absorption) gives rise to nonlinear optical effects in the same spectral range. Therefore, in practice, the SPR effect may be enhanced by raising the

Metal-Polymer Nanocomposites, Edited by Luigi Nicolais and Gianfranco Carotenuto
ISBN 0-471-47131-3

nanoparticle concentration in the composite—that is, by increasing the volume fraction of the metal phase (filling factor) in the insulator. Systems with a higher filling factor offer a higher nonlinear cubic susceptibility, when all other things being the same [1].

Metal particles may be embedded in a polymer matrix in a variety of ways. These are chemical synthesis in an organic solvent [2], vacuum deposition on viscous-flow polymers [3], plasma polymerization combined with metal evaporation [4], and so on. However, they all suffer from disadvantages, such as a low filling factor or a great spread in size and shape of particles synthesized, which offsets the good optical properties of composites.

One more promising method is ion implantation [5], which provides controllable synthesis of MNPs at various depths under the surface and unlimitedly high-impurity doses. Despite the intensive study of metal nanoparticle synthesis by ion implantation in dielectrics, such as nonorganic glasses and crystals, which was started in 1973 by Davenas et al. with Na and K ions [6] and in 1975 by Arnold with Au ions [7], the formation of nanoparticles in organic matrices was realized only at the beginning of the 1980 by Koon et al. in their experiments on implantation of Fe ions into polymers [8]. By implantation, one can produce almost any metal–insulator (specifically, metal–polymer) composites, as follows from Table 8.1, which gives a comprehensive list of references [9–40], MNP shapes, and implantation conditions for various organic matrices. Note that noble metals exhibit the most pronounced SPR effect and, hence, the highest nonlinearity of the MNP optical properties in insulators [2]. This chapter focuses on polymer-based optical materials containing metallic nanoparticles (Ag, Au, etc.) characterized by SPR. Only recently were such materials fabricated by Ag implantation into epoxy resins [38] and polymethylmethacrylate (PMMA) [18, 33–35]. However (see, e.g., references 33 and 34), an unusually weak plasmon resonance absorption was detected in the case of the Ag–PMMA composite despite the high concentration of the synthesised Ag nanoparticles. The aim of this work is therefore to study the SPR-related linear optical properties of MNPs introduced into a polymer matrix by implantation. We compare experimental optical spectra for silver nanoparticles implantation-synthesized in PMMA with model spectra calculated based on the Mie classical electrodynamic theory [2, 41].

2. EXPERIMENTAL

As substrates, 1.2-mm-thick PMMA plates, which are optically transparent in a wide spectral range (400–1000 nm), were used [42]. Substrates were implanted by 30-keV Ag^+ ions with doses in the range from 3.1×10^{15} to $7.5 \times$

TABLE 8.1. Conditions for MNP Synthesis by Ion Implantation into Polyvinylidene Fluoride (PVF_2), Polyimide (PI), Polymethylmethacrylate (PMMA), Polymethylmethacrylate with Phosphorus-Containing Fragments (PMMA + PF), Polyethylene (PE), Poly(ethyleneterephthalate) (PET), Silicone Polymer (Phenylmethyl-silane Resin with Tin Diethyldicaprilate), Epoxy Polycarbonate (PC), and Polyetherimide (PEI)[a]

Metal of Particles	Matrix	Ion Beam Energy (keV)	Ion Dose (ion/cm²)	Ion Current Density (μA/cm²)	Substrate Temperature (K)	Shape of Particles	Study Methods	Authors
Ti	PET	40	2.0×10^{17}	4.5			TEM XRD	Wu et al. 2001 [9]
Cr	PET	40	1.0×10^{17}	4.5			TEM XRD	Wu et al. 2000 [10]
Fe	PVF_2	25	$0.1–1.0 \times 10^{17}$	—	300		FMR	Koon et al. 1984 [8]
Fe	PM	100–150	$0.1–1.0 \times 10^{17}$	0.1–5	300	Spherical	TEM	Ogawa 1988 [11]
Fe	PM	40	$0.25–1.2 \times 10^{17}$	4, 8, 12	300	Spherical and their aggregates	TEM	Khaibullin R.I. et al. 2002 [12]
							FMR	Popok et al. 2002 [13]
Fe	PMMA	100–150	$0.1–1.0 \times 10^{17}$	0.1–5	300	Spherical	TEM	Ogawa 1988 [11]
Fe	PMMA	40	$0.1–6.0 \times 10^{17}$	1–10	300	Spherical and their aggregates	TEM FMR	Petukhov et al. 1993 [14] 1995 [15] 1996 [16] 2001 [17]
								Bazarov et al. 1995 [18]
Fe	PMMA–PF	40	$1.0–3.0 \times 10^{16}$	1–6	300	Spherical and their aggregates	TEM FMR	Petukhov et al. 2001 [17]

TABLE 8.1. Continued

Metal of Particles	Matrix	Ion Beam Energy (keV)	Ion Dose (ion/cm^2)	Ion Current Density (μA/cm^2)	Substrate Temperature (K)	Shape of Particles	Study Methods	Authors
Fe	PE	25	$0.1–1.0 \times 10^{17}$	—	300		FMR	Koon et al. 1984 [8]
Fe	PET	100–150	$0.1–1.0 \times 10^{17}$	0.1–5	300	Spherical	TEM	Ogawa 1988 [11]
Fe	PET	40	$1.0–3.0 \times 10^{16}$	1–6	300		TEM FMR	Petukhov et al. 2001 [17]
Fe	Silicone polymer	40	$0.3–1.8 \times 10^{17}$	4	300	Spherical and their aggregates	TEM	Khaibullin R.I. et al. 1998 [19] 1999 [20] 2000 [21] Rameev et al. 2000 [22]
Co	PM	100–150	$0.1–1.0 \times 10^{17}$	0.1–5	300	Spherical	TEM	Ogawa 1988 [11]
Co	PM	40	$0.25–1.2 \times 10^{17}$	4, 8, 12	300	Spherical and their aggregates	TEM	Popok et al. 2002 [13]
Co	Silicone polymer	40	$0.3–1.8 \times 10^{17}$	4	300	Spherical and their aggregates	TEM FMR	Khaibullin R.I. et al. 2000 [21] Rameev et al. 2000 [22]
Co	Epoxy	40	$0.3–2.5 \times 10^{17}$	2–8	300	Spherical, needle-like, drop-like, cubic-faced, and so on.	TEM	Stepanov et al. 1994 [23] 1995 [24] Abdulin et al. 1996 [25] 1996 [26] 1998 [27] Khaibullin R.I. et al. 1996 [28]

								Khaibullin I.B. et al. 1997 [29]
Cu	PM	150	$0.5–1.0 \times 10^{17}$	1–5	<360	Spherical	TEM	Yushida and Iwaki 1987 [30]
Cu	PM	80 100	5.0×10^{16}	0.1	<630	Spherical	TEM	Kobayashi et al. 2001 [31]
Cu	PMMA	40	$0.1–6.0 \times 10^{17}$	1–6	360	Spherical	TEM	Petukhov et al. 2001 [17]
Cu	PMMA–PF	40	$1.0–3.0 \times 10^{16}$	1–6	360	Spherical	TEM	Petukhov et al. 2001 [17]
Cu	PET	40	$0.5–2.0 \times 10^{17}$	4–5	360	Spherical	TEM AFM XRD	Wu et al. 2000 [10] 2001 [32]
Cu	PET	40	$1.0–3.0 \times 10^{16}$	1–6	360	Spherical	TEM	Petukhov et al. 2001 [17]
Zn	PM	150	5.0×10^{17}	1–5	<360	Spherical	TEM	Yoshida and Iwaki 1987 [30]
Pd	PM	100	$0.1–1.0 \times 10^{17}$	0.1	<630	Spherical	TEM	Kobayashi et al. 2001 [31]
Ag	PM	130	$0.1–5.0 \times 10^{17}$	1–3	<630	Spherical	TEM	Kobayashi et al. 2001 [31]
Ag	PMMA	30	$1.0–7.5 \times 10^{16}$	4	300	Spherical	TEM OS	Stepanov et al. 1994 [33] 2000 [34] 2002 [35] Bazarov et al. 1995 [18]

TABLE 8.1. Continued

Metal of Particles	Matrix	Ion Beam Energy (keV)	Ion Dose (ion/cm^2)	Ion Current Density (μA/cm^2)	Substrate Temperature (k)	Shape of Particles	Study Methods	Authors
Ag	PET	79	$0.5–2.0 \times 10^{17}$	4.5	—	Spherical	TEM	Wu et al. 2000 [10] 2001 [9] 2002 [36]
Ag	Silicone polymer	30	$0.6–1.8 \times 10^{17}$	4	300	Spherical and their aggregates	TEM OS	Khaibullin R.I. et al. 1998 [19] 1999 [20] Stepanov et al. 1998 [37]
Ag	Epoxy	30	$0.22–7.5 \times 10^{17}$	4	300	Spherical	TEM OS	Stepanov et al. 1995 [38] 1997 [39] Khaibullin I.B. et al. 1997 [29]
Pt	PC	106	1.0×10^{17}	—	—	Spherical	TEM	Rao et al. 1994 [40]
Pt	PEI	106	1.0×10^{17}	—	—	Spherical	TEM	Rao et al. 1994 [40]

[a] FMR, ferromagnetic resonance; TEM, transmission electron microscopy; AFM, atomic-force microscopy; OS, optical spectroscopy; and XRD, X-ray diffraction.

10^{16} ion/cm^2 at ion current density of 4 μA/cm^2 in a vacuum of 10^{-6} torr using an ILU-3 ion implanter. In a control experiment, Xe-ion implantation into PMMA at the same conditions was performed. Spectra of optical density were measured from 300 to 900 nm at room temperature in air using a dual-beam spectrophotometer Hitachi 330. All spectra were recorded in a standard differential mode in order to normalize substrate effects. The samples obtained were examined with transmittance electron microscopy (TEM) with a Tesla BM-500 microscope and with Rutherford backscattering (RBS) using a beam of 2-MeV $^4He^+$ ions with van de Graff accelerator.

Optical spectra of spherical MNPs embedded in various dielectric media were simulated in terms of the Mie electromagnetic theory [41], which allows one to estimate the extinction cross section σ_{ext} for a wave incident on a particle. This value is related to the light intensity loss ΔI_{ext} of an incident light beam I_0 that passes through a transparent particle-containing dielectric medium due to absorption σ_{abs} and elastic scattering σ_{sca}, where $\sigma_{ext} - \sigma_{abs} + \sigma_{sca}$. Following the Lambert–Beer law, it is obseve

$$\Delta I_{ext} = I_0\left(1 - e^{-\#\sigma_{ext}h}\right) \tag{1}$$

where h is the thickness of the optical layer and # is the density of nanoparticles in a sample. The extinction cross section is connected to the extinction constant γ as $\gamma = \#\sigma_{ext}$.

Experimental spectral dependencies of optical density (OD) are given by

$$OD = -\lg(I/I_0) = \gamma \cdot \lg(e) \cdot h \tag{2}$$

Hence, for samples with electromagnetically noninteracting nanoparticles, it is possible to put $OD \sim \sigma_{ext}$. Therefore, experimental OD spectra are compared with model spectral dependences that are expressed through σ_{ext} found from the Mie theory.

3. ION SYNTHESIS OF METAL NANOPARTICLES

Ion implantation is an effective tool for introducing single impurities into the surface layer to a depth of several micrometers [5]. The surface modification of the material depends on its properties, as well as on ion implantation parameters (ion type and energy, ion current density, target temperature, etc.). A critical implantation parameter is ion dose F_0, which determines the implant amount. Depending on the modification of an insulating target (polymers, inorganic glasses, ionic crystals, minerals, etc.), ion implantation may be conven-

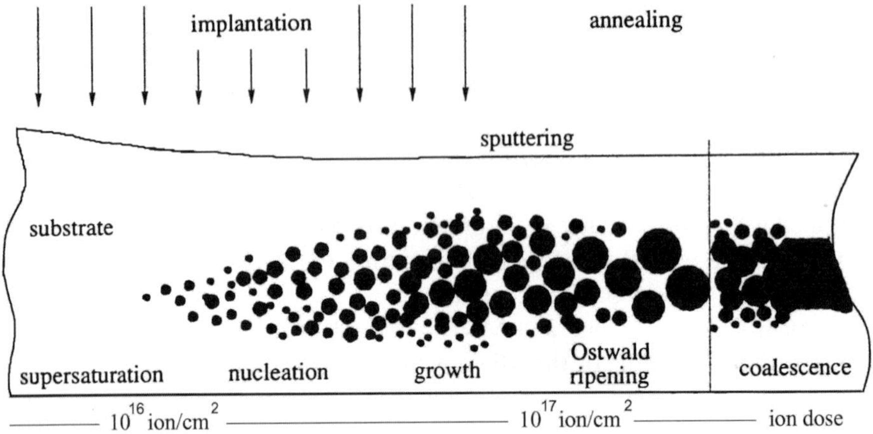

Figure 8.1. Basic physical stages of nanoparticle synthesis by ion implantation versus ion dose.

tionally divided into low-dose and high-dose implantation (Figure 8.1). In the former case ($F_0 \leq 5 \times 10^{14}$ ion/cm^2), the stopped ions are disperse (isolated from one another) in the insulating matrix. The energy of ions implanted is transferred to the matrix through the excitation of electronic shells (ionization) and nuclear collisions. This causes radiation-induced defects, which, in turn, may reversibly or irreversibly modify the material structure [5]. Various types of polymer structure damage have been observed [43]: breaking of covalent bonds in macromolecules, generation of free radicals, cross-linkage, oxidation and carbonization of irradiated layers, formation of new chemical bonds between atoms of the insulator or between ions implanted, and so on. In addition, ion implantation may be accompanied by the intense sputtering of the surface exposed [36, 44] or, sometimes, by the swelling of the polymer [45].

High-dose implantation may also be divided into dose (or time) stages (Figure 8.1). At F_0 between 10^{15} and 10^{16} ion/cm^2, the equilibrium solubility of metallic implants in insulators (in particular, polymers) is usually exceeded, causing the nucleation and growth of MNPs. The dose threshold value depends on the type of the insulator and implant. For 25 keV silver ions implanted into $LiNbO_3$, the threshold dose was found to be $F_0 \approx 5.0 \times 10^{15}$ ion/cm^2 [46]; for 30-keV silver ions implanted into epoxy resin, $F_0 \approx 10^{16}$ ion/cm^2 [38].

At the next stage of high-dose implantation, starting from $F_0 \geq 10^{17}$ ion/cm^2, the existing MNPs coalesce to form MNP aggregates or quasi-continuous films in the surface layer (Figure 8.1). For example, the irradiation of epoxy resin by 49-keV cobalt ions at higher-than-threshold doses favors the formation of thin labyrinth structures [25, 27]. The MNP distribution established in the insulator after coalescence or Ostwald ripening may be disturbed by postimplantation thermal or laser annealing.

In this work, composites where MNPs are disperse and isolated from one another—that is, synthesized at ion doses of 10^{15}–10^{17} ion/cm^2, were studied. In the present case of implantation by heavy Ag^+ ions at relatively low energy 30 keV, nuclear collisions prevail in ion–insulator interaction. They displace atoms in the polymer matrix and break some of chemical bonds in it. Along with this, target atoms effectively lose electrons and the implanted Ag^+ ions deionize with the formation of neutral silver atoms (Ag^0). In principle, Ag atoms may combine with arising organic radicals and polymer ions or take part in the oxidation reaction. However, because of a great difference in Gibbs free energy between Ag atoms and atoms of PMMA elements, Ag–Ag bonding is energetically more favorable.

The formation of MNPs proceeds in several stages: the accumulation of and subsequent supersaturation by Ag^0 atoms in a local surface region of the polymer, the formation of nuclei consisting of several atoms, and the growth of silver particles from the nuclei. Assuming that the nanoparticles nucleate and grow via the successive attachment of silver atoms (which are neutralized embedded Ag^+ ions), one may conclude that this process is governed simultaneously by the diffusion coefficient and local concentration of silver atoms—that is, depends on the matrix temperature. In this work, ion implantation was performed under identical conditions; specifically, the polymer during irradiation was kept at room temperature.

As follows from electron microscopy data, silver ion implantation under the experimental conditions considered will result in the formation of silver nanoparticles. For example, the cross-sectional micrograph in Figure 8.2 (a dose of 5.0×10^{16} ion/cm^2) shows dark spherical nanoparticles against the bright field (polymer). The irradiation of PMMA by xenon ions did not result in such patterns. Microdiffraction patterns demonstrate that the nanoparticles have the fcc structure of metallic silver. The patterns consist of thin rings (corresponding to polycrystalline nanoparticles) imposed on wide diffuse faint rings from the amorphous polymer matrix. By comparing the experimental diffraction patterns with standard ASTM data, it possible to conclude that implantation does not form any chemical compounds involving silver ions.

From RBS spectra (Figure 8.3), it is seen that the silver implantation layer (i.e., the depth where the nanoparticles are located) is almost independent of the ion dose. The dose dependence is observed only for the silver peak RBS intensities, with the widths and positions of the lines remaining unchanged. It is known that the implantation depth of an ion depends largely on its energy (accelerating voltage) [4], provided that the chemical constitution of the surface

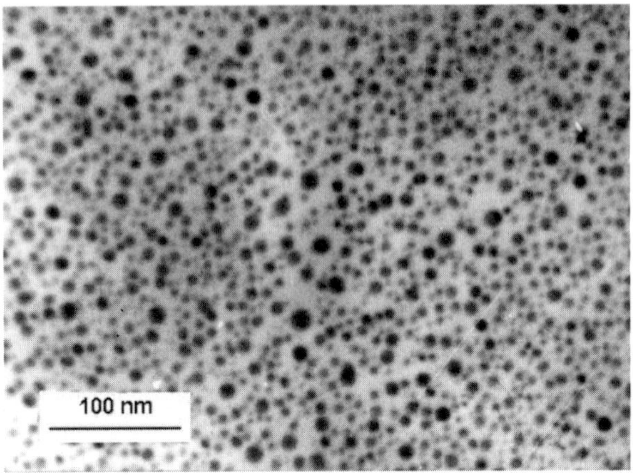

Figure 8.2. Micrograph of silver nanoparticles produced by Ag^+ implantation into PMMA at a dose of 5×10^{16} ion/cm^2.

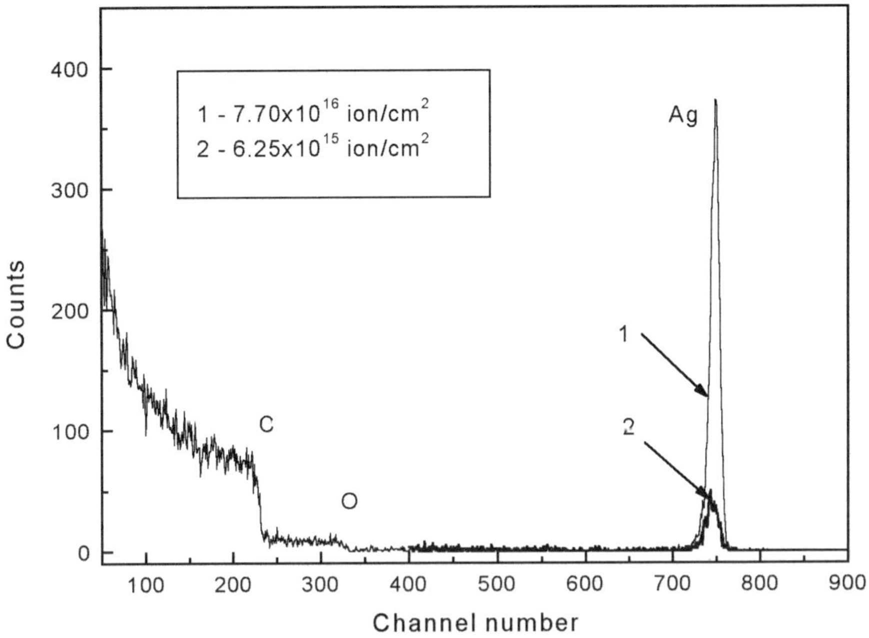

Figure 8.3. RBS spectra from PMMA irradiated by silver ions for doses of (1) 7.7×10^{16} and (2) 6.25×10^{15} ion/cm^2.

irradiated does not change dramatically [44]. The similarity of the RBS spectra shown in Figure 8.3 implies that the arising MNPs do not restrict the penetration depth of silver ions at the higher dose. Thus, in PMMA, the implantation dose, being responsible for the amount of the implant, influences directly the MNP size but does not affect the implant distribution profile at present conditions. As was noted above, the particles nucleate at a dose of ~10^{16} ion/cm^2 (low-dose implantation). For silver in PMMA, this dose provides MNPs with a size of about 2 nm. However, at a dose of 5.0×10^{16} ion/cm^2 (Figure 8.2), the particles grow to 10 nm.

Experimental optical absorption spectra for PMMA irradiated by xenon and silver ions at various doses are shown in Figure 8.4. It is evident that the xenon

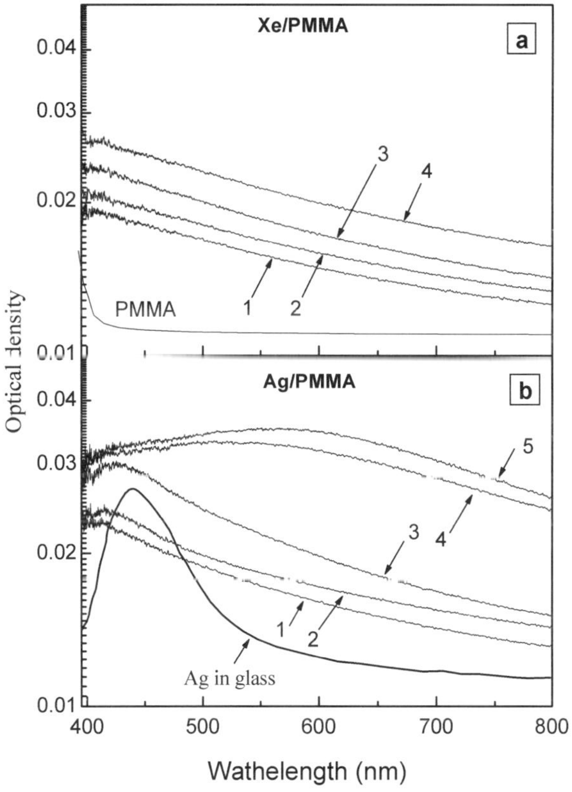

Figure 8.4. Optical density spectra from PMMA irradiated by (a) xenon and (b) silver ions for doses of (1) 0.3×10^{16}, (2) 0.6×10^{16}, (3) 2.5×10^{16}, (4) 5.0×10^{16}, and (5) 7.5×10^{16} ion/cm^2. The spectrum taken from silica glass irradiated by silver ions (5.0×10^{16} ion/cm^2) [44] is shown for comparison.

irradiation of PMMA does not produce nanoparticles, as also follows from the micrographs (Figure 8.2). In Figure 8.4a, as the xenon ion dose increases, the absorption of the polymer in the visible (especially in the close-to-UV) range also increases monotonically. This indicates the presence of radiation-induced structure defects in the PMMA. The absence of absorption bands in these spectral curves is noteworthy. The implantation of silver not only generates radiation-induced defects but also causes the nucleation and growth of MNPs. Therefore, along with the absorption intensity variation as in Figure 8.4a, an absorption band associated with silver nanoparticles is observed (Figure 8.4b). For the lowest dose, the peak of this band is near 420 nm and shifts to longer waves (up to 600 nm) with dose, with the band broadening significantly. The peak of this band is not high, although it is related to the SPR effect in the silver nanoparticles. The very low intensity of SPR absorption is atypical of silver nanoparticles in PMMA and cannot be explained by the polymer environment of the particles. When silver particles were synthesized in PMMA by the convection technique [47], the SPR intensity was very intense, unlike our experiment. Nor can the weak SPR absorption be explained by any features of the implantation process. For comparison, Figure 8.4 shows the optical density spectrum for inorganic silica glass (SiO_2) irradiated by silver ions under the conditions similar to the ion synthesis conditions used in this work (silica glass has the refractive index $n \approx 1.5$ close to that of PMMA) [44]. It is seen that the absorption of Ag nanoparticles in the glass (Figure 8.4b) is much more intense (even in view of the background absorption due to matrix structure imperfections) than the absorption of the particles in the polymer. Note that the particle size distributions in the glass and PMMA are nearly the same. Below, the optical properties of the Ag–PMMA composite are simulated, and we discuss various effects that may clarify the SPR absorption of Ag nanoparticles synthesized in PMMA by ion implantation.

4. SIMULATION OF OPTICAL EXTINCTION

4.1. Extinction of Silver Particles in Dependence of Surrounding Matrix

The attenuation (extinction) of an optical wave propagating in a medium with MNPs depends on the amount of the SPR effect and the light scattering efficiency. The wavelength of optical radiation, the particle size, and the properties of the environment are governing factors in this process. Within the framework of classical electrodynamics (the Maxwell equations), the problem of interaction between a plane electromagnetic wave and a single spherical particle was exactly solved in terms of optical constants of the interacting objects by Mie [2, 41]. According to the Mie theory, the extinction and scattering cross sections are expressed as an infinite sum of spherically symmetric partial electric and magnetic waves that generate fields similar to those generated by the particle when it is viewed as an excited multipole. In the general case, the Mie electromagnetic theory imposes no limitations on the wavelength of optical radiation. Therefore, the operation on the optical constants of the particles and matrix results in extinction spectra, so-called Mie resonance bands [2], which agree well with experiment. However, the Mie theory, which relies on the spectral dependence of the optical constants, does not allow one to penetrate deep into the physics of Mie optical peaks exhibited by the particles. Yet, independent investigations [2] into the behavior of silver nanoparticles showed that Mie resonances are due to the SPR effect, so that analytical Mie spectra may be compared with experimental data.

Simulated extinction spectra for Ag nanoparticles embedded in a polymer matrix to compare with experimental data shown in Figure 8.4. In theoretical calculations, we used the complex value of the optical constant ε_{Ag} in the visible range [48] that was obtained by measurements on a set of fine silver particles. Such an approach [48] takes into account limitations imposed on the electron free path in particles of different size and electron scattering at the particle–insulator interface [49] and thus yields a more exact value of ε_{Ag} than does the procedure of correcting optical constants for bulk silver [50]. The complex value of ε_{PMMA} for the polymer matrix was found elsewhere [42]. The extinction was calculated for particles of size between 1 and 10 nm (according to the MNP sizes in Figure 8.2).

At the early stage of simulation, consider the simple case where Ag nanoparticles are incorporated into the PMMA matrix. Associated extinction spectra for different metal particle sizes are shown in Figure 8.5. These spectra feature a wide band, which covers the entire spectral range. In the given range of particle sizes, the position of the SPR absorption maximum (near 440 nm) is almost independent of the particle size. However, the extinction band intensity grows while the band itself somewhat narrows with increasing particle size. Comparing the analytical and experimental spectra, it is seen that, to the greatest extent, Figure 8.5 refers to the situation where PMMA is irradiated by silver ions with doses between 0.33×10^{16} and 2.5×10^{16} ion/cm^2 (Figure 8.4b, curves 1–3). This dose range corresponds to the early stage of MNP nucleation and growth in the *OD* spectral band with a maximum between 420 and 440 nm. Thus, one may conclude that ion implantation in this dose range results in the formation of Ag nanoparticles, as also revealed microscopically. It may be supposed that radiation-induced defects in the PMMA have an insignificant effect on the MNP optical properties in this case. However, at higher implantation doses, the recorded *OD* spectra and the analytical spectra shown in Figure 8.5 diverge; hence, the structure of the metal–polymer composite should be considered.

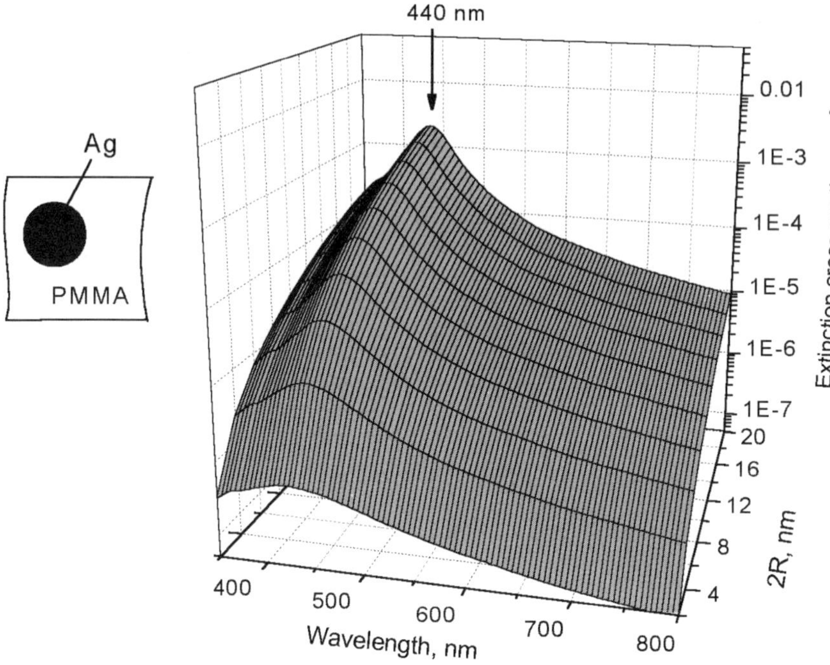

Figure 8.5. Analytical optical extinction spectra for silver nanoparticles embedded in PMMA versus particle size.

To explain the experimental dependences corresponding to high-dose silver implantation into PMMA, we will first elucidate the difference between implantation into polymers and inorganic insulators (silicate glasses, single crystals, minerals, etc.). The most important distinction is that as the absorbed dose grows, so does the number of dangling chemical bonds along the track of an ion. Because of this, gaseous hydrogen, low-molecular hydrocarbons (e.g., acetylene), CO, and CO_2 evolve from the matrix [43]. In particular, ion-irradiated PMMA loses HCOOCH3 methoxy groups [51]. The evolution of several organic fractions leads to the accumulation of carbon in the polymer layer irradiated, and radiation-induced chemical processes may cause chain linking. Eventually, an amorphous hydrogenated carbon layer is produced. Polymer carbonization starts with the formation of polycyclic compounds (in essence, primary carboniferous clusters) and, at higher doses, ends up with the formation of the well-developed carbonized phase via carbon cluster linking.

In view of the specific phase structure of the polymer irradiated, it is of interest to analyze the optical properties (extinction) of Ag nanoparticles embedded in the amorphous carbon matrix (C matrix). For this system, the extinction cross-section spectra versus particle size dependence (Figure 8.6)

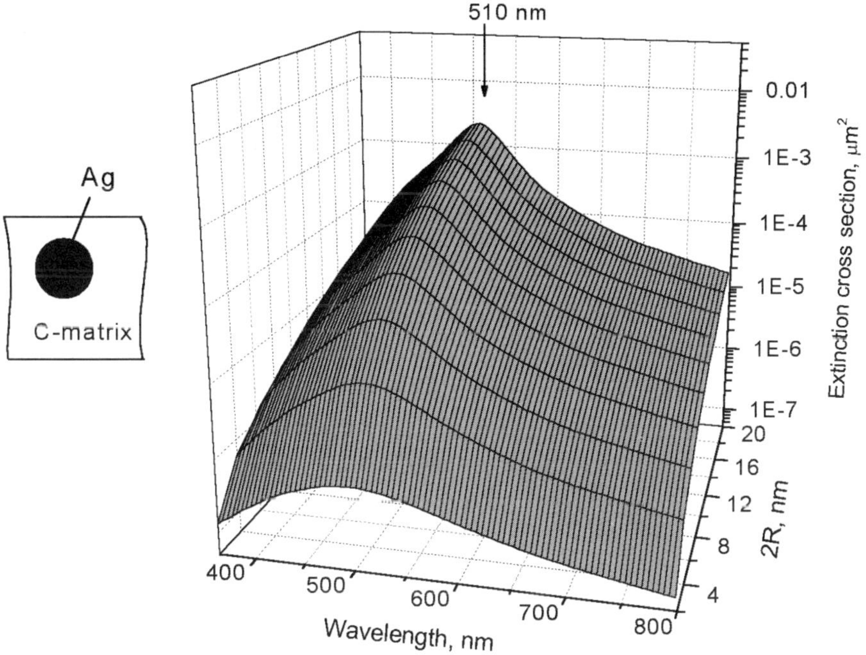

Figure 8.6. Analytical optical extinction spectra for silver nanoparticles embedded in the C matrix versus particle size.

was simulated in the same way as for the MNP–PMMA system—that is, by using complex optical constants ε_C for amorphous carbon, which were taken from reference 52. As before (Figure 8.5), throughout the particle size interval, the extinction spectra exhibit a single broad band, which covers the visible range, with a peak at longer waves (510 nm). The calculated longwave shift of the peak, which is observed upon changing the matrix, may be assigned to a longer wave *OD* band in the experimental spectra for the PMMA, which arises when the Ag ion dose exceeds 2.5×10^{16} ion/cm^2 (Figure 8.4b; curves 3 and 4). It seems that this spectral shift may be associated with the fact that the pure polymeric environment of the Ag nanoparticles turns into the amorphous carbon as the implantation dose rises. The broader extinction bands in the C matrix (Figure 8.6) compared with the PMMA (Figure 8.5) also count in favor of this supposition, since the broadening of the extinction bands is observed in the experiments as well (Figure 8.4b). In a number of experiments, however, the carbonization of the polymer surface layer depended on the type of the polymer and ion, as well as on the process parameters, and completed at doses of $(0.5–5.0) \times 10^{16}$ ion/cm^2 but the entire material was not carbonized. The carbon clusters may reach several tens of nanometers in size [43]. Thus, the assumption that the polymer irradiated is completely carbonized, which was used in the simulation (Figure 8.6), does not become a reality when the process lasts for a long time. Below, the variation of the extinction spectra with amount of carbon in the PMMA layer is analyzed in terms of a model that considers the optical properties of silver MNPs covered by the amorphous carbon sheath.

4.2. Extinction of Silver Particles with Carbon Shell

Extinction spectra for nanoparticles represented as a silver core covered by a carbon sheath in an insulating matrix (PMMA) will be analyzed in terms of the Mie relationships for sheathed cores [53, 54]. Here, an additional interface for which electrodynamic boundary conditions must be set up arises. Plasmon–polariton modes may be excited in both the core and the sheath. These modes, interacting through the inner interface, are responsible for the resulting extinction spectrum.

Optical extinction spectra for a Ag nanoparticle with a fixed size of the core (4 nm) and a varying thickness of the carbon sheath (from 0 to 5 nm) are shown in Figure 8.7. The maximum of the SPR bands of the particles is seen to shift

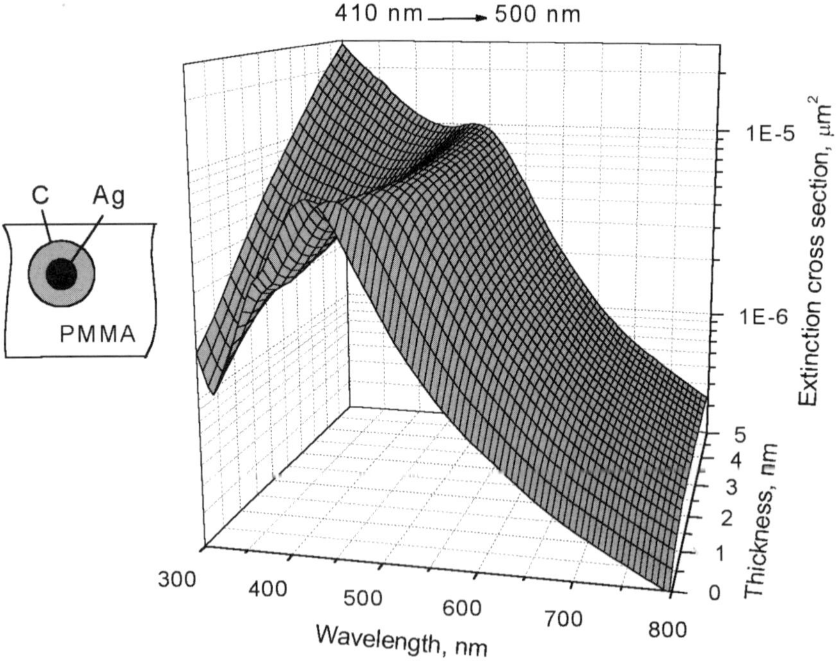

Figure 8.7. Analytical optical extinction spectra for 4-nm silver nanoparticles with the carbon sheath that are placed in the PMMA matrix versus sheath thickness.

from 410 nm (uncovered particle, Figure 8.5) to approximately 510 nm. Simultaneously, the SPR band intensity decreases, while the UV absorption increases, so that the absorption intensity at 300 nm and a sheath thickness of 5 nm exceeds the SPR absorption of the particles. Both effects (namely, the shift of the SPR band to longer waves and the increased absorption in the near ultraviolet) agree qualitatively with the variation of the experimental optical density spectra (Figure 8.4b) when the implantation dose exceeds 2.5×10^{16} ion/cm^2. Thus, our assumption that the increase in the carbonized phase fraction with implantation dose and the variation of the optical density spectra (Figure 8.4b) go in parallel is sustained by the simulation of the optical extinction for complex particles (sheathed cores, Figure 8.7).

In spite of the fact that the model dependences on the carbon sheath thickness and the experimental dose dependences agree qualitatively, discrepancies still exist, particularly in the position of the long-wave maximum in the optical density spectra and in the breadths of the simulated and experimental spectra. Possible reasons for such quantitative discrepancies are discussed below.

4.3. Effects Arising at the Silver Core–Carbon Sheath Interface

Interest in carbon-based composites with MNPs goes back a long way. Examples are the studies of magnetic properties of cobalt particles [55], electric and optical properties of layers with copper [56] or silver [57, 58] nanoparticles, and so on. It was found in optical absorption experiments that copper and silver nanoparticles [56, 58] dispersed in carbon matrices exhibit a weak SPR effect as in our work (Figures 8.4b, 8.6, and 8.7).

When analyzing the optical properties of nanoparticles embedded in a medium, one should take into account effects arising at the particle–matrix interface, such as the static and dynamic redistributions of charges between electronic states in the particles and the environment in view of their chemical constitution [59].

Consider first the charge static redistribution. When an atom is deposited (adsorbed) on the MNP surface, the energy levels of this atom ε_a change their positions compared with those in the free state [59, 60] (Figure 8.8).When the number of the adsorbed matrix atoms becomes significant, their contact generates a wide distribution of density of states. Most frequently, the adsorbed atoms are separated from surface atoms of the metal by a tunnel barrier. The gap between the energy positions ε_a of the adsorbed atoms and the Fermi level ε_F of the particles depends on the type of the adsorbate (Figure 8.8). In addition, the overlap between the energy positions of the matrix atoms and the energy positions of the silver surface atoms depends on the rate with which the electrons tunnel through the barrier. Accordingly, the conduction electron density in the particles embedded will change compared with that in the particles placed in a vacuum (without adsorbates): It decreases if the electrons tunnel toward the adsorbed atoms, or it increases when the electrons tunnel in the reverse direction. Eventually, equilibrium between the particle and the matrix sets in; that is, a constant electrical charge (Coulomb barrier) forms at the nanoparticle surface.

Such a charge static redistribution due to the deposition of an adsorbate on the particle surface and the respective change in the electron concentration in the MNPs were also observed in the SPR absorption spectra [2, 59]. In metals (silver, sodium, aluminum, etc.), where free conduction electrons dominate, the SPR spectral maximum $h\omega_{max}$ depends on the concentrations of electrons, N, in nanoparticles as

$$h\omega_{\max} \approx \left[N/(\varepsilon_0 m_{\mathrm{eff}})\right]^{1/2}\left[2\varepsilon_m + 1 + \chi_1^{\mathrm{inter}}\right]^{1/2} \tag{3}$$

where ε_m is the permittivity of the matrix, specifying the contribution of the real part of the susceptibility of interband optical transitions in a metal, and m_{eff} is the effective mass of an electron.

It was shown [61] that the incorporation of Ag nanoparticles into the carbon matrix of C_{60} fullerene (or the deposition of carbon on the nanoparticle surface) reduces the concentration of 5*sp* electrons in the particle roughly by 20%, since they are trapped by matrix molecules. According to (equation 3), the decrease in N is bound to shift the MNP extinction spectrum toward longer waves, as also demonstrated by comparing the experimental spectra of the particles in free space (without an adsorbate) with those of the particles in the C_{60} matrix [61]. Samples studied in reference 61 were similar to those obtained by ion implantation in our work (a carbonized layer near silver particles implanted into the polymer). Thus, the shift of the SPR extinction band into the longer wave range with implantation dose in this experiment (Figure 8.4) may be explained by the formation of a carbon sheath around silver nanoparticles. This sheath traps conduction electrons of the particles. The simulation (Figure 8.7) also demonstrates the shift of the SPR maximum. However, the effect of charge dynamic redistribution is disregarded in the Mie theory. Therefore, the long-wave shift of the SPR band due to the charge static redistribution at the particle–matrix interface is an additional reason why the experimental spectra are observed at longer waves than the model ones (Figures 8.6 and 8.7).

Along with the charge static redistribution at the interface, the charge at the same interface may also change dynamically—that is, with a high rate [59]. After the static state of the charge has been established and the Fermi level at the interface has been stabilized, the MNP electrons optically excited above the Fermi level (hot electrons) may tunnel (by fluctuations) to the matrix over or through the static barrier (Figure 8.8). Levels occupied by the electrons in the

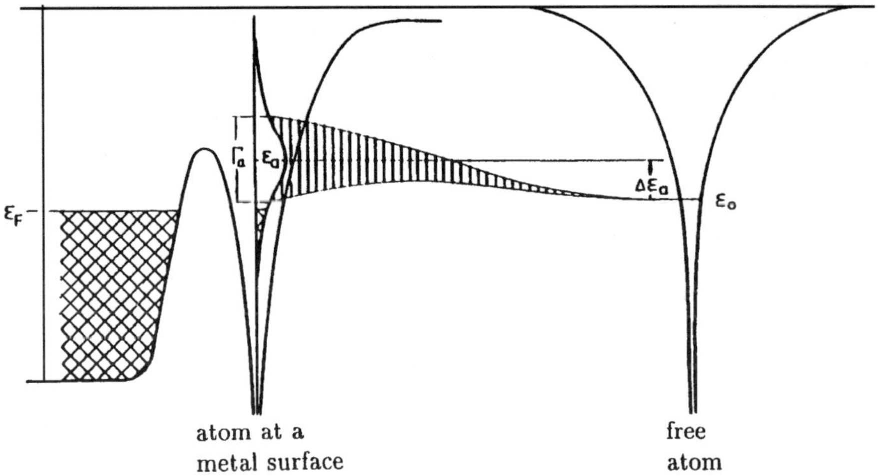

Figure 8.8. Electron energy levels in an atom adsorbed on a metal surface [60]. A free atom (*right*) reaches the surface (*left*). Γ_a is the spread of energy levels ε_a. Electron levels in the conduction band of the metal are occupied up to the Fermi level ε_F.

intermediate (between the particle and the matrix) state depend on the chemical constitution of the materials. Within a residence lifetime, the electrons may tunnel again from the acceptor levels of the matrix to the particle, and this process may occur over and over.

The charge dynamic variation in time at the particle–matrix interface causes the electron concentration in the particle to fluctuate. Fluctuation influences directly the SPR relaxation. The lifetime of excited conduction electrons in the particle defines the SPR spectral width. Here, the contribution from electron scattering by the interface (because of restrictions imposed on the electron free path [4]) adds up with the charge dynamic variation at the interface. Thus, the temporal capture of conduction electrons from the particle broadens the SPR-related extinction spectra. This was demonstrated with a set of silver nanoparticles embedded in the C_{60} matrix [61]. Silver nanoparticles in the carbon matrix exhibit the much broader SPR band than in free space. We may therefore suppose that as the dose rises, the charge dynamic redistribution may broaden the SPR spectra of silver nanoparticles synthesized by ion implantation in PMMA. This is because implantation carbonizes the irradiated layer with increasing absorbed dose and raises the amount of acceptor levels on the MNP surface, which change the relaxation time of electrons excited. Since the classical Mie theory disregards the charge dynamic redistribution, the model spectra (Figure 8.7) must be narrower than the experimental spectra, which is the case.

5. SUMMARY

In this chapter, we studied the formation of silver nanoparticles in PMMA by ion implantation and optical density spectra associated with the SPR effect in the particles. Ion implantation into polymers carbonizes the surface layer irradiated. Based on the Mie classical electrodynamic theory, optical extinction spectra for silver nanoparticles in the polymeric or carbon environment, as well as for sheathed particles (silver core + carbon sheath) placed in PMMA, as a function of the implantation dose are simulated. The analytical and experimental spectra are in qualitative agreement. At low doses, simple monatomic silver particles are produced; at higher doses, sheathed particles appear. The quantitative discrepancy between the experimental spectra and analytical spectra obtained in terms of the Mie theory is explained by the fact that the Mie theory disregards the charge static and dynamic redistributions at the particle–matrix interface. The influence of the charge redistribution on the experimental optical spectra taken from the silver–polymer composite at high doses, which cause the carbonization of the irradiated polymer, is discussed. Table 8.1, which summarizes available data for ion synthesis of MNPs in a polymeric matrix, and the references cited therein may be helpful in practice.

ACKNOWLEDGMENTS

I wish to thank the Alexander Humboldt Foundation for the financial support of the investigations made in Germany; the Austrian Scientific Foundation under the auspices of the Lisa Meitner program; and the State Foundation in Support of Leading Scientific Schools of the Russian Federation (grant no. SS 1904.2003.2). My special thanks go to S. N. Abdullin and V. I. Nuzhdin for the assistance in carrying out an implantation and electron microscopy study, and I also thank D. E. Hole (University of Sussex, United Kingdom) for taking RBS spectra.

REFERENCES

1. C. Flytzanis, F. Hache, M. C. Klein, D. Ricard, and P. Rousignol, *Nonlinear Optics in Composite Materials*, Elsevier Science, Amsterdam (1991).
2. U. Kreibig and M. Vollmer, *Optical Properties of Metal Clusters*, Springer, Berlin (1995).
3. S. N. Abdullin, A. L. Stepanov, Yu. N. Osin, and I. B. Khaibullin, *Surf. Sci.* **395**, L242 (1998).
4. M. Quinten, A. Heilmann, and A. Kiesow, *Appl. Phys. B* **68**, 707 (1999).

5. P. T. Townsend, P. J. Chandler, and L. Zhang, *Optical Effects of Ion Implantation*, Cambridge University Press, Cambridge (1994).
6. J. Davenas, A. Perez, P. Thevenard, and C. H. S. Dupuy, *Phys. Stat. Sol. A* **19**, 679 (1973).
7. G. W. Arnold, *J. Appl. Phys.* **46**, 4466 (1975).
8. N. C. Koon, D. Weber, P. Pehrsson, and A. I. Sindler, *Mater. Res. Soc. Symp. Proc.* **27**, 445 (1984).
9. Y. Wu, T. Zhang, Y. Zhang, H. Zhang, H. Zhang, and G. Zhou, *Nucl. Instr. Meth. B* **173**, 292 (2001).
10. Y. Wu, T. Zhang, H. Zhang, X. Zhang, Z. Deng, and G. Zhou, *Nucl. Instr. Meth. B* **169**, 89 (2000).
11. K. Ogawa, U.S. Patent 4,751,100 (1988).
12. R. I. Khaibullin, V. N. Popok, V. V. Bazarov, E. P. Zheglov, B. Z. Rameev, C. Okay, L. R. Tagirov, and B. Aktas, *Nucl. Instr. Meth. B* **191**, 810 (2002).
13. V. N. Popok, R. I. Khaibullin, V. V. Bazarov, V. F. Valeev, V. Hnatowicz, A. Mackova, and V. B. Odzhaev, *Nucl. Instr. Meth. B* **191**, 695 (2002).
14. V. Yu. Petukhov, V. A. Zhikharev, N. R. Khabibullina, and I. B. Khaibullin, *Vysokochist. Veshchestva* **3**, 45 (1993) [Russian].
15. V. Yu. Petukhov, V. A. Zhikharev, I. F. Makovskii, Yu. N. Osin, M. A. Mitryaikina, I. B. Khaibullin, and S. N. Abdullin, *Poverhnost* **4**, 27 (1995) [Russian].
16. V. Petukhov, V. Zhikharev, M. Ibragimova, E. Zheglov, V. Bazarov, and I. Khaibullin, *Solid State Commun.* **97**, 361 (1996).
17. V. Yu. Petukhov, M. I. Ibragimova, N. P. Khabibullina, S. V. Shulyndin, Yu. N. Osin, E. P. Zheglov, T. A. Vakhonina, and I. B. Khaibullin, *Polym. Sci. A* **43**, 1154 (2001). Translated from *Vysokomol. Soedin. A* **43**, 1973 (2001) [Russian]
18. V. V. Bazarov, V. Yu. Petukhov, V. A. Zhikharev, and I. B. Khaibullin, *Mater. Res. Soc. Symp. Proc.* **388**, 417 (1995).
19. R. I. Khaibullin, Yu. N. Osin, A. L. Stepanov, and I. B. Khaibullin, *Vacuum* **51**, 289 (1998).
20. R. I. Khaibullin, Yu. N. Osin, A. L. Stepanov, and I. B. Khaibullin, *Nucl. Instr. Meth. B* **148**, 1023 (1999).
21. R. I. Khaibullin, V. A. Zhikharev, Yu. N. Osin, E. P. Zheglov, I. B. Khaibullin, B. Z. Rameev, and B. Aktas, *Nucl. Instr. Meth. B* **166–167**, 897 (2000).
22. B. Z. Rameev, B. Aktas, R. I. Khaibullin, V. A. Zhikharev, Yu. N. Osin, and I. B. Khaibullin, *Vacuum* **58**, 551 (2000).
23. A. L. Stepanov, R. I. Khaibullin, S. N. Abdullin, Yu. N. Osin, and I. B. Khaibullin, *Mater. Res. Soc. Symp. Proc.* **343**, 161 (1994).
24. A. L. Stepanov, R. I. Khaibullin, S. N. Abdullin, Yu. N. Osin, V. F. Valeev, and I. B. Khaibullin, *Pros. Inst. Phys. Conf. Ser.* **147**, 357 (1995).
25. S. N. Abdullin, A. L. Stepanov, R. I. Khaibullin, F. V. Valeev, Yu. N. Osin, and I. B. Khaibullin, *Phys. Solid State* **38**, 1412 (1996) Translated from *Fiz. Tverdogo Tela* **38**, 2574 (1996) [Russian].

26. S. N. Abdullin, A. L. Stepanov, R. I. Khaibullin, and I. B. Khaibullin, Russian Federation Patent 2096835 (1996).
27. S. N. Abdullin, A. L. Stepanov, Yu. N. Osin, R. I. Khaibullin, and I. B. Khaibullin, *Surf. Coat. Techn.* **106**, 214 (1998).
28. R. I. Khaibullin, C. N. Abdullin, A. L. Stepanov, Yu. N. Osin, and I. B. Khaibullin, *Tech. Phys. Lett.* **22**, 112 (1996). Translated from *Pisma Zh. Techni. Fiz.* **22**, 48 (1996) [Russian].
29. I. B. Khaibullin, R. I. Khaibullin, S. N. Abdullin, A. L. Stepanov, Yu. N. Osin, V. V. Bazarov, and S. P. Kurzin, *Nucl. Instr. Methods B* **127–128**, 685 (1997).
30. K. Yoshida and M. Iwaki, *Nucl. Instr. Methods B* **19–20**, 878 (1987).
31. T. Kobayashi, T. Iwata, Y. Doi, and M. Iwaki, *Nucl. Instr. Methods B* **175–177**, 548 (2001).
32. Y. Wu, T. Zhang, Y. Zhang, G. Zhou, H. Zhang, and X. Zhang, *Surf. Coat. Techol.* **148**, 221 (2001).
33. A. L. Stepanov, S. N. Abdullin, R. I. Khaibullin, Yu. N. Osin, and I. B. Khaibullin, *Proc. R. Micr. Soc.* **29**, 226 (1994).
34. A. L. Stepanov, S. N. Abdullin, V. Yu. Petukhov, Yu. N. Osin, R. I. Khaibullin, and I. B. Khaibullin, *Philos. Mag. B* **80**, 23 (2000).
35. A. L. Stepanov, V. N. Popok, I. B. Khaibullin, and U. Kreibig, *Nucl. Instr. Meth. B* **191**, 473 (2002).
36. Y. Wu, T. Zhang, A. Liu, and G. Zhou, *Surf. Coat. Techol.* **157**, 262 (2002).
37. A. L. Stepanov, R. I. Khaibullin, and I. B. Khaibullin, *Philos. Mag. Lett.* **77**, 261 (1998).
38. A. L. Stepanov, S. N. Abdullin, R. I. Khaibullin, V. F. Valeev, Yu. N. Osin, V. V. Bazarov, and I. B. Khaibullin, *Mater. Res. Soc. Symp. Proc.* **392**, 267 (1995).
39. A. L. Stepanov, S. N. Abdullin, R. I. Khaibullin, and I. B. Khaibullin, Russian Federation Patent 97 109708 (010137) (1997).
40. G. R. Rao, K. Monar, E. H. Lee, and J. R. Treglio, *Surf. Coat. Technol.* **64**, 69 (1994).
41. G. Mie, *Ann. Phys. (Leipzig)* **25**, 377 (1908).
42. M. A. Khashan and A. Y. Nassif, *Opt. Commun.* **188**, 129 (2001).
43. D. V. Sviridov, *Russian Chem. Rev.* **71**, 315 (2002).
44. A. L. Stepanov and D. E. Hole, *Recent Res. Dev. Appl. Phys.* **5**, 1 (2002).
45. V. B. Odzhaev, I. P. Kozlov, V. N. Popok, and D. V. Sviridov, *Ion Implantation into Polymers*, Belarusk. Gos. University, Minsk (1998) [Russian].
46. S. Deying, Y. Saito, and S. Suganomata, *Jpn. J. Appl. Phys.* **33**, L966 (1994).
47. W. Scheunemann and H. Jäger, *Z. Phys.* **265**, 441 (1973).
48. M. Quinten, *Z. Phys. B* **101**, 211 (1996).
49. U. Kreibig, *J. Phys. F 4*, **999** (1974).
50. P. B. Johnson and R. W. Christy, *Phys. Rev. B* **6**, 4370 (1972).
51. B. Pignataro, M. E. Fragala, and O. Puglisi, *Nucl. Instr. Methods B* **131**, 141 (1997).

52. E. D. Palik, *Handbook of Optical Constants of Solids*, Academic, London (1997).
53. A. Aden and M. Kerker, *J. Appl. Phys.* **22**, 1242 (1951).
54. J. Sinzig and M. Quinten, *Z. Phys. D* **26**, 242 (1993).
55. H. Wang, S. P. Wong, W. Y. Cheung, N. Ke, M. F. Chiah, H. Liu, and X. X. Zhang, *J. Appl. Phys.* **88**, 2063 (2000).
56. V. I. Ivanov-Omskii, A. V. Tolmatchev, and S. G. Yastrebov, *Philos. Mag. B* **73**, 715 (1996).
57. H. Biederman, Z. Chmel, A. Fejfar, M. Misina, and J. Pesicka, *Vacuum* **40**, 377 (1990).
58. O. Stenzel, H. Kupfer, T. Pfeifer, A. Lebedev, and S. Schulze, *Opt. Mater.* **15**, 159 (2000).
59. U. Kreibig, *Handbook of Optical Properties*, Vol. 2: *Optics of Small Particles, Interfaces, and Surfaces*, edited by R. E. Hummel and P. Wissmann, CRC, London (1997).
60. J. Hölzl, F. Schulte, and H. Wagner, *Solid Surface Physics*, Springer, Berlin (1979).
61. U. Kreibig, M. Gartz, and A. Hilger, *Ber. Bunssenges. Phys. Chem.* **101**, 1593 (1997).

9

OPTICALLY ANISOTROPIC METAL–POLYMER NANOCOMPOSITES

W. Caseri

Department of Materials, Institute of Polymers, ETH Zentrum, Zürich, Switzerland

1. INTRODUCTION

In composites based on polymer matrices with incorporated inorganic nanoparticles, a random dispersion of the particles is usually attempted. This is of particular importance with respect to the nanocomposites' optical properties. That is, if the particle sizes are far below the wavelength of visible light, scattering of visible light is strongly suppressed [1] and can typically be neglected in the resulting materials. This property renders nanocomposites attractive as materials with uncommon optical properties, which may be of use, for instance, in the area of photoconductivity [2, 3], nonlinear optics [2, 3], transparent magnetic materials [2, 3], transparent UV-absorbing layers [4–6], or extreme refractive indices [7–12]. However, nanocomposites with an ordered arrangement of nanoparticles have also been described, for example, in very thin films [13–19]. In fact, nanocomposites containing particles that form a regular lattice can show remarkable optical properties such as iridescence [20, 21]. In this chapter, however, anisotropic nanocomposites containing uniaxially oriented arrays of metal nanoparticles and their dichroic behavior are the main focus of our atten-

Metal-Polymer Nanocomposites, Edited by Luigi Nicolais and Gianfranco Carotenuto
ISBN 0-471-47131-3

tion. Dichroism, which is a phenomenon that has been studied for a long time in colored anisotropic crystals [22], can be recognized with simple methods: If a dichroic sample is placed before or behind a polarizer, the color of the sample changes upon rotation of the sample or of the polarizer. In ambient light, these colors are usually mixed and therefore not observed individually (unless the sample is in an appropriate position to naturally polarized light—for example, present in certain locations in the sky [23]). The color differences observed with polarized light emerge as a result of absorption differences of the two components of the light which are polarized perpendicular to each other.

The appearance of a color is a prerequisite for dichroism in the visible wavelength region, and this color can be introduced by metal nanoparticles. It is well established that metal nanoparticles are frequently colored, with the color depending on the type of metal and its particle size. For very small particles, the color is caused by absorption rather than by scattering, as demonstrated by Mie and Steubing about 100 years ago [24–27]. They found in the example of spheric gold particles with diameters below 50 nm that the intensity loss of transmitted light originated predominately from absorption [24–27]; considerable scattering arose only for gold particles with diameters above ~50 nm [28, 29]. In particle arrays such as present in the dichroic nanocomposites referred to below, it must be considered that the color of nanoparticles can depend not only on the size of the particles but also on the distance between the particles [30, 31]. This was realized about 100 years ago when Kirchner and Zsigmondy reported that nanocomposites of gelatin and colloidal silver [32] or gold [33] reversibly changed their color from blue to red upon swelling with water. They suggested that the color of nanocomposites must therefore be influenced by the distance between the embedded particles [32, 33], which was also substantiated around the same time by theoretical analyses of Maxwell Garnett [34, 35]. Hence, polarized light may interact with a particle array in a different way for parallel and perpendicular orientation of the polarization plane of the light with respect to the long axis of the particle array since cooperative optical effects that emerge in adjacent metal particles under the influence of electromagnetic radiation are more pronounced for parallel orientation (provided that the short axis of the particles is small enough).

2. DICHROISM IN NATURAL POLYMERS CONTAINING METAL NANOPARTICLES

Many fibers composed of natural polymers contain uniaxially oriented anisotropic hollow spaces in which metal particles can crystallize under appropriate conditions. As a consequence, the impregnated fibers contain anisotropic assemblies of metal nanoparticles that induce anisotropic optical properties such as dichroism. In 1896, Ambronn prepared related materials. He immersed fibers or thin sections of natural polymers (leaches, cotton, and fir) for some time in the dark in an aqueous solution containing 1–2% silver nitrate [23]. Thereafter, he removed the specimen from the solution, let it dry in the air, and subsequently exposed it to ambient light for ~2 days whereupon the silver ions were reduced to elemental silver. Upon observation through a polarizer, the specimen appeared bright yellow or bright greenish for parallel orientation of the long axis of the fiber and the polarization plane of the light and appeared blue, green, red, or violet for perpendicular orientation. Dichroic samples could also be prepared upon *in situ* reduction of fine silver nitrate powder that had been strewed on dry plant fibers. This method was also successful for the preparation of dichroic animal sinews, white human hair, or gelatin that had been previously stretched (without stretching, dichroism in gelatin samples was not observed). Dichroic samples with stretched gelatin were also obtained after treatment with a silver-nitrate-saturated 80% ethanolic solution for several days. Moreover, upon immersion of natural fibers in aqueous or alcoholic solution of 1–2% gold chloride for some time followed by drying in the air and subsequent exposure to ambient light for 2–3 days, elemental gold was formed and a red color was found for parallel and blue-green or green for perpendicular orientation of the polarization plane of incident light and the long axis of the fiber. A related example prepared by Ambronn but displayed by Braun in 1905 [36] is shown in Figure 9.1. Reduction of the incorporated gold ions with formic acid was not successful for the manufacture of dichroic fibers, in contrast to the results mentioned below.

Only a short time after Ambronn's publication, Apáthy reported in 1897 that he had been busy for many years with the impregnation of animal fibers with gold salts that were subsequently converted to gold colloids by exposure to light (besides colloidal gold, Apáthy also considered the formation of AuO, according to a concept of the prominent chemist Berzelius) [37]. Apáthy studied the influence of a number of parameters on the reduction process and the quality of the coloration of the fibers. In the absence of light or organic matter, the characteristic colors that appear after reduction of gold salts to gold colloids were not observed, and the extent of the dichroism in the impregnated fibers depended on parameters such as the chemical composition ($H[AuCl_4]\cdot xH_2O$, where x was supposed to be 4, or $AuCl_3\cdot xH_2O$, where x was supposed to be 0 or 2) and the commercial source of the gold salt, the amount of gold salt that was incorporated in the fiber, the temperature during the conversion into gold colloids, the presence of organic acids (formic acid, acetic acid, and citric acid), and the light intensity. The best results were obtained when the samples were immersed in a 1% $H[AuCl_4]\cdot xH_2O$ solution for 2 to ~12 hr in the dark followed by immersion of the wet samples into 1% formic acid. The reduction to gold colloids was then performed advantageously by subsequent exposure of the

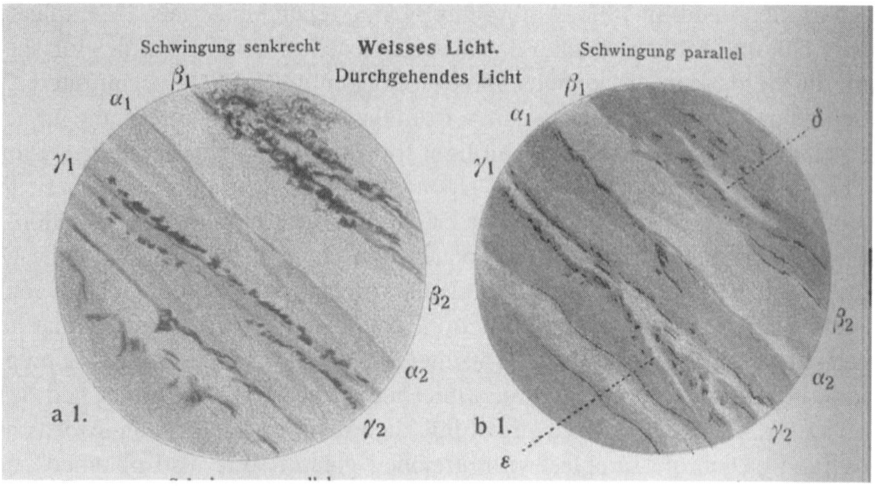

Figure 9.1. Microscope image of a dichroic nanocomposite consisting of gold particles embedded in wood (probably fir), published in 1905 [36]. The polarization plane of the light is parallel to the orientation axis of the wood in the left image and perpendicular in the right image. See color insert.

impregnated specimen to the light of a clear winter day during at least 8 hr without interruption at 10–15°C; however, at elevated temperatures in summer, exposure to diffuse daylight during at least 6 hr without interruption was recommended. The sizes of the formed gold particles were recognized to be below the resolution of an optical microscope, and the impregnated specimen adopted colors that varied from a bright pink to dark violet. Dichroism was observed at this type of specimen, with the colors changing for instance from cherry red at parallel orientation of the polarization plane to the long axis of the fibrils to blue-black at perpendicular orientation.

Dichroic samples in natural fibers had been prepared not only by reduction of metal salts in the fibers themselves. In the nineteenth century, dichroic specimen of gold or silver nanoparticles in gelatin were obtained by exposure of gelatin to dispersions of corresponding metal colloids followed by stretching of the swollen gelatin and subsequent drying [38]. A change in the position between the polarization plane of the incident light and the fiber direction from parallel to perpendicular orientation manifested in a color change in the case of gold from bright red or yellowish red to dark blue violet and in the case of silver from bright brown to dark brown or from yellow to olive green.

The elemental state of incorporated metal colloids in dichroic fibers was experimentally proven by early X-ray studies [39, 40]. The diffraction patterns revealed that the crystal lattice of the silver and gold particles in ramie fibers did not differ from that of the the corresponding bulk metals [53, 60]. In the same era, a number of dichroic nanonomposites consisting of anisotropic natural polymers and gold [36, 39–48], silver [36, 39–41, 46–49], platinum [47], palladium [36, 47], rhodium [47], copper [47], bismuth [47], mercury [39, 44, 47, 48], or silver amalgam [48] were reported. The metal content in the fibers was determined only in relatively few cases. As an example, a silver content in cotton fibers of 0.02–0.2% w/w [46] and in ramie fibers of 0.1% and 5% w/w [46]—respectively 0.45% and 1.2% w/w (0.07% and 0.18% v/v) [49]—was found. In samples of gold in ramie fibers, metal contents between 0.6% and 1.6% w/w were measured [46]. For silver contents of as low as 0.1% w/w and gold contents of only 0.7% w/w, dichroism was weak but still just visible [46].

The dichroism in natural fibers with incorporated metal nanoparticles was found to depend strongly not only on the element employed [50] but also on the particle size [51], which was calculated from the full-width at half-maximum of X-ray diffraction patterns, yielding values between 5 and 14 nm in ramie, hemp, bamboo, silk, viscose silk, acetate rayon, and wool fibers [51]. As an example of a particle-size-dependent dichroism, a color transition between perpendicular and parallel orientation of polarization plane and long fiber axis in ramie fibers appeared from straw yellow to indigo blue for gold particles of 8.5-nm diameter and from claret red to green at 12.3-nm diameter [51].

It was suggested from the beginning that the dichroism in the nanocomposites composed of natural fibers and metal colloids must originate in an anisotropy in the context with the embedded metallic particles. Ambronn assumed the dichroism to be caused most likely by the formation of anisotropic, labile crystal modifications of the elemental metals [52], while Braun attributed the origin of the dichroism to metallic rods that are distributed in the matrix at appropriate distances for acting as a wire grid polarizer [36]. X-ray patterns of ramie fibers with incorporated gold particles showed well-formed rings stemming from the lattice of the metal atoms, demonstrating that the crystal axes of the metal particles were randomly oriented in the fibers, and it was concluded that the dichroism in those fibers was not due to an oriented anisotropic modification of silver or gold primary particles [40, 51] but was, instead, due to a uniaxially oriented linear arrangement of isotropic metal crystallites as a consequence of their growth in oriented, anisotropic spaces of the fibers (Figure 9.2) [50, 51, 53]. Hence, the particles themselves were considered not to be dichroic, and the dichroism was supposed to originate exclusively as a consequence of the anisotropic shape of the particle-filled space between oriented fibrils [50, 53]. It was also reported, however, that the X-ray diffraction patterns of silver in ramie fibers did not consist of completely homogeneous

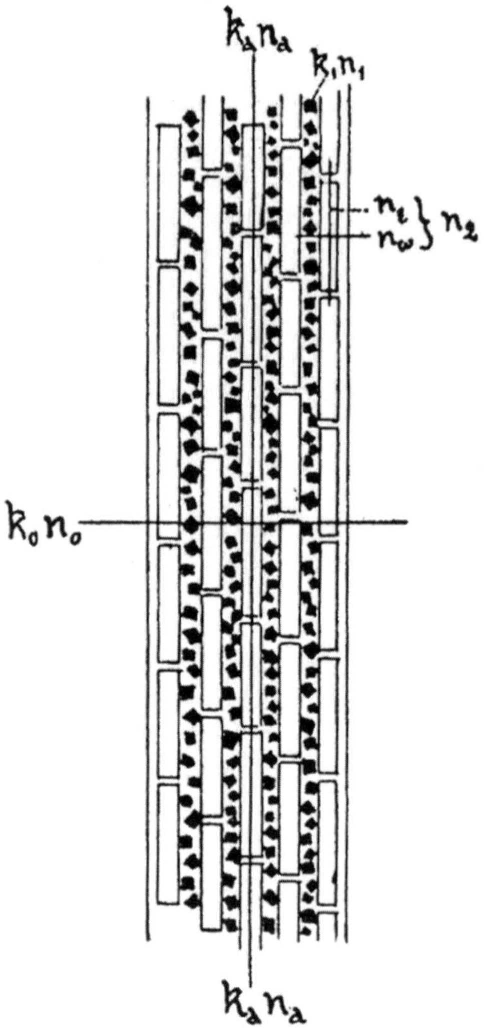

Figure 9.2. Schematic illustration of Frey's explanation of the dichroism in anisotropic natural fibers containing metal colloids, proposed in 1927 [50, 53].

rings but that they displayed some symmetric deviations [39], and it was therefore suggested that a fraction of the crystal axes of the metal particles was oriented, but another report considered these deviations in the X-ray diffraction patterns not to be significant [51]. Finally, observations in an ultramicroscope in combination with polarized light led to the conclusion that gold and silver particles in plant fibers (probably ramie) indeed were arranged in parallel elongated aggregates that caused the dichroic behavior of the samples [48].

3. DICHROIC FILMS OF POLY(VINYL ALCOHOL) AND GOLD NANOPARTICLES

For the preparation of dichroic nanocomposites based on poly(vinyl alcohol) (PVAL) as the matrix, colloids of elemental gold were synthesized *in situ* by reduction of sodium tetrachloroaurate(III) with trisodium citrate dihydrate in water [54, 55]. The colloidal gold dispersions thus obtained were added to aqueous PVAL solutions. The resulting homogeneous mixtures were poured in a flat dish; and after water evaporation, nanocomposite films were left. These films, which still contained reaction side products stemming from the reduction of tetrachloroaurate(III) with trisodium citrate, contained 3.4% w/w gold particles of 9.5-nm diameter. The incorporated gold particles were randomly dispersed in the PVAL, as evident from transmission electron microscopy (TEM), but agglomerates consisting of few particles might have been present, too. The absorption maximum in UV–Vis spectra, which appeared at 527 nm in aqueous dispersions of the gold particles, became broad in the isotropic nanocomposites and shifted to 564 nm (Figure 9.3).

The poly(vinyl alcohol)–gold nanocomposites were finally drawn at 120°C up to a draw ratio (DR), defined as the ratio of the length before and after drawing, of 5. It is well known that the polymer molecules orient upon drawing, and it was evident from TEM images of thin sections of the drawn materials

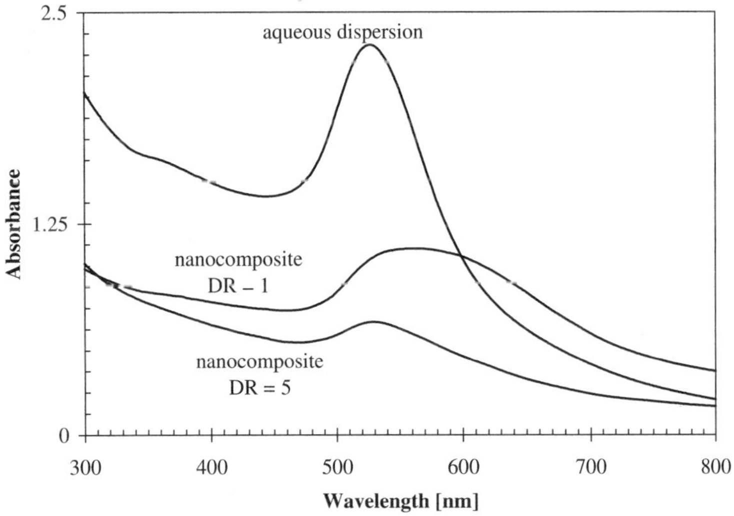

Figure 9.3. UV–Vis spectra of gold particles prepared *in situ* in water and of the related particles embedded in poly(vinyl alcohol) before drawing (DR = 1) and after drawing (DR = 5).

that the drawing also resulted in an alignment of a part of the gold particles into arrays that were oriented in the direction of the drawing axis. This alignment might arise as a consequence of the exclusion of particles between orienting polymer chains, thus maximizing the intermolecular forces between adjacent polymer molecules. The UV–Vis spectrum of the drawn samples recorded with nonpolarized light resembled to the spectrum of the gold colloids in aqueous dispersion (Figure 9.3; note that the absorbances in this figure have to be dealt with qualitatively since the amounts of particles in the optical path are not known), and the absorption maximum at 529 nm was close to that reported above in aqueous dispersion. This indicates that the size of the particles remained in the same order of magnitude upon preparation and processing of the composites as also evident from TEM images. The absorption maximum became sharper upon drawing, maybe as a consequence of a disruption or deformation of small particle agglomerates under the action of shear forces.

As in the case of the above-described natural fibers comprising metal nanoparticles, the color of the PVAL–gold nanocomposites changed upon observation in linearly polarized light as the angle φ between the polarization plane and the drawing axis was turned, as a consequence of the uniaxially oriented arrays of metal colloids. As an example, a specimen appeared blue for parallel ($\varphi = 0°$) and red for perpendicular ($\varphi = 90°$) orientation between the polarization plane of the incident light and the drawing axis. The color shifts of the nanocomposites upon variation of φ also manifested in UV–Vis spectra. The wavelength of maximum absorption (λ_{max}) shifted subsequently to shorter wavelengths as φ increased with λ_{max} at $\varphi = 0°$ at 528 nm and at $\varphi = 90°$ at 570 nm (Figure 9.4). Between 400 and 800 nm, the absorbance was higher at parallel than at perpendicular orientation of the polarization direction with respect to the drawing direction almost in the entire region, but between ~490 and 520 nm the absorbance was rather independent of φ.

4. DICHROIC FILMS OF POLY(ETHYLENE) AND GOLD OR SILVER NANOPARTICLES

Poly(ethylene) (PE) is a material that can be drawn to a higher extent than poly(vinyl alcohol) (the latter was the focus of Section 3). However, PE is insoluble in water; therefore, aqueous dispersions of *in situ*-prepared metal nanoparticles as described above are not suited for the preparation of PE–metal nanocomposites. Corresponding materials could be gained, however, by using precipitated metal nanoparticles surrounded by a layer of surface-bound organic molecules [54–59]. The organic surface layer prevented the formation of strongly connected agglomerates of metal particles in the isolated powder since the surface layer markedly diminishes the attractive interactions between metal

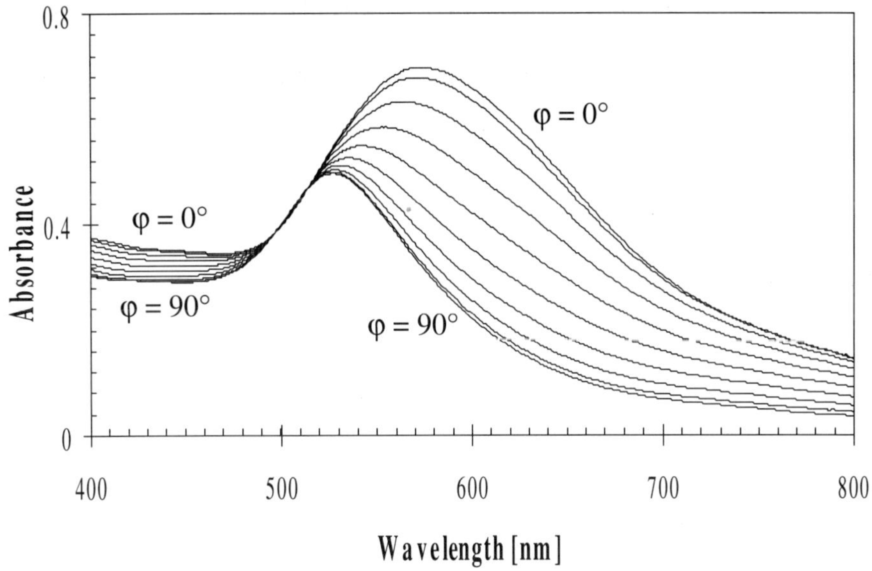

Figure 9.4. UV–Vis spectra of a drawn poly(vinyl alcohol)–gold nanocomposite in polarized light at different angles φ representing the angle between the polarization plane of the light and the drawing direction of the nanocomposite. The angle φ varies from 0° to 90° in steps of 10°, and a pure poly(vinyl alcohol) sample with similar thickness as the nanocomposite was inserted in the reference beam of the spectrometer.

particles; note that in absence of a surface layer the interactions between metal particles are so strong that the agglomerates usually cannot be disrupted during the processing steps for nanocomposite manufacture. For the creation of nanocomposites with PE, gold or silver particles coated with a layer of dodecanethiol were synthesized on the basis of previously described recipes [60, 61]. Gold or silver salts were reduced with sodium borohydride in a two-phase toluene–water mixture containing 1-dodecanethiol and tetraoctylammonium bromide as a phase transfer catalyst. Upon adsorption of the 1-dodecanethiol molecules on the *in situ*-formed metal particles, the metal colloids became well-soluble in toluene, and hence the organic phase adopted the color of the nanoparticles while the aqueous phase became completely colorless. Thereafter, the toluene phase was separated and the surface-modified colloids were precipitated by the addition of ethanol. Transmission electron microscopy (TEM) revealed average particle diameters of gold of 2–3 nm and of silver of 4–5 nm [54, 56, 57, 59].

Basically, surface-modified particles can be mixed with polymers present in the molten or in the dissolved state. However, when working with PE, it has to be considered that some grades of PE are difficult to dissolve. A suited solvent for all PE grades is *p*-xylene at 130°C, and hence the dodecanethiol-coated particles were usually dispersed in *p*-xylene–PE solutions at 130°C followed by casting of the resulting mixture and solvent evaporation [54–59]. As an alternative route in the case of silver, the particles were also mixed with molten PE in an extruder at 180°C [57]. In order to improve the homogeneity of the nanocomposites, the solvent-casted and the extruded samples were compression-molded at 150–180°C; molding periods of 20 and 120 min resulted in materials with similar optical properties (cf. Table 9.1). The color of the surface-modified gold particles dispersed in *p*-xylene was retained in the nanocomposites prepared from the corresponding dispersions, and the absorption maxima (λ_{max}) in UV–Vis spectra of the *p*-xylene dispersions and the corresponding nanocomposites differed only slightly (0–20 nm, depending on the annealing period of the particles in xylene as described below), and the UV–Vis spectrum

TABLE 9.1. Absorption Maxima of Drawn Poly(ethylene)–Gold Nanocomposites (Draw Ratio 15) at Different Reaction Conditions and Different Angles φ Between the Polarization Plane of the Incident Linearly Polarized Light and the Drawing Axis of the Specimen[a]

C [% w/w]	t_h [min]	t_c [min]	$\lambda_{max}(0°)$ [nm]	$\lambda_{max}(45°)$ [nm]	$\lambda_{max}(90°)$ [nm]	$\Delta\lambda_{max}$ [nm]
3.75	90	20	542	534	527	15
		120	560	544	536	24
3.36	120	20	618	584	553	65
		120	619	575	548	71
3.75	180	20	614	570	546	68
		120	618	562	539	79
4.00	240	20	641	578	548	93
		120	648	571	541	107
3.96	300	20	690	580	534	156
		120	682	580	542	140
3.91	330	20	690	657	631	59
		120	678	640	618	60
4.18	480	120	676	641	614	62
3.85	810	20	653	625	607	46
		120	650	616	596	54

[a]Here c is the ratio of gold (including 1-dodecanethiol layer) and poly(ethylene), t_h is the heating period of the gold dispersion in a poly(ethylene) solution in *p*-xylene, and t_c is the compression molding period of the isotropic nanocomposites at 150°C. The absorption maxima at $\varphi = 0°$, 45°, and 90° are designated as $\lambda_{max}(0°)$, $\lambda_{max}(45°)$, and $\lambda_{max}(90°)$, respectively, and $\Delta\lambda_{max}$ is the difference between $\lambda_{max}(0°)$ and $\lambda_{max}(90°)$.

of the silver particles was essentially the same for the dispersion in *p*-xylene and the resulting PE–silver nanocomposites. However, in contrast to the initial gold dispersions in xylene and the resulting PE nanocomposites, aggregates of silver particles seemed to be present in the freshly prepared xylene dispersions and in the nanocomposites.

Sections of the nanocomposite sheets obtained after compression molding were drawn on a hot stage at 120°C up to maximum draw ratios of 22–60 depending on the grade of PE (high-density PE or ultrahigh-molecular-weight PE) [54–59]. Under appropriate conditions (which are described below), the drawing induced the formation of particle arrays which were oriented in the drawing direction (Figure 9.5), as evident from transmission electron microscope (TEM) and scanning electron microscope (SEM) images (an SEM picture displaying oriented arrays of gold particles is shown in Figure 9.6). The formation of oriented arrays of metal particles was connected with a dichroism in the resulting materials. Noteworthy, the extent of the dichroism of the drawn PE–gold nanocomposites strongly depended on an annealing period of the gold dispersions in the PE solutions used for nanocomposite preparation [54, 55, 59]. After drawing, most pronounced dichroism was observed when the gold particles had been kept in the PE–xylene dispersion at 130°C for 300–330 min before solvent evaporation (cf. Table 9.1). While the color of the mixture in xylene did not change significantly during heating for 1 hr, further annealing resulted in a color change from red to blue, and accordingly the absorption in UV–Vis spectra shifted to higher wavelengths (Figure 9.7). This effect might be due to an increase in the diameter of the primary gold particles [62, 63] or, as mentioned in the Introduction, to the formation of agglomerates. TEM images did not allow an unambiguous decision if the size of the primary gold particles changed during the treatment in *p*-xylene at 130°C. However, agglomerates were clearly evident in TEM images of samples that had been preannealed for 300 min [54, 59]. The cross section of these agglomerates ranged from a few up to ~10,000 primary particles. Agglomeration was not pronounced in TEM images without annealing of the gold dispersions in PE solutions in *p*-xylene or after annealing for 20 min. The agglomeration might be caused by partial desorption of the surface-bound dodecanethiol molecules, thus enhancing

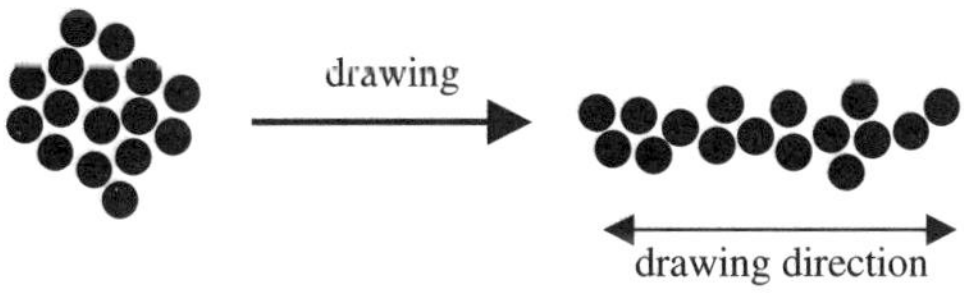

Figure 9.5. Schematic illustration of the deformation of spheric aggregates of metal particles into linear aggregates upon drawing.

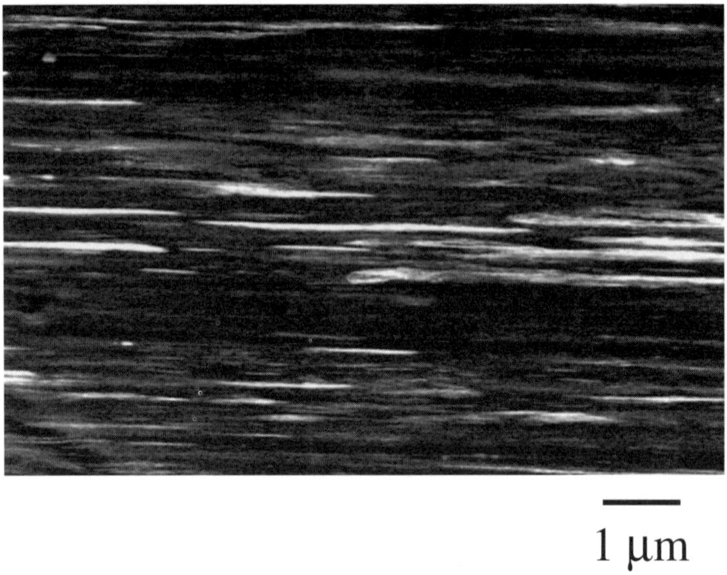

Figure 9.6. Scanning electron micrograph (detection of backscattered electrons) of a poly(ethylene)–gold nanocomposite. The bright lines represent arrays of gold nanoparticles that are not resolved individually.

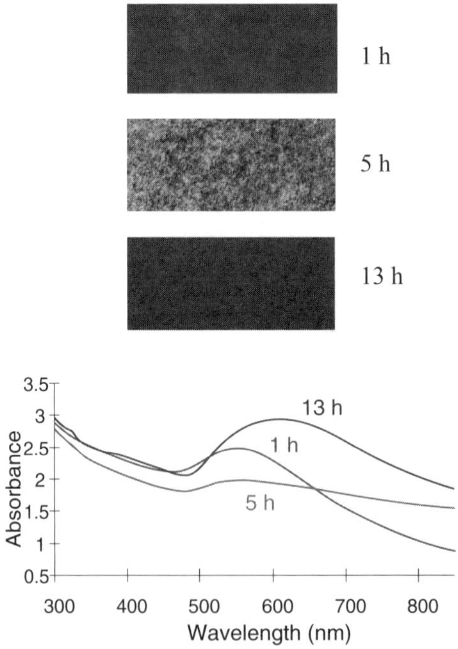

Figure 9.7. Colors and UV–Vis spectra of dispersions of gold particles (average diameter 2.2 nm, covered with a layer of 1-dodecanethiol) kept in a poly(ethylene) solution in *p*-xylene at 130°C for 1 hr, 5 hr, and 13 hr, respectively. See color insert.

the interactions between adjacent gold particles. Obviously, the aggregates of gold particles in the undrawn films were deformed during solid-state drawing of the nanocomposites, resulting in arrays of particles that were oriented in the drawing direction. Since the agglomeration of the gold particles required some time and thus the conditions became favorable for the formation of particle arrays only after a certain annealing period, only a weak dichroism was observed in drawn nanocomposites after annealing for 90 min—for example, with $\Delta\lambda_{max}$ (difference of the absorption maxima at $\varphi = 0°$ and $\varphi = 90°$) of ~15–25 nm compared to ~140–160 nm in nanocomposites prepared in the same way with the same grade of PE but with an annealing period of the gold dispersion in *p*-xylene of 300 min (Table 9.1). It has to be noted that dichroism also became less remarkable when the gold dispersion was kept in xylene at 130°C for periods significantly longer than 300 min—for example, with $\Delta\lambda_{max}$ of ~45–55 nm after 810 min annealing (Table 9.1)—probably as a result of an increase in the adhesion between the gold particles due to the desorption of a larger quantity of dodecanethiol molecules.

As already implied above, the UV–Vis spectra of drawn PE–gold nanocomposites gradually changed with φ in the UV and the visible wavelength range except at isosbestic points that emerged in all samples (an example of a UV–Vis spectrum is shown in Figure 9.8) [54, 55, 59]. The absorption maximum at parallel orientation of the polarization plane of the incident light ($\varphi = 0°$) appeared at higher wavelengths (typically between 670 and 690 nm, depending on the PE grade and the detailed experimental parameters) than at perpendicular orientation ($\varphi = 90°$, λ_{max} typically between 545 and 590 nm). Accordingly, the color of the drawn nanocomposites changed from blue at $\varphi = 0°$ to red at $\varphi = 90°$ (Figure 9.9).

Besides the annealing period of the gold particles dispersed in PE solutions in xylene, the influence of the amount of gold present in PE and the draw ratio

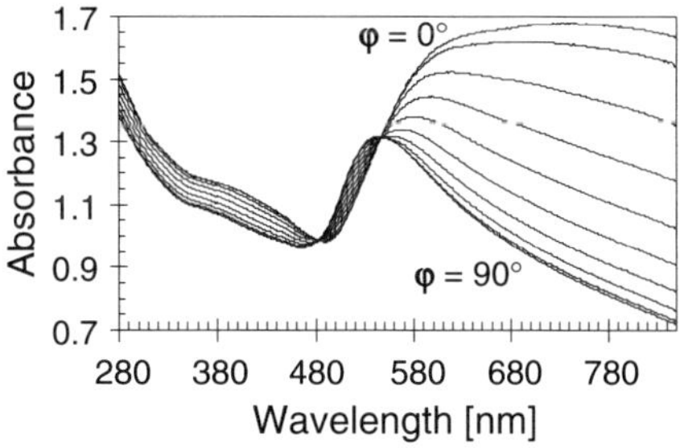

Figure 9.8. UV–Vis spectra of a drawn poly(ethylene)–gold nanocomposite in polarized light at different angles φ representing the angle between the polarization plane of the incident light and the drawing direction of the nanocomposite. The angle φ varies from 0° to 90° in steps of 10°.

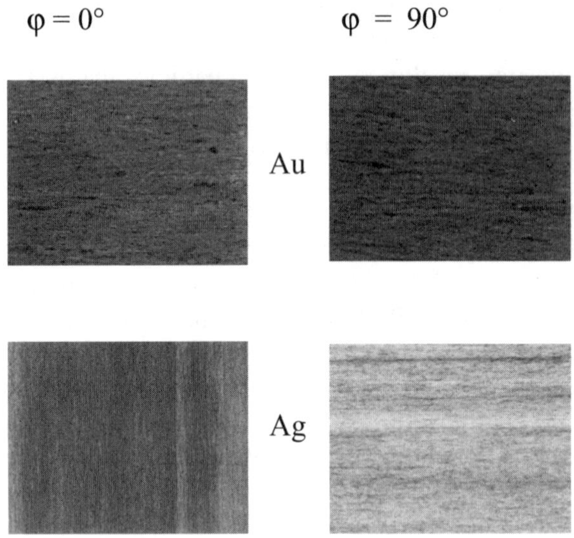

Figure 9.9. Colors of a drawn poly(ethylene)–gold and a poly(ethylene)–silver nanocomposite in polarized light with the polarization plane of the incident light and the drawing direction of the nanocomposite parallel (φ = 0°) and perpendicular (φ = 0°), respectively. See color insert.

on the dichroic properties of the composites was also investigated [54]. When the gold content was varied between 0.9% and 7.4% w/w, most pronounced shifts of the absorption maxima between $\varphi = 0°$ and $\varphi = 90°$ were observed in drawn samples with gold fractions of 2–4% w/w. The difference between the absorption maxima at $\varphi = 0°$ and $\varphi = 90°$ did not change exceedingly at draw ratios above 6. For example, a PE–gold specimen showed a $\Delta\lambda_{max}$ of 65 nm at a draw ratio of 6 and a $\Delta\lambda_{max}$ of 85 nm at a draw ratio of 18; and for oriented nanocomposites with another grade of PE, a $\Delta\lambda_{max}$ of 45 nm and 100 nm resulted for draw ratios of 10 and 45, respectively

TEM micrographs of undrawn nanocomposites of PE and silver (Figure 9.10) revealed that the silver particles were agglomerated, to a certain extent, in the polymer matrix also without annealing of the silver dispersions in *p*-xylene containing dissolved PE, in contrast to the above-described behavior of the systems comprising gold particles. Aggregation of silver particles was also indicated by X-ray scattering patterns of undrawn nanocomposites that revealed lattice planes with a spacing of 3.5 nm [57]. As far as could be concluded from the individual silver nanoparticles that still were visible in TEM images of the nanocomposites, the size and size distribution of the primary particles did not change within the experimental precision upon preparation and processing of the nanocomposites. After drawing, an X-ray diffraction pattern with individ-

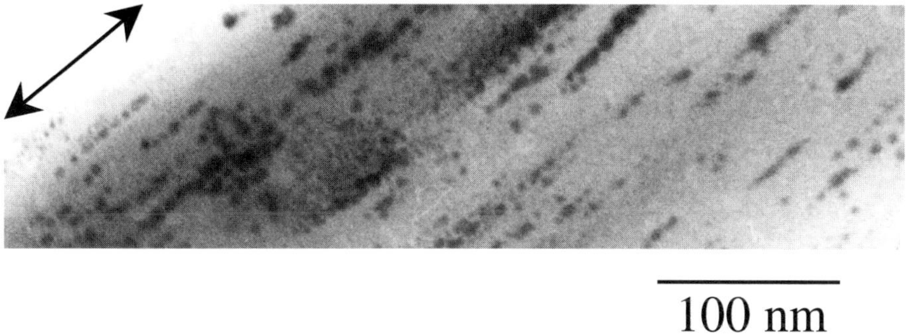

Figure 9.10. Transmission electron micrograph of a section of a drawn poly (ethylene)–silver nanocomposite showing uniaxially oriented arrays of silver particles. The doble arrow in the left upper corner indicates the drawing direction.

ual spots instead of rings was observed; the spots corresponded to lattice planes separated by 3.4 nm and indicated an orientation in the system. As described above for the gold particles, spheric aggregates formed by surface-modified silver particles in the undrawn materials also aligned into linear structures under the tensile deformation forces acting during solid-state drawing.

As in the case of the PE–gold nanocomposites described above, the UV–Vis spectra of the drawn PE–silver samples strongly depended on φ (an example is shown in Figure 9.11) [55–57]. Light with the polarization plane parallel to the drawing axis of the nanocomposites was absorbed at higher wavelengths than at the perpendicular orientation. As with drawn materials containing gold, isosbestic points also appeared in the UV–Vis spectra of drawn PE–silver nanocomposites. The PE–silver nanocomposites became dichroic at moderate draw ratios; in an example, $\Delta\lambda_{max}$ at $\varphi = 0°$ and $\varphi = 90°$ remained essentially constant (90 nm) between draw ratios of 6 and 22. The shift in the absorption maximum between $\varphi = 0°$ and $\varphi = 90°$ in the PE–silver samples was virtually independent of the procedure used for nanocomposite manufacture—that is, solution casting or melt processing. As implied by the UV–Vis spectra in polarized light at different φ, the color of the nanocomposites changed upon variation of φ, and the nanocomposites appeared red and yellow at φ of 0° and 90°, respectively (Figure 9.9). Annealing of PE–silver samples at 180°C for 15 hr resulted in an increase of the average particle size from 4.5 to 10 nm. The increase in particle diameter led to a shift of the absorption maximum in UV–Vis spectra from 435 nm to 463 nm. Accordingly, after drawing, the color turned from mauve or purple-red (depending on the sample) at $\varphi = 0°$ to amber at $\varphi = 90°$.

Dichroic nanocomposites with pronounced color changes are potentially useful in liquid crystal display (LCD) applications. The setup of a traditional

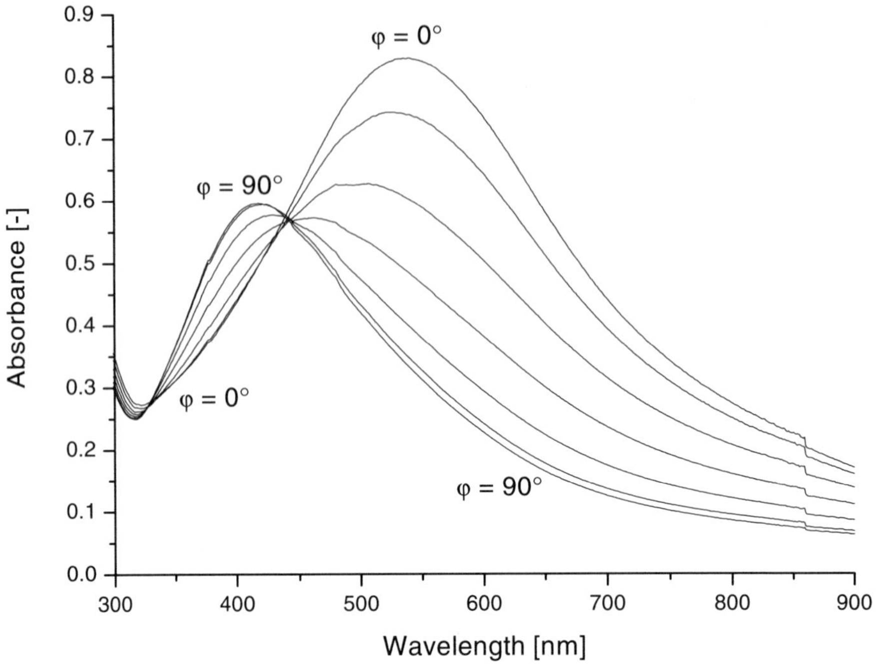

Figure 9.11. UV–Vis spectra of a drawn poly(ethylene)–silver nanocomposite in polarized light at different angles φ representing the angle between the polarization plane of the incident light and the drawing direction of the nanocomposite. The angle φ varies from 0° to 90° in steps of 15°.

LCD is shown in Figure 9.12. The incoming light passes two crossed polarizers that enclose an electrooptical cell containing liquid-crystalline molecules [64]. In the absence of a voltage, the liquid-crystalline molecules are arranged in a twisted-nematic structure that turns the direction of the polarization plane of the incoming light by 90°; that is, the light passes the second polarizer. In the presence of a voltage, however, the liquid-crystalline molecules orient parallel to the electric field and therefore the polarization plane of the incoming light does not change anymore; that is, this light is absorbed when arriving at the second polarizer. If, however, the second polarizer is replaced by a dichroic nanocomposite as described above with the drawing axis either parallel or perpendicular to the remaining polarizer, light is transmitted both in the absence and the presence of a voltage but in different colors (Figure 9.12), which results in a bicolored display [56, 59] instead of a common grayish/black display; an example is shown in Figure 9.13.

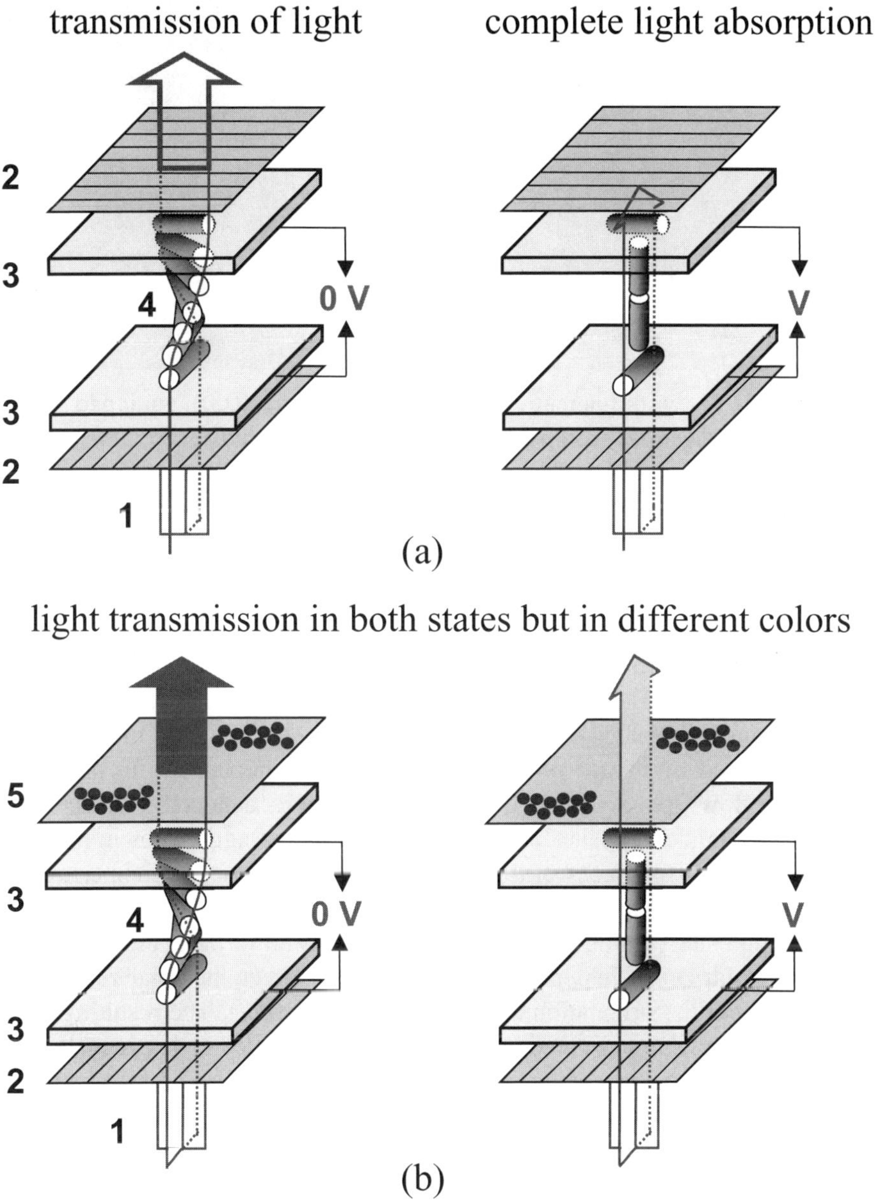

Figure 9.12. Schematic representation of **(a)** a standard twisted-nematic liquid crystal display and **(b)** a related display equipped with a drawn nanocomposite. 1, incoming light (unpolarized); 2, polarizer; 3, glass plate coated with an electrode layer and an orientation layer for the liquid-crystalline molecules; 4, liquid-crystalline molecules forming a 90° helical twist in absence of a voltage or a linear array parallel to an electric field in presence of an electric field; 5, nanocomposite with oriented arrays of metal particles. See color insert.

Figure 9.13. A twisted-nematic liquid crystal display (LCD) equipped with a poly(ethylene)–silver nanocomposite that had been annealed at 180°C for 15 hr and subsequently drawn as described in the text. The drawing axis of the nanocomposite is oriented parallel to the polarizer in the left image and perpendicular in the right image. See color insert.

6. CONCLUSIONS

Dichroic nanocomposites based on polymers with incorporated metal colloids can be prepared by *in situ* synthesis of the metal nanoparticles in fibers with oriented hollow spaces or by drawing of isotropic nanocomposites. In the former case, nanocomposites form uniaxially oriented aggregates in the originally present hollow spaces of the fibers; in the latter case, particle arrays oriented in the drawing direction are established as a result of the drawing process. In such composites, the absorption and hence the color of the samples observed in linearly polarized light depend on the angle φ between the polarization plane of the light and the orientation axis of the particle arrays. The resulting colors depend on the metal and the size of the primary particles and aggregates. The optical properties of such nanocomposites are similar to those of systems with wire-type metal entities [65–71]. Dichroic nanocomposites may find application, for example, in liquid crystal displays (LCD) that appear bicolored instead of grayish and black in corresponding standard displays.

ACKNOWLEDGMENTS

I express my thanks to my colleagues who have invaluably contributed to the results presented from our laboratory (including Figures 9.3 to 9.13), namely, Cees Bastiaansen, Cyril Darribère, Yvo Dirix, Wilbert Heffels, and Paul Smith.

REFERENCES

1. B. M. Novak, *Adv. Mater.* **5**, 422 (1993).
2. L. L. Beecroft and C. K. Ober, *Chem. Mater.* **9**, 1302 (1997).
3. D. Yu. Godovski, *Adv. Polym. Sci.* **119**, 81 (1995).
4. T. Kyprianidou-Leodidou, P. Margraf, W. Caseri, U. W. Suter, and P. Walther, *Polym. Adv. Technol.* **8**, 505 (1997).
5. R. J. Nussbaumer, W. R. Caseri, P. Smith, and T. Tervoort, *Macromol. Mater. Eng.* **288**, 44 (2003).
6. R. J. Nussbaumer, W. Caseri, T. Tervoort, and P. Smith, *J. Nanoparticle Res.* **4**, 319 (2002).
7. M. Weibel, W. Caseri, U. W. Suter, H. Kiess, and E. Wehrli, *Polym. Adv. Technol.* **2**, 75 (1991).
8. L. Zimmermann, M. Weibel, W. Caseri, U. W. Suter, and P. Walther, *Polym. Adv. Technol.* **4**, 1 (1993).
9. L. Zimmermann, M. Weibel, W. Caseri, and U. W. Suter, *J. Mater. Res.* **8**, 1742 (1993).
10. F. Papadimitrakopoulos, P. Wisniecki, and D. E. Bhagwagar, *Chem. Mater.* **9**, 2928 (1997).
11. T. Kyprianidou-Leodidou, W. Caseri, and U. W. Suter, *J. Phys. Chem.* **98**, 8992 (1994).
12. T. Kyprianidou-Leodidou, H.-J. Althaus, Y. Wyser, D. Vetter, M. Büchler, W. Caseri, and U. W. Suter, *J. Mater. Res.* **12**, 2198 (1997).
13. J. F. Ciebien, R. T. Clay, B. H. Sohn, and R. E. Cohen, *New J. Chem.*, 685 (1998).
14. V. Sankaran, C. C. Cummins, R. R. Schrock, R. E. Cohen, and R. J. Silbey, *J. Am. Chem. Soc.* **112**, 6858 (1990).
15. J. P. Spatz, A. Roescher, and M. Möller, *Adv. Mater.* **8**, 337 (1996).
16. M. Möller, J. P. Spatz, A. Roescher, S. Mössmer, S. T. Selvan, and H.-A. Klok, *Macromol. Symp.* **117**, 207 (1997).
17. S. T. Selvan, J. P. Spatz, H.-A. Klok, and M. Möller, *Adv. Mater.* **10**, 132 (1998).
18. J. P. Spatz, S. Mössmer, and M. Möller, *Chem. Eur. J.* **2**, 1552 (1996).
19. J. P. Spatz, S. Mössmer, M. Möller, T. Herzog, A. Plettl, and P. Ziemann, *J. Luminescence* **76&77**, 168 (1998).
20. H. B. Sunkara, J. M. Jethmalani, and W. T. Ford, *Chem. Mater.* **6**, 362 (1994).
21. J. M. Jethmalani and W. T. Ford, *Chem. Mater.* **8**, 2138 (1996).
22. H. Ambronn, *Anleitung zur Benutzung des Polarisationsmikroskops bei histologischen Untersuchungen*, J. H. Robolsky, Leipzig (1892).
23. H. Ambronn, *Kgl. Sächs. Ges. Wiss.* **8**, 613 (1896).
24. G. Mie, *Ber. Deutsch. Phys. Ges.* 492 (1907).
25. G. Mie, *Phys. Z.* **8**, 769 (1907).
26. G. Mie, *Z. Chem. Ind. Kolloide* **2**, 129 (1907).

27. W. Steubing, *Ann. Phys., vierte Folge (Drude's Ann.)* **26**, 329 (1908).
28. G. Mie, *Ann. Phys., vierte Folge (Drude's Ann.)* **25**, 377 (1908).
29. J. Turkevich, G. Garton, and P. C. Stevenson, *J. Colloid Sci. Suppl. 1* **9**, 26 (1954).
30. R. Zsigmondy, *Kolloidchemie, 1. Allgemeiner Teil*, Otto Spamer, Leipzig (1925).
31. M. Quinten and U. Kreibig, *Surf. Sci.* **172**, 557 (1986).
32. F. Kirchner, *Ber. Königl. Sächs. Ges. Wiss., Math.-Phys. Klasse* **54**, 261 (1902).
33. F. Kirchner and R. Zsigmondy, *Ann. Phys., vierte Folge (Drude's Ann.)* **15**, 573 (1904).
34. J. C. Maxwell Garnett, *Philos. Trans. R. Soc. London A* **205**, 237 (1906).
35. J. C. Maxwell Garnett, *Philos. Trans. R. Soc. London A* **203**, 385 (1904).
36. F. Braun, *Ann. Phys., vierte Folge (Drude's Ann.)* **16**, 238 (1905).
37. S. Apáthy, *Mitt. Zool. Stat. Neapel* **12**, 495 (1897).
38. H. Ambronn and R. Zsigmondy, *Ber. Sächs. Ges. Wiss.* **51,** 13 (1899).
39. F. Bion, *Helv. Phys. Acta* **1**, 165 (1928).
40. S. Berkman, J. Böhm, and H. Zocher, *Z. Phys. Chem.* **124**, 83 (1926).
41. H. Siedentopf, *Verh. Dtsch. Phys. Ges.* **12**, 6 (1910).
42. H. O. E. Patteri, *Z. Zellforsch. Mikrosk. Anat.* **16**, 723 (1932).
43. W. J. Schmidt, "Submikroskopischer Bau und Färbung des Chitins," in *3. Wanderversammlung Deutscher Entomologen in Giessen*, edited by F. van Emden and W. Horn, Dahlem, Berlin (1929), p. 100.
44. F. Fox, *Z. Farben-Textil-Ind.* **4**, 257 (1905).
45. W. J. Schmidt, *Z. Wiss. Mikrosk.* **52**, 8 (1935).
46. E. Kolbe, *Ueber die Färbung von Pflanzenfasern mit Silber- und Goldsalzen*, Ph.D. Thesis, Universitätsdruckrei von G. Neuenhahn, Jena, 1912.
47. A. Frey, *Z. Wiss. Mikrosk.* **42**, 421 (1925).
48. A. Frey-Wyssling, *Protoplasma* **27**, 563 (1937).
49. A. Frey-Wyssling and O. Wälchli, *J. Polym. Sci.* **1**, 266 (1946).
50. A. Frey, *Jahrbücher Wiss. Botanik* **67**, 597 (1927).
51. A. Frey-Wyssling, *Protoplasma* **27**, 372 (1937).
52. H. Ambronn, *Z. Wiss. Mikrosk.* **22**, 349 (1905).
53. A. Frey, *Das Wesen der Chlorzinkjodidreaktion und das Problem des Faserdichroismus (Ein Beitrag zur Theorie der Färbungen)*, ETH Habilitationsschrift, Zürich (1927).
54. W. Heffels, J. Friedrich, C. Darribère, J. Teisen, K. Interewicz, C. Bastiaansen, W. Caseri, and P. Smith, *Recent Res. Dev., Macromol. Res.* **2**, 143 (1997).
55. C. Bastiaansen, W. Caseri, C. Darribère, S. Dellsperger, W. Heffels, A. Montali, C. Sarwa, P. Smith, and C. Weder, *Chimia* **52**, 591 (1998).
56. Y. Dirix, C. Bastiaansen, W. Caseri, and P. Smith, *Adv. Mater.* **11**, 223 (1999).
57. Y. Dirix, C. Bastiaansen, W. Caseri, and P. Smith, *J. Mater. Sci.* **34**, 3859 (1999).

58. Y. Dirix, C. Bastiaansen, W. Caseri, and P. Smith, *Mater. Res. Soc. Symp. Proc.* **559**, 147 (1999).
59. Y. Dirix, C. Darribère, W. Heffels, C. Bastiaansen, W. Caseri, and P. Smith, *Appl. Optics* **38**, 6581 (1999).
60. J. R. Heath, C. M. Knobler, and D. V. Leff, *J. Phys. Chem. B* **101**, 187 (1997).
61. M. Brust, M. Walker, D. Bethell, D. J. Schiffrin, and R. Whyman, *J. Chem. Soc., Chem. Commun.* 801 (1994).
62. M. Ohtaki, Y. Ohshima, K. Eguchi, and H. Arai, *Chem. Lett.*, 2201 (1992).
63. C. A. Foss, G. L. Hornyak, J. A. Stockert, and C. R. Martin, *J. Phys. Chem.*, **98**, 2963 (1994).
64. M. Schadt and W. Helfrich, *Appl. Phys. Lett.* **18**, 127 (1971).
65. G. Chumanov., K. Sokolov, and T. M. Cotton, *J. Phys. Chem.*, **100**, 5166 (1996).
66. W. Gotschy, K. Vonmentz, A. Leitner, and F. R. Aussenegg, *Opt. Lett.*, **21**, 1099 (1996).
67. M. Takakuwa, K. Baba, and M. Miyagi, *Opt. Lett.*, **21**, 1195 (1996).
68. K. Baba, J. Katsu, and M. Miyagi, *Opt. Lett.*, **17**, 622 (1992).
69. K. Baba and M. Miyagi, *Opt Lett.*, **16**, 964 (1991).
70. A. H. Lu, G. H. Lu, A. M. Kessinger, and C. A. Foss, Jr., *J. Phys. Chem. B* **101**, 9139 (1997).
71. A. G. de León, Y. Dirix, Y. Staedler, K. Feldman, G. Hähner, W. R. Caseri, and P. Smith, *Appl. Opt.* **39**, 4847 (2000).

INDEX

t refers to tables, f refers to figures.

Metal-Polymer Nanocomposites, Edited by Luigi Nicolais and Gianfranco Carotenuto
ISBN 0-471-47131-3